Water Environment Pollution and Control, Volume II

Water Environment Pollution and Control, Volume II

Editors

Weiying Feng
Fang Yang
Jing Liu

MDPI

Basel • Beijing • Wuhan • Barcelona • Belgrade • Novi Sad • Cluj • Manchester

Editors
Weiying Feng
School of Materials Science and Engineering
Beihang University
Beijing
China

Fang Yang
State Key Laboratory of Environmental Criteria and Risk Assessment
Chinese Research Academy of Environmental Sciences
Beijing
China

Jing Liu
Environment Research Institute
Shandong University
Shandong
China

Editorial Office
MDPI AG
Grosspeteranlage 5
4052 Basel, Switzerland

This is a reprint of articles from the Special Issue published online in the open access journal *Water* (ISSN 2073-4441) (available at: https://www.mdpi.com/journal/water/special_issues/C2LN7T3RX0).

For citation purposes, cite each article independently as indicated on the article page online and as indicated below:

Lastname, A.A.; Lastname, B.B. Article Title. *Journal Name* **Year**, *Volume Number*, Page Range.

ISBN 978-3-7258-1568-5 (Hbk)
ISBN 978-3-7258-1567-8 (PDF)
doi.org/10.3390/books978-3-7258-1567-8

Contents

About the Editors

Weiying Feng

Weiying Feng is an associate professor at the School of Materials Science and Engineering, Beihang University, China. She completed her Ph.D. (2016) at the College of Water Science, Beijing Normal University, and went to the Chinese Research Academy of Environmental Sciences as a postdoc from 2016 to 2019. Her research focuses on water environmental pollution and control, microplastic pollution and treatment, biological carbon sequestration, etc. She has presided over and participated in more than 10 scientific research projects, such as those of the National Natural Science Foundation of China and National Key Research and Development Projects. More than 100 academic papers have been published, and 30 invention patents from China and Australia have been authorized. She has won many honors, such as "Outstanding Young Scholar" and 'Young Top Talent and Excellent Mentor" from Beihang University. She has served on the youth Editorial Board of international journals such as *Carbon Research* and *Reviews of Environmental Contamination and Toxicology*.

Fang Yang

Yang Fang, an associate professor, primarily focuses on researching the nutrient cycle and pollution migration mechanisms in eutrophic lakes during the ice period. Additionally, she specializes in basin pollution tracing technology based on isotopes and spectral fingerprints, basin pollution prevention, and ecological restoration. She has successfully presided over the National Natural Science Foundation, published over 40 academic papers, and been granted more than 10 invention patents.

Jing Liu

Jing Liu is an associate professor at Shandong University. Her primary research focuses on the interactions between nanoparticles and heavy metals, the application of nanomaterials in agriculture and environmental remediation, the environmental behaviors of emerging pollutants, etc. She has led projects funded by the Young Future Scholar Program at Shandong University, the National Natural Science Foundation of China, and the Shandong Provincial Natural Science Foundation. She has published 27 journal articles in prestigious journals such as *Environmental Science & Technology* (EST), etc.

water

MDPI

Editorial

Sources, Risks, and Remediation Technologies of Pollutants in Aquatic Environments

Jing Liu [1], Weiying Feng [2,*] and Fang Yang [3,*]

1 Environment Research Institute, Shandong University, Qingdao 266237, China; liu_jing@email.sdu.edu.cn
2 School of Materials Science and Engineering, Beihang University, Beijing 100191, China
3 State Key Laboratory of Environmental Criteria and Risk Assessment, Chinese Research Academy of Environmental Sciences, Beijing 100012, China
* Correspondence: fengweiying@buaa.edu.cn (W.F.); yang.fang@craes.org.cn (F.Y.)

1. Introduction

Water, the lifeblood of our planet, is encountering unprecedented challenges stemming from a diverse array of pollution sources, including industrial wastewater, agricultural runoff, and urban domestic sewage [1,2]. Water pollution poses significant threats to aquatic ecosystems, water quality, and ultimately the health of our planet and its inhabitants [3]. With urbanization and industrialization rapidly developing, conventional and new pollutants pose additional potential risks to the environment and human health [4,5]. Substantial information is required to stay informed on pollutant sources and their potential risks in aquatic environments, as well as to develop innovative technologies for pollution control and remediation, to establish sustainable solutions for maintaining the health of ecology and humans [6–8].

2. Overview of This Special Issue

The Special Issue entitled "*Water Environment Pollution and Control, Volume II*" contains 12 contributions, including nine original articles on diverse topics within the scope, two reviews, and one communication. These articles focus on both organic (organophosphate esters, phenolic compounds, polycyclic aromatic hydrocarbons, fulvic acid, rhodamine B dye, etc.) and inorganic pollutants (mercury, strontium, salinity, etc.). They also discuss contaminants of emerging concern in groundwater, irrigation water, wastewater, drinking water, and seawater, as well as those environmental compartments closely related to the aquatic environment. Innovative technologies have also been proposed to treat the polluted water resources, such as alkaline prehydrolysis, sonochemical catalysis, and photocatalysis. Further, the potential functional applications of seagrass, a promising resource with significant potential for industrial applications and environmental remediation, are explored.

Cao et al. provide valuable insights into the spatiotemporal variations, source determination, and potential ecological risks of organophosphate esters (OPEs) in water and sediment samples collected from 19 coastal tourist resorts in the Shandong Peninsula, China (contribution 1). OPEs are common flame retardants and plasticizers in various consumer products. This study contributes to the knowledge on the environmental fate and impacts of OPEs by highlighting the widespread presence of OPEs in coastal tourist resorts.

Gu et al. investigated alkaline prehydrolysis as a means to enhance phenolic compound absorption in oil wastewater (contribution 2). Phenolic compounds are notoriously toxic and difficult to degrade, but the research showed that hydrolysis in alkaline conditions (pH 12) for 12 h rendered them more susceptible to destruction. Prehydrolysis plays a crucial role in reducing chemical oxygen demand (COD). This research provides valuable insight into the effective treatment of phenolic compounds in oil wastewater and their environmental implications.

Citation: Liu, J.; Feng, W.; Yang, F. Sources, Risks, and Remediation Technologies of Pollutants in Aquatic Environments. *Water* **2024**, *16*, 1532. https://doi.org/10.3390/w16111532

Received: 21 May 2024
Accepted: 24 May 2024
Published: 26 May 2024

Hu et al. investigated the impact of soil salinity distribution, groundwater depth, and irrigation management on agricultural sustainability with a case study in the Hetao Irrigation District in China (contribution 3). They found that groundwater depth and irrigation practices significantly affected soil salinity. Salinized farmland adversely affects crop yield and water resource utilization. The dynamic relationship between soil and groundwater necessitates further research on the temporal variability of soil salinity, in order to implement effective measures for enhancing resource utilization efficiency and agricultural production.

The study by Malov focused on strontium pollution in drinking groundwater and the associated health risks in North-West Russia (contribution 4). Key findings indicate that groundwater with strontium concentrations exceeding the maximum permissible limits exhibited an increased correlation between strontium concentrations, total dissolved solids, and the saturation indices of celestite and gypsum. The research highlights the formation of strontium pollution in drinking groundwater, its potential health implications, and the need for continued monitoring and management to safeguard public health.

Mulenga et al. provide a comprehensive review of mercury (Hg) pollution in aquatic environments resulting from artisanal and small-scale gold mining in Sub-Saharan Africa (contribution 5). The article highlights that this gold mining, which is prevalent in many regions of Sub-Saharan Africa, often uses mercury to extract gold from ore. This process results in significant mercury pollution, which enters aquatic environments through runoff, leaching, and direct discharge. The impacts of mercury pollution on aquatic ecosystems and potential interventions to address it are also explored.

Feng et al.'s systematic review comprehensively explores the global research trends on contaminants of emerging concern (CECs) over the past decade (contribution 6). The authors provide an in-depth analysis of CECs, which covers their occurrence in water, soil, and air. They also delve into the various sources of these contaminants, including industrial discharges, agricultural runoff, and urban waste. This comprehensive coverage enhances understanding of CECs' scope and scale. The risks that CECs pose to ecosystems and human health are particularly concerning. The study highlights the potential for CECs to bioaccumulate in organisms, affecting food chains and ultimately human health. The implications for water quality are also discussed, emphasizing the need for effective monitoring and management strategies to ensure safe drinking water supplies.

Tian et al. studied the remediation of saline–alkali farmland by drainage with subsurface pipes in typical arid and semi-arid agricultural areas, i.e., the Hetao Irrigation District in China (contribution 7). Water and salt migration in amended heavy saline soil was investigated with field experiments on growing sunflowers. The DRAINMOD model and drainmod equation were used to calculate the water and salt migration with the support of field studies. The results can help in developing strategies for desalination and salt control in arid and semi-arid soil with the optimal layout of subsurface pipes. This method is highly significant in effectively managing soil and water resources and improving agricultural production efficiency in these relatively infertile areas. With continuous technological advancements and in-depth research, more innovative solutions are anticipated to emerge, bringing greater breakthroughs to soil and water resource management, as well as the sustainable development of agricultural production.

Tsenter et al. investigated the synergistic inactivation of bacteria during water disinfection by ultrasound coupled with sonocatalysts, including persulfate, nano- and micro-zinc oxide (contribution 8). In terms of environmental remediation, sonochemical reactors have demonstrated strong potential in advanced oxidation processes, especially in the remediation of microbial hazards in water and wastewater treatment. Research has indicated that combining ultrasound and different catalysts can significantly improve the degradation efficiency of organic pollutants, providing new effective means for environmental remediation. Meanwhile, nanotechnology has also driven breakthroughs in environmental remediation, such as the synergistic effect of nanocrystals and ultrasound.

Wu et al. investigated the concentration, composition, source, and ecological and health risks of polycyclic aromatic hydrocarbons (PAHs) in the soil surrounding oil wells in the Yellow River Delta, China (contribution 9). Their comprehensive analyses revealed the pollution status and sources of PAHs in the region. Although the current risks of PAHs to the ecology and human health are low, potential threats still exist in the long run. Future work is proposed to strengthen environmental monitoring and assessment for pollution treatment and remediation.

Xiao et al. examined the dynamics of sedimentary fulvic acid (FA) fractions from an urban river affected by anthropogenic activities (contribution 10). Utilizing advanced spectroscopic techniques, including excitation–emission matrix (EEM) fluorescence spectroscopy with parallel factor (PARAFAC) analysis and two-dimensional correlation spectroscopy (2D-COS), the researchers tracked changes in the composition, spatial distribution, sources, and transformation mechanism of FA fractions from different reaches of the Baitapu River in Shenyang, north-east China. Sediment samples were collected along gradients with varying human activities, including rural, town, and urban sections. The findings help explain how anthropogenic activities alter the composition and behavior of organic matter in aquatic sediments, which is crucial for water quality management and environmental protection.

Photocatalysis has shown significant potential in environmental pollution control and energy conversion in recent years. Zhang et al. compared the degradation effects of rhodamine B (RhB) dye by graphite carbon nitride (g-C_3N_4) synthesized with different precursors under natural sunlight and LED lights, and found that natural light achieves better degradation results due to its diverse spectral components (contribution 11). Meanwhile, g-C_3N_4 synthesized with urea as a precursor has better photocatalytic performance in degrading RhB dye. These research findings not only broaden the application scope of photocatalysis but also strongly support addressing energy crises and environmental pollution issues. In the future, photocatalysis technology is poised to become an important pathway for solving energy shortages and environmental pollution problems, contributing to the sustainable development of human society.

In recent years, seaweed has attracted widespread attention for its commercial potential and biological activity. Lee et al. suggested that seaweed has broad applications as biofuels, functional food ingredients, and sources of biologically active polysaccharides (contribution 12). Specifically, *Sargassum* from Mexico has demonstrated tremendous potential in the food, agricultural, and environmental protection industries, due to its nutritional value, eco-friendly characteristics, and growth-stimulating effects. They also improved the extraction methods of organic components from *Sargassum* for safe application in commercial products.

3. Conclusions

Collectively, these papers showcase cutting-edge research spanning multiple disciplines on water environment pollution and control. Our intention with this Special Issue is to spark further interest in this crucial area of research, as it is desperately needed. While admiring this impressive compilation of work, we still need to remember that water environmental protection is a cornerstone in our efforts to tackle the global environmental crisis. We must collaborate in many battles ahead to achieve sustainable development within our global community.

Author Contributions: J.L.: writing—original draft preparation; W.F. and F.Y.: writing—review and editing. All authors have read and agreed to the published version of the manuscript.

Funding: This research received no external funding.

Conflicts of Interest: The authors declare no competing financial interests.

List of Contributions

1. Cao, X.; Wang, B.; Liu, X.; Cheng, J.; Wang, S. Study of the Spatiotemporal Variations, Source Determination, and Potential Ecological Risk of Organophosphate Esters in Typical Coastal Tourist Resorts in China. *Water* **2023**, *15*, 3976.
2. Gu, C.; Yan, K.; Bo, L.; Zhou, X.; He, Y.; Feng, J.; Qin, J. Alkaline Prehydrolysis Prolongs Resin Life and Enhances the Adsorption of Phenolic Compounds. *Water* **2023**, *15*, 2566.
3. Hu, Z.; Miao, Q.; Shi, H.; Feng, W.; Hou, C.; Yu, C.; Mu, Y. Spatial Variations and Distribution Patterns of Soil Salinity at the Canal Scale in the Hetao Irrigation District. *Water* **2023**, *15*, 3342.
4. Malov, A.I. Features of the Formation of Strontium Pollution of Drinking Groundwater and Associated Health Risks in the North-West of Russia. *Water* **2023**, *15*, 3846.
5. Mulenga, M.; Ouma, K.O.; Monde, C.; Syampungani, S. Aquatic Mercury Pollution from Artisanal and Small-Scale Gold Mining in Sub-Saharan Africa: Status, Impacts, and Interventions. *Water* **2024**, *16*, 756.
6. Feng, W.; Deng, Y.; Yang, F.; Miao, Q.; Ngien, S.K. Systematic Review of Contaminants of Emerging Concern (CECs): Distribution, Risks, and Implications for Water Quality and Health. *Water* **2023**, *15*, 3922.
7. Tian, F.; Miao, Q.; Shi, H.; Li, R.; Dou, X.; Duan, J.; Liu, J.; Feng, W. Study on Water and Salt Transport under Different Subsurface Pipe Arrangement Conditions in Severe Saline–Alkali Land in Hetao Irrigation District with DRAINMOD Model. *Water* **2023**, *15*, 3001.
8. Tsenter, I.; Kobunova, E.; Matafonova, G.; Batoev, V. Synergistic Piezo-Catalytic Inactivation of Bacteria by Dual-Frequency Ultrasound (120 + 1700 kHz) Using Persulfate and ZnO Nano- and Microparticles. *Water* **2023**, *15*, 2937.
9. Wu, Y.; Zhao, Y.; Qi, Y.; Li, J.; Hou, Y.; Hao, H.; Xiao, N.; Zhi, Q. Characteristics, Source and Risk Assessment of Soil Polycyclic Aromatic Hydrocarbons around Oil Wells in the Yellow River Delta, China. *Water* **2023**, *15*, 3324.
10. Xiao, Y.; Yu, H.; Song, Y. Influences of Anthropogenic Pollution on the Dynamics of Sedimentary Fulvic Acid Fractions as Revealed via Spectroscopic Techniques Combined with Two-Dimensional Correlation Spectroscopy. *Water* **2023**, *15*, 2256.
11. Zhang, M.; Xing, M.; Dong, B.; Zhang, H.; Sun, X.; Li, Q.; Lu, X.; Mo, J.; Zhu, H. Thermal Polymerisation Synthesis of g-C3N4 for Photocatalytic Degradation of Rhodamine B Dye under Natural Sunlight. *Water* **2023**, *15*, 2903.
12. Lee, M.C.; Yeh, H.Y.; Chang, C.M.; Liou, Y.F.; Nan, F.H.; Wungen-Sani, J. Tracking and Utilizing Sargassum, an Abundant Resource from the Caribbean Sea. *Water* **2023**, *15*, 2694.

References

1. Kuang, X.; Liu, J.; Scanlon, B.R.; Jiao, J.J.; Jasechko, S.; Lancia, M.; Biskaborn, B.K.; Wada, Y.; Li, H.; Zeng, Z.; et al. The Changing Nature of Groundwater in the Global Water Cycle. *Science* **2024**, *383*, 3976. [CrossRef]
2. Gu, B.; Zhang, X.; Lam, S.K.; Yu, Y.; Van Grinsven, H.J.; Zhang, S.; Wang, X.; Bodirsky, B.L.; Wang, S.; Duan, J.; et al. Cost-effective Mitigation of Nitrogen Pollution from Global Croplands. *Nature* **2023**, *613*, 77–84. [CrossRef] [PubMed]
3. Zhang, H.; Cao, X.; Huo, S.; Ma, C.; Li, W.; Liu, Y.; Tong, Y.; Wu, F. Changes in China's river water quality since 1980: Management implications from sustainable development. *npj Clean Water* **2023**, *6*, 45. [CrossRef]
4. Ateia, M.; Wei, H.; Andreescu, S. Sensors for Emerging Water Contaminants: Overcoming Roadblocks to Innovation. *Environ. Sci. Technol.* **2024**, *58*, 2636–2651. [CrossRef] [PubMed]
5. Li, P.; Liu, J. Micro(nano)plastics in the Human Body: Sources, Occurrences, Fates, and Health Risks. *Environ. Sci. Technol.* **2024**, *58*, 3065–3078. [CrossRef] [PubMed]
6. Sultana, F.; McAllister, T.; Katyaini, S.; Blackstock, M.D. How to Achieve Safe Water Access for All: Work with Local Communities. *Nature* **2024**, *627*, 732–734. [CrossRef] [PubMed]

7. He, J.; Ma, W.; Han, L.; Chen, L.; Xu, E.G.; Xing, B.; Yang, Z. Unraveling the role of natural and pyrogenic dissolved organic matter in photodegradation of biodegradable microplastics in freshwater. *Carbon Res* **2023**, *2*, 18. [CrossRef]
8. Wu, A.; Yang, C.; Zhao, X.; Wang, J.; Liang, W.; Wang, X.; Zhou, L.; Teng, M.; Hou, G.; Niu, L.; et al. Heteroaggregation and sedimentation of natural goethite and artificial Fe3O4 nanoparticles with polystyrene nanoplastics in water. *Carbon Res.* **2024**, *3*, 38. [CrossRef]

water

MDPI

Review

Aquatic Mercury Pollution from Artisanal and Small-Scale Gold Mining in Sub-Saharan Africa: Status, Impacts, and Interventions

Mary Mulenga [1,2], Kennedy O. Ouma [3,4,*], Concillia Monde [2,3] and Stephen Syampungani [2,5,6]

1 Department of Biological Sciences, School of Mathematics and Natural Sciences, Copperbelt University, Kitwe P.O. Box 21692, Zambia; marymulenga452@gmail.com
2 Oliver R Tambo Africa Research Chair Initiative (ORTARChI), Chair of Environment and Development, Copperbelt University, Kitwe P.O. Box 21692, Zambia; concillia.monde@cbu.ac.zm (C.M.); ssyampungani@yahoo.com (S.S.)
3 Department of Zoology and Aquatic Sciences, School of Natural Resources, Copperbelt University, Kitwe P.O. Box 21692, Zambia
4 Africa Centre of Excellence for Sustainable Mining (CBU-ACESM), Copperbelt University, Kitwe P.O. Box 21692, Zambia
5 Department of Plant and Environmental Sciences, School of Natural Resources, Copperbelt University, Kitwe P.O. Box 21692, Zambia
6 Forest Science Postgraduate Program, Department of Plant and Soil Sciences, Plant Sciences Complex, University of Pretoria, Private Bag x20, Hatfield, Pretoria 0002, South Africa
* Correspondence: kennedy.ouma@cbu.ac.zm

Abstract: Mercury (Hg) pollution remains an environmental global concern due to its non-degradable and toxic nature. Natural and anthropogenic sources of Hg adversely affect the functioning of aquatic ecosystems and biological processes. In sub-Saharan Africa (SSA), unregulated artisanal and small-scale gold mining (ASGM) contributes up to 20% of global gold production and uses 205–496 tonnes/yr of Hg. Despite being a vital economic driver for 20–30 million people, ASGM threatens the health of aquatic systems from Hg pollution, presenting a complex challenge that demands urgent interventions. This review seeks to (1) establish the current status of aquatic Hg pollution, (2) explore the environmental impacts of aquatic Hg, and (3) highlight the proposed interventions for aquatic Hg pollution in SSA. We examined publications and institutional reports between 2000 and 2023 addressing aquatic Hg pollution, impacts, and interventions in the ASGM of SSA. Results indicate a rise in aquatic Hg pollution due to the expansion and intensification of ASGM. West Africa remained the highest contributor (50.2%), followed by Central Africa (39.6%), Southern Africa (9.6%), and Eastern Africa (<1%). Contamination of freshwater ecosystems, toxicity to aquatic biota, and environmental health risks to humans were evident. Alternative Hg-free ASGM technologies, including physical, metallurgical, and pyrometallurgical, were investigated from case studies and recommended for adoption.

Keywords: aquatic mercury pollution; Minamata Convention; mercury-free technologies; artisanal and small-scale mining (ASGM); sub-Saharan Africa; Global Mercury Assessment (GMA); Minamata Initial Assessment (MIA)

Citation: Mulenga, M.; Ouma, K.O.; Monde, C.; Syampungani, S. Aquatic Mercury Pollution from Artisanal and Small-Scale Gold Mining in Sub-Saharan Africa: Status, Impacts, and Interventions. *Water* **2024**, *16*, 756. https://doi.org/10.3390/w16050756

Academic Editors: Weiying Feng, Fang Yang and Jing Liu

Received: 20 November 2023
Revised: 5 December 2023
Accepted: 12 December 2023
Published: 2 March 2024

1. Introduction

Mercury (Hg) pollution and toxicity remain an environmentally relevant global concern across the different biospheric matrices due to its ubiquitous and non-degradable nature [1]. Studies have identified the impact of natural and anthropogenic sources of Hg on the functioning of ecosystems and biological processes [2]. Chemically, Hg occurs in three major forms: elemental or metallic (Hg0), organic (OrgHg), predominantly methyl mercury (MeHg), and inorganic (IHg), mainly as mercuric chloride ($HgCl_2$) [3,4]. These different forms are cumulatively called total mercury (THg) [5,6]. Hg, a non-essential element, is liquid at ambient temperature and highly volatilises into the atmosphere. Atmospheric Hg can be transported and deposited in terrestrial and aquatic environments [1,7,8]. MeHg is among the top 10 highly toxic contaminants that negatively impact aquatic biota, accounting for up to 50% of THg pollution [9,10].

In the aquatic ecosystem, under favourable environmental conditions, microbially-mediated biogeochemical (e.g., sulphate-reducing bacteria) and abiotic processes (e.g., transmethylation) transform the IHg into the most bioavailable and toxic MeHg [11–13]. Mercury methylation in aquatic environments results in the bioaccumulation and biomagnification of MeHg, which increases the toxicity risk to higher trophic-level biota [14,15] and humans [4].

The fourth Global Mercury Assessment report [16] catalogued the trends of Hg emissions from key sectors in 2015 (Figure 1). The contribution of atmospheric THg increased by 450% to 4400 tonnes/yr, and almost half of this (2500 tonnes/yr) came mainly from anthropogenic emissions from industrial and artisanal mining and mineral processing, energy production, and losses from Hg-based products and processes. Approximately 600 tonnes entered the freshwater ecosystems, at least 30% of which were conveyed to the marine environment. Natural Hg, releases from volcanic processes, soils, and vegetation burning, accounted for 2100 tonnes of Hg in the atmosphere, while the oceans contributed 250% of Hg (3400 tonnes) in emissions. In the terrestrial environment, natural sources, including organic and mineral soils, retained 750,000 tonnes/yr. The oceanic Hg cycle retained about 2600 tonnes/yr in surface waters, 310,000 tonnes/yr (37%) of the land, and atmospheric THg deposits in the intermediate and deep waters. Net losses of THg from the marine environment contributed about 45–55% of THg to the atmosphere, both transported to land and freshwater or redeposited in the oceans.

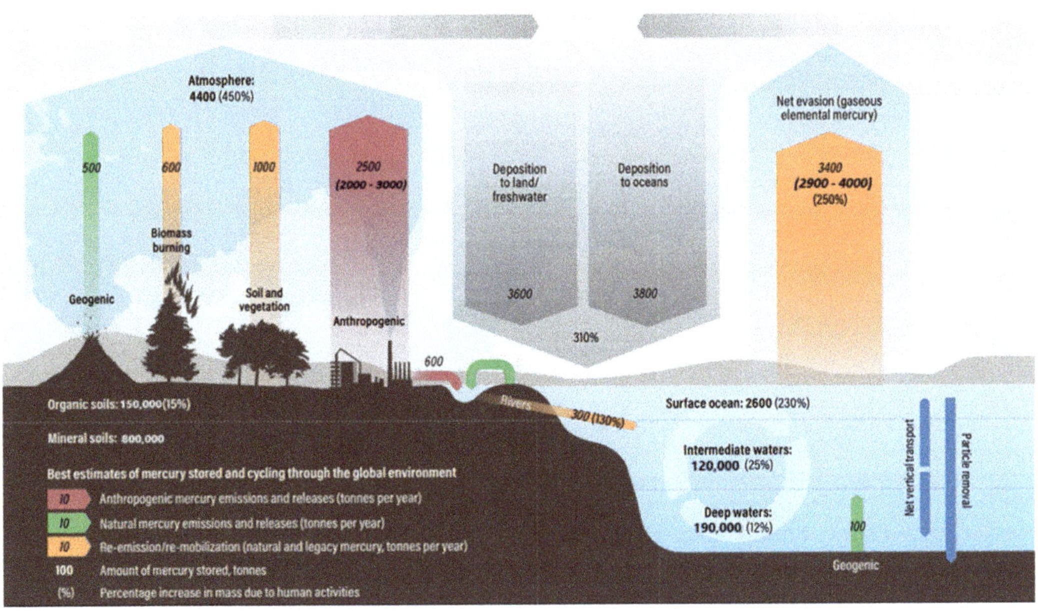

Figure 1. The global Hg budget for emissions, anthropogenic influence on the mercury cycle, and temporary storage in the atmosphere, soils, and aquatic environments [16].

Amotaey and Baawain [17], in a global review of the impact of metal pollution on aquatic biota, observed Hg bioaccumulation in fish muscles, eggs, ovaries, and zooplankton biomass. La Colla et al. [14] reiterated the significant health risk of THg accumulation in fish muscles along an anthropogenically impacted southeast Argentina's Bahía Blanca estuarine ecosystem. Aquatic MeHg concentrations and biomagnification above acceptable limits in fish communities (732–918 ng/g), frogs (1.075 ng/g), and riparian spiders (347–1140 ng/g) in tropical highland aquatic systems in southwest Colombia have also been reported [18–20]. In addition, Lino et al. [21] observed that artisanal and small-scale gold mining (ASGM) and deforestation were the primary sources of THg and MeHg in Brazil's Tapajós River basin. In the Gambia River flowing through the Kedougou region in eastern Senegal, pollution of stream sediments was in the magnitude of 2–6 mg/kg THg [22] but recently reduced to 1.16 mg/kg THg and 3.2 ng/g MeHg [23].

Streets et al. [24] globally observed a 1.4 Tg cumulative release of Hg between the year 1510 and 2010 period, of which 23% was atmospheric and 77% dissipated into aquatic and terrestrial environments. Furthermore, the authors noted that North America and Europe contributed a cumulative 413 Gg and 427 Gg in this period. With the increased ASGM activities in Africa and the Middle East, about 72% of cumulative atmospheric Hg0 was released [25]. Additionally, ASGM was the largest source of global Hg emissions (775 Mg) in 2015. The aquatic and terrestrial environments that accumulated over 83% of the atmospheric THg were Africa, the Middle East, and Oceania, which contributed 77%, while up to 89% came from South America. Streets et al. [25] further documented an increase in annual Hg emissions from 2188 Mg to 2390 Mg (+1.8%) between 2010 and 2015, where emissions increased in Eastern Africa (>4%), South Asia (>4.6%), and Central America (>5.4%).

In 2011, the ASGM industry employed over 6 million miners globally, who extracted 380–450 tonnes of gold, and by 2018, over 44.67 million people worked in the ASGM sector [16,26]. In sub-Saharan Africa, approximately 1.322 million people work in the ASGM sector, where 27,200 kg of gold was produced in 2014 [16,27]. According to UNEP [16], global Hg emission sources ranged from 40 kg in the ferrous metal industries to 52 kg from biomass burning, 162 kg from waste products, 313 kg from non-ferrous metal production, and 838 kg from ASGM. In 2015, ASGM contributed nearly 1220 tonnes of Hg to global water and terrestrial ecosystems from mining activities in sub-Saharan Africa (8%), East and Southeast Asia (36%), and South America (53%) [16]. The increased use of Hg in ASGM has contaminated aquatic ecosystems. Mercury transforms into highly toxic MeHg, which bioaccumulates in aquatic biota, threatening aquatic productivity and the health of aquatic resource users [28,29]. Furthermore, the ASGM communities risk exposure to inhalation of gaseous Hg, ingestion of contaminated water and food, and dermal contact, resulting in health problems, including neurological disorders [30,31]. For example, communities near ASGM sites in countries like Indonesia [32–34] and Ghana [35–37] have reported both environmental impacts and health issues associated with mercury exposure.

Additionally, Hg used in ASGM enters the global Hg cycle following atmospheric release during gold amalgams, contributing to long-range transport and deposition thus impacting ecosystems globally [38,39]. In aquatic ecosystems, Hg bioaccumulation threatens aquatic biodiversity, particularly in ASGM-intensive areas [40]. For instance, studies in the Amazon basin reported high Hg levels in fish, impacting aquatic ecosystems and human health [41,42]. Although ASGM is often a source of livelihood for marginalised communities, it poses environmental and health risks associated with Hg use, which further undermines the health and productivity of ASGM communities.

In October 2013, the UNEP-led Minamata Convention on Mercury, enforced in August 2017, targeted, among other interventions, the elimination of Hg use in ASGM [43]. Parties to the Convention were obliged to conduct enabling assessments under the Minamata Initial Assessments (MIA) and generate country profiles on Hg status [43]. According to Anan and Toda [44], analysis of country priorities from MIA reports shows that most African countries prioritised three key areas for reduction of Hg release into the environment (Article 8), including Hg waste management, eliminating Hg in products such as cosmetics and dental amalgam (Article 4), and adopting Hg-free gold processing in the ASGM industry (Article 7) (Figure 2). The identified priority area later formed the basis for developing National Action Plans (NAP) to enforce the Minamanta Convention. Furthermore, Anan and Toda [44] identified 16 SSA countries that had prioritised the phasing out of Hg-added products and the safe handling of Hg wastes to reduce environmental contamination. Similarly, 14 country MIA reports highlighted total elimination or reduction of Hg use in ASGM. Waste incineration, open burning of mercury wastes, and uncontrolled waste damping release approximately 132,776kg (23.2%) and 229,681 kg (40.1%) Hg, respectively. The application, use, and disposal of Hg-added products and dental amalgam contribute 54,521 kg (10%) of Hg, whereas the ASGM sector, dominated by the use of Hg amalgamation during gold production accounts for 156,350 kg (27.3%) of Hg releases. Cumulatively, the implementation of these action areas would reduce approximately 573,328 kg of Africa's contribution to global Hg emissions [44].

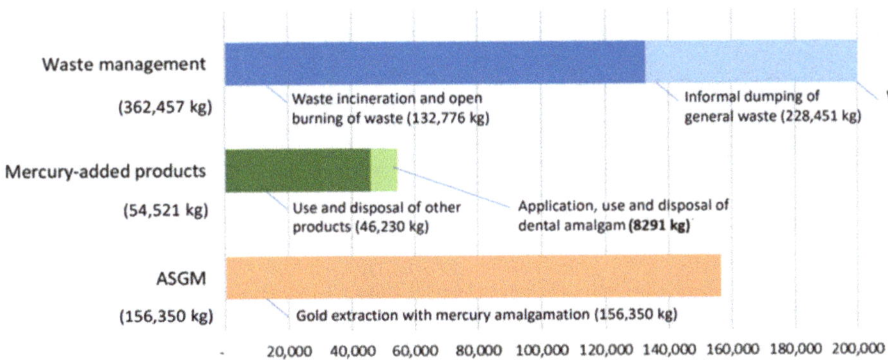

Figure 2. Mercury releases from ASGM, mercury-added products, and waste management sectors from 17 MIA reports from countries in Africa [44].

From the above statistics, it remains clear that the environmental presence and use of Hg in the ASGM sector and industrial products exposes the biosphere to toxicity from Hg interaction with air, land, and water and consumption of Hg-contaminated food, moreso in SSA aquatic systems [16,44]. To address the Hg intoxication problem, there is a need to establish the present status, environmental impacts, and existing interventions to address this global and continental problem, particularly aquatic Hg pollution. The ASGM sector remains a critical contributor to aquatic Hg in SSA [5,45,46]. However, few studies have attempted to establish the magnitude and impact of aquatic Hg pollution from the SSA perspective but have not entirely focused on the contribution of the ASGM sector [47–49]. Other studies have reviewed the impacts of Hg contamination of aquatic environments from ASGM at the country level, including in South Africa [50], the Lake Victoria basin, Eastern Africa [51], and Ghana [35]. However, regional comparative studies on Hg aquatic pollution and related environmental impacts and interventions across and from sub-regional ASGM activities across SSA that provide regional interventions are missing. Consequently, the review focuses on (1) establishing the current status of aquatic Hg pollution, (2) exploring the environmental impacts of Hg pollution on aquatic ecosystems in sub-Saharan Africa, and (3) highlighting proposed interventions and identifying gaps for further research on the management of aquatic Hg pollution from ASGM in SSA.

2. Methodological Approach

We considered primary research and reviews published in English between 2000 and 2023 and the latest annual reports from relevant institutions, such as the UNEP Minamata Convention on Mercury (https://minamataconvention.org (accessed on 18 October 2023)), country-based Minamata Initial Assessment (MIA) reports (https://minamataconvention.org/en/parties/minamata-initial-assessments (accessed on 21 October 2023)), the UNEP 2018 Global Mercury Assessment (GMA) report (https://wedocs.unep.org/handle/20.500.11822/29830 (accessed on 27 October 2023)), UN Environment Global Mercury

Partnership (UNGMP) (https://www.unep.org/globalmercurypartnership (accessed on 4 November 2023)), and the Global Environmental Facility (GEF) (https://www.thegef.org (accessed on 8 November 2023)). The articles were retrieved from SCOPUS, Google Scholar, and institutional databases. We further conducted a citation search on the documents retrieved to identify additional relevant studies. Different combinations of keywords were used to access the relevant literature, including "environmental", "impacts", "aquatic", "ecosystems", "small-scale", "artisanal", "gold mining", "mercury", "pollution", and "country" ("country name"). Publications from South America and Asia were also accessed for a comparative assessment of aquatic ASGM pollution in the southern hemisphere. We retrieved 151 documents, 119 primary research articles, and 32 from secondary data (Figure 3).

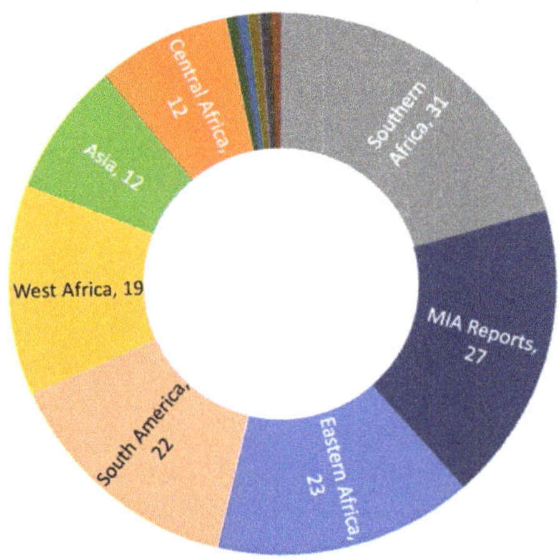

Figure 3. Number of publications between 2000 and 2023 per region on aquatic Hg pollution from ASGM and processing.

3. Mercury Pollution in Sub-Saharan Africa: Trends and Impacts

ASGM has recently intensified in SSA, driven by its potential to provide livelihoods and alleviate poverty in resource-rich but economically disadvantaged localities [47,52]. However, the expansion of ASGM in SSA has accelerated Hg pollution with severe environmental impacts on air, water, land, and human health. The use of Hg in gold processing, often informal and unregulated, severely threatens the aquatic ecosystems of SSA. The release of this toxic element into aquatic systems not only endangers the health of aquatic biota but also the well-being of the riparian community at local and global scales [16]. Several studies have documented the environmental impacts of aquatic Hg pollution across the sub-regions of SSA: Eastern, Central, Western, and Southern, as discussed below.

3.1. Mercury Pollution to Air, Land, and Aquatic Ecosystems: MIA Synthesis

This section presents an account of Hg pollution in air, terrestrial, and aquatic environments in Eastern, Southern, Central, and West Africa from the three key sources, which are primary metal production (PMP), production of other materials and metals (OMM), and extraction and use of fuel/energy sources (FES) listed in the MIA reports (https://minamataconvention.org/en/parties/minamata-initial-assessments (accessed on 21 October 2023). Twenty-seven MIA reports from sub-Saharan African countries were synthesised (Figure 3). Generally, THg emissions were 962,827 kg/yr, of which a total of 261,765 kg/yr (27.2%) was emitted to air, 150,908 kg/yr (15.7%) to water, and 550,154 kg/yr (57.1%) was emitted to land from three major contributors, namely, primary metal production, minerals and other materials, and emissions from the extraction and use of fuel/energy (Figure 4) (MIA, 2017–2023). Burkina Faso and Sudan were the main contributors of THg, with 335,698 (34.9%) and 320,038 (33.2%) kg/yr, respectively. Ghana, South Africa, Senegal, and Zambia contributed between 30,665 (3.18%) and 65,249 (6.78%) kg/yr THg, while the remaining countries accounted for 114,728 kg/yr (11.92%) THg releases in SSA (MIA, 2017–2023).

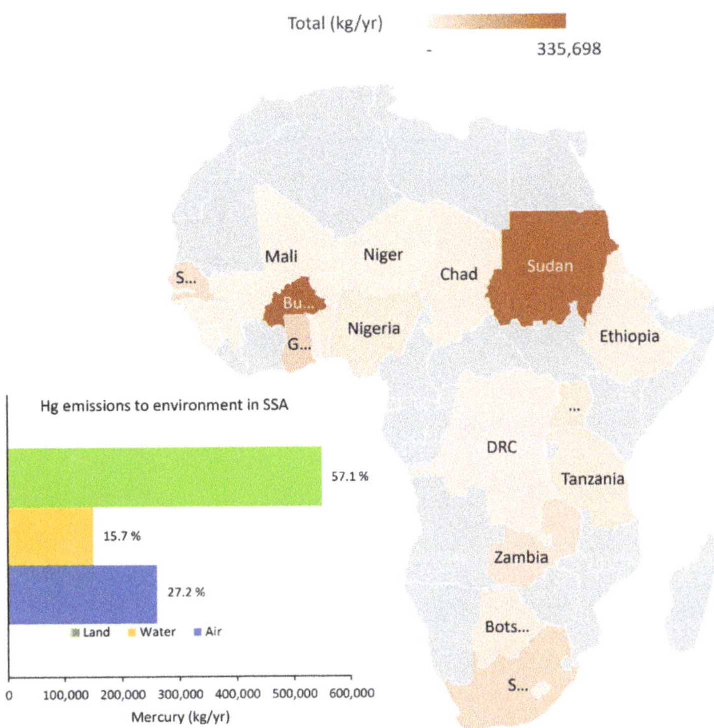

Figure 4. Contribution to total Hg pollution from extraction and consumption of fuel and energy sources and exploitation of primary metals, other minerals, and raw materials in sub-Saharan Africa (data source: MIA, 2017–2023 reports).

3.1.1. Eastern Africa

According to the MIA, between 2017 and 2023, atmospheric Hg emissions in Eastern African countries (kg/yr) increased in the order Eritrea (45.1) < Rwanda (77.4) < Sudan (237) < Tanzania (970) < Ethiopia (5739) < Uganda (12,138) (Figure 5). Annual land THg pollution from the extraction of core metals, such as Au and diamond, from ASGM in the region was predominant in Tanzania (9363), Rwanda (13,193), Eritrea (4106 kg/yr), and Uganda (3333 kg/yr). The extraction of OMM contributed to atmospheric THg emissions of 31.4 kg/yr (Rwanda) and 645 kg/yr (Tanzania), totalling 1646.25 kg/yr. Primary metal production was a significant source of aquatic THg in Ethiopia (375 kg/yr) and Sudan (103,622 kg/yr).

3.1.2. Southern Africa

Only three country MIA reports were available in Southern Africa: Botswana, South Africa, and Zambia; the latter are major sub-Saharan African mining countries. Primary metal production accounted for 65% (19,781 kg/yr) of THg in South Africa, 49% (14,705 kg/yr) in Zambia, and 8% (69 kg/yr) of atmospheric THg emissions. Additionally, 8275 kg/yr of THg was released to aquatic environments in South Africa, 1719 kg/yr in Zambia, and only 44.4 kg/yr in Botswana. Zambia was the highest emitter of THg on land (139,976 kg/yr), followed by Botswana (10,598 kg/yr) and South Africa (2183 kg/yr) (Figure 6).

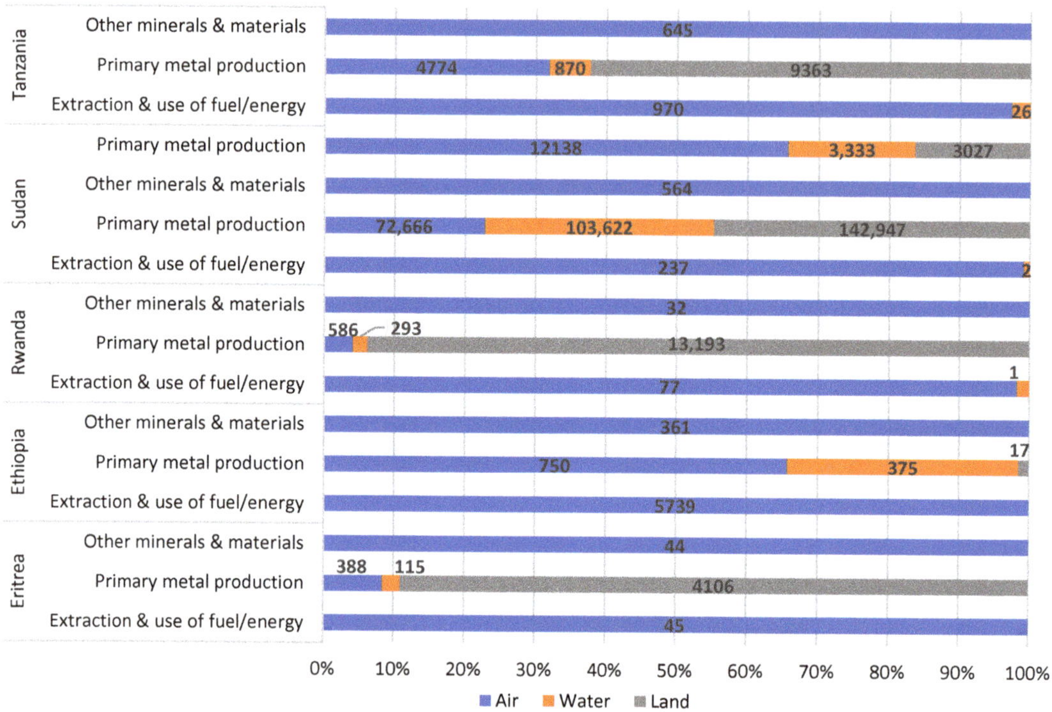

Figure 5. Total Hg emissions to air, water, and land in Eastern Africa from the three main contributors (data source: MIA, 2017–2023 reports).

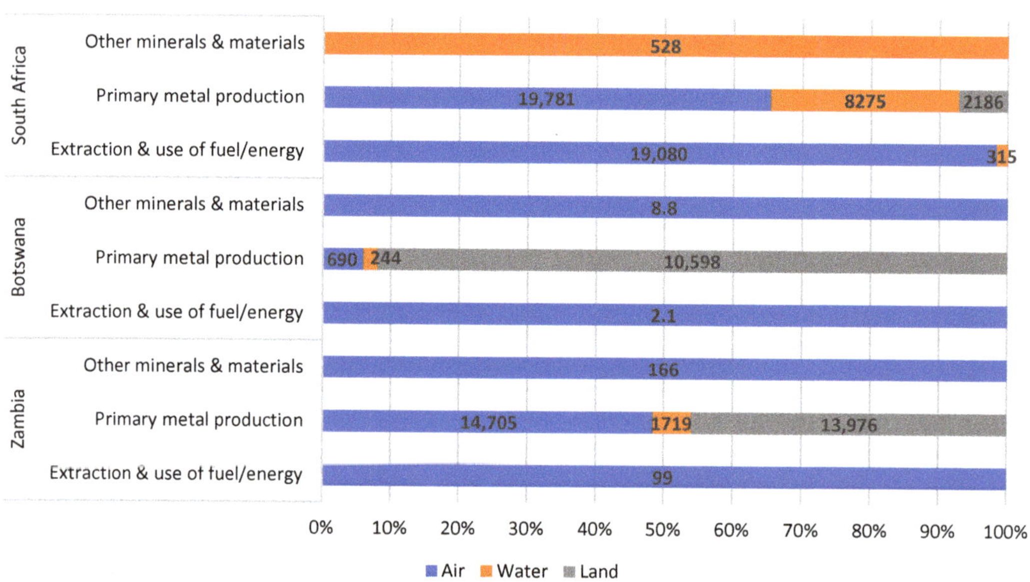

Figure 6. Emissions of THg to air, water, and land in Southern Africa from the production of primary metals, other minerals and materials, and extraction and use of fuel/energy (data source: MIA, 2017–2023 reports).

3.1.3. West Africa

From Figure 7, atmospheric THg pollution was mainly from the extraction of fuel and other energy sources from Nigeria (2370 kg/yr), Ghana (1228 kg/yr), Burkina Faso (286 kg/yr), Senegal (260 kg/yr), Togo (144.2 kg/yr), and Sierra Leone (118 kg/yr). Primary metal processing in West Africa contributed significantly to atmospheric THg, particularly from Burkina Faso (36,996 kg/yr), Ghana (33,344 kg/yr), Nigeria (4396 kg/yr), and Cameroon (2000 kg/yr). The PMP sector also released large amounts of THg to aquatic systems in Benin (3286 kg/yr), Ghana (6527 kg/yr), and Senegal (7266 kg/yr). The West African terrestrial ecosystems equally received alarming quantities of THg from Burkina Faso (287,881 kg/yr), Ghana (6075 kg/yr), Nigeria (6730 kg/yr), and Senegal (21,752 kg/yr).

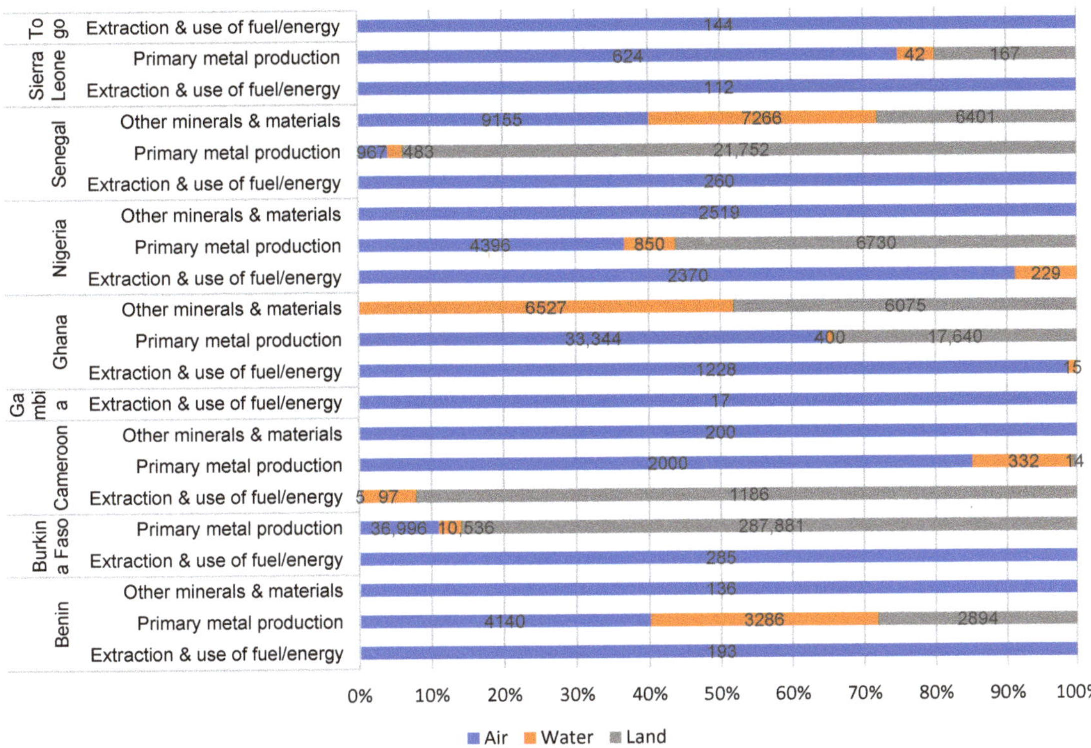

Figure 7. Mercury emissions to air, water, and land in West Africa from the extraction and processing of minerals and fuel/energy resources (data source: MIA, 2017–2023 reports).

Extraction and use of fuel and other energy sources contributed a cumulative 1683 kg/yr to atmospheric THg, approximately 1186 kg/yr to land, and another 326 kg/yr to water bodies in the sub-region.

3.2. *Regional Patterns of Aquatic Hg Pollution from ASGM*

In SSA, the western Africa region is richly endowed with mineral and natural fuel energy resources compared to the rest. Consequently, as depicted in Figure 8, based on THg emissions from three key sectors, primary metal production, mining and processing of secondary metals and other mineral resources, and extraction of fuels and other energy sources, West Africa was the highest contributor to environmental THg pollution (50.2%), followed by Central Africa (39.6%), Southern Africa (9.6%) and Eastern Africa (0.5%). The implications of these emissions to the aquatic environmental and potential health risks from the utilisation of water and water resources (e.g., fish) in the different sub-regional locations are described in Section 3.2.1 below and summarised in Table 1 for clarity.

3.2.1. Eastern Africa

Uganda

In the Okame River basin of Eastern Uganda, River Omanyi was the most contaminated from ASGM by Hg (0.0191 mg/L), followed by Nankuke (0.0163 mg/L) and Nabewo (0.0158 mg/L) [53]. Omara et al. [54] investigated THg contamination in water, sediments, and fish from the Namukombe stream in Uganda's Busia gold mining district. THg concentrations in water were up to 1.21 mg/L; fish muscle contained 0.11 µg/g; and sediments had 0.14 µg/g. In the Western Uganda Rwizi River ecosystem, sediments had 0.01–0.4 µg/g, while surface water contained 0.01–0.3 µg/L of Hg traced to ASGM [54]. However, the health risk from drinking water in the Rwizi River was considerably low [55]. In the Napoleon and Winam Gulfs of Uganda and Kenya, respectively, THg ranged from 1.7 to 5.8 ng/L, while MeHg levels ranged between 0.2 and 1.0 ng/L in the L. Victoria waters and were considerably higher than Hg levels in temperate lakes [56].

Tanzania

Earlier studies by Ikingura and Akagi [57] and Ikingura et al. [58] in the L. Victoria goldfields of Tanzania revealed low Hg in fish (7.0 µg/g), sediments (4.19 µg/g) and soils (3.39 µg/g) within the ASGM locations. However, Hg concentrations in urine samples (2.6 ng/mL) and human hair (947 ug/g) from miners exposed to Hg vapour during amalgamation were significantly higher [57,58]. In the Mugusu mine, Hg concentrations in stream sediments varied from 0.5 to 6.0 mg/kg, while tailings from the Mwagamaza mine had 165–332 mg/kg THg [59]. In Nungwe Bay, however, the low Hg concentrations (2–23 µg/kg) in fish were linked to background levels. Furthermore, the tailings released high MeHg concentrations (629–710 ng/g) [56]. Chibunda [60], in a bioassay to determine the bio-uptake and toxicity of mercuric chloride ($HgCl_2$) from ASGM activities on aquatic biota of L. Victoria basin goldfields, observed that assay concentrations ranging from 8 µg/L in the freshwater shrimps (*Caridina nilotica*) to 68 µg/L in freshwater snails (*Bulinus forskalii*) and 162 µg/L in fish (*Hypochromis nubilus*) exhibited biotoxic effects. Taylor et al. [61] comprehensively assessed the environmental impact of Hg release from ASGM in the Rwagamaze gold mine region of the Geita district in Tanzania. In that study, Hg concentration increased from 0.01 to 1.6 mg/kg in stream sediments and 0.006–3.5 mg/kg in fish from Lake Victoria; these values exceeded the WHO limits (0.5 mg/kg) for human consumption and the WHO maximum allowable limit (0.2 mg/kg) for vulnerable groups.

Kenya

The environmental contamination from ASGM Hg release in the Migori Gold Belt in the L. Victoria basin, Kenya, has been extensively documented. According to Ogola et al. [62], stream sediments at the gold sites carried 0.28–348 mg/kg, and tailings at the panning sites released 0.46–1920 mg/kg Hg. Findings from that study estimated that about 150–200 kg Hg/month was used for Au processing, of which 40% and 60% were lost during panning and amalgamation, respectively. Odumo et al. [63] investigated Hg concentrations in gold ores and sediment from four ASGM areas within the Gold Belt (Osiri A and B, Mikei, and Makalda) and reported 16.1–149.9 mg/kg Hg. Ngure et al. [64] also observed significant Hg concentrations in water (0.36–52.1 µg/L), soil (0.51–1830 mg/kg), and fish (0.26–355 mg/kg), which were above the WHO/FAO maximum acceptable limits. In the same period in the Migori-Transmara ASGM region, high Hg was detected in sediment (430 µg/kg), soil (140 µg/kg), and tailings (8900 µg/kg) and was associated with wet and dry depositions [65]. In recent investigations, Tampushi et al. [66] reported 0.66 mg/L in surface water and 24.63 mg/kg Hg in sediments, positively correlated with ASGM ore and tailings contamination sources in the Lolgorian mining region of southwestern Kenya. In Kakamega and Vihiga counties, soil samples contained Hg levels about 49% above the

USEPA standards with 1–72% bioaccessibility. Metal concentrations in 25% of water sources exceeded the WHO (10 μg/L) limit, which increased the risk of non-carcinogenic health effects in humans [67].

Sudan

In sub-Saharan Africa, Sudan was ranked as the largest Hg importer between 2008 and 2015, accounting for an average of 43.25 tonnes of Hg annually, mainly used for gold extraction in ASGM [68]. The use of Hg in gold extraction, particularly when miners employ rudimentary techniques, contaminates aquatic ecosystems. For instance, improper handling and disposal of Hg-laden tailings releases this toxic element that ultimately leaches into aquatic ecosystems [69]. Ahmed et al. [70] assessed the consequences of Hg use in ASGM processes on the surface water quality of the Alebedia area in the River Nile State, Sudan. In the study, the average Hg concentration in water ranged between 0.001 and 0.005 mg/L, attributed to ASGM activities that used mercury amalgamation for gold processing. Hg levels in stream sediments from five regions of the Dar-Mali gold mining area in Sudan's Nile State were lowest in the eastern part (0.001 mg/kg) and increased in the northern (0.002 mg/kg), southern (0.004 mg/kg), and western (0.005 mg/kg) parts [71]. Another study by Mubarak and Ali [72] investigated Hg content among other metals in two fish species, Nile perch (*Lates niloticus*) and Nile tilapia (*Oreochromis niloticus*) of Lake Nubia, Sudan. It was noted that Hg levels ranged from 0.017 to 0.094 mg/kg in *O. niloticus* and 0.085–0.172 mg/kg in *L. niloticus*, indicating bioaccumulation in fish, thus posing health risks to consumers.

3.2.2. Southern Africa

South Africa

South Africa is a significant coal and Au producer in SSA and an important source of Hg pollution globally, contributing about 46.4 tonnes of Hg in 2018 [73,74]. For instance, atmospheric Hg emissions and deposition to the aquatic environment are significantly elevated in South Africa from activities such as coal combustion, ASGM, and industrial gold mining [47]. South Africa is the second-highest emitter of atmospheric Hg from coal-powered energy production plants, which supply ~64% of energy, releasing 0.02–0.16 g Hg per ton of coal burned [75]. South Africa is also the second largest producer of Au, and 70–80% of Hg emissions emanate from Au mining, far higher than coal combustion, while the remaining 20–30% is deposited onto terrestrial and aquatic environments [76]. Additionally, 50 ng/g THg and 1.3 ng/g MeHg were reported in sediment cores and surface sediments from South Africa's Berg River [77], up to 68 ng/g in the Olifants River [74], and up to 0.036 μg/L in the Ga-Selati River [78].

Hg contamination of aquatic systems by coal-fired power generation in the Upper Vaal and Olifants and ASGM in the Inkomati catchment has been observed from fish, water, and sediment analyses [50]. According to Lusilao-Makiese et al. [79], Hg released from ASGM activities in abandoned gold mines in Randfontein, west Johannesburg, release up to 10 μg/g to sediments, 2 ng/L in surface water, and 223 ng/L in groundwater, negatively impacting the Krugersdorp Game Reserve ecosystem. Mine tailings and sediments have been reported to release significant Hg (837–867 μg/kg) to surface waters, of which 90% occur as toxic MeHg [80]. In another study, the variability in MeHg concentration (~13 μg/kg) and IHg (~8480 μg/kg) in sediments in low flows was influenced by seasonality [81]. Verhaert et al. [82] investigated THg in water, surface sediments, aquatic biota, and the trophic transfer dynamics of Hg in the Olifants River basin. The biomagnificative property of THg was noted in fish that accumulated between 0.1 and 6.1 μg/g and trophic magnification factor (TMF) > 1, mostly desorbed from sediments (0.001–0.078 μg/g THg), with potential risk from consumption from fish in the basin.

Zimbabwe

van Straaten [45] compared Hg contamination from ASGM activities in Zimbabwe and Tanzania, where 70–80% was lost to the atmosphere and 20–30% to terrestrial and aquatic ecosystems. While the THg released in Tanzania's Lake Victoria goldfields ranged between 3 and 4 tonnes and above 3 tonnes in Zimbabwe, approximately 1.2–1.5 g of Hg was also released to the environment during Au processing, corresponding to the global 1–2 g Hg loss per gram of Au produced [83]. Green et al. [84] reported aquatic sediment Hg levels between 6 and 154 mg/kg, which positively correlated with organic content but negatively correlated from the source along a four-stream system draining the Farvic Gold Mine area, southeast of Bulawayo, Zimbabwe. In a study to determine the ecological impacts of illegal ASGM on freshwater crabs (*Potamonautes* sp.) and quantify Hg levels in 49 river sampling sites across three land use categories in Chimanimani, Eastern Zimbabwe, Dalu et al. [85] reported Hg levels in stream water from 0 to 0.1 mg/L in communal lands, up to 0.3 mg/L in the national parks, and up to 0.06 mg/L in the timber plantations. Furthermore, the national parks had strong-to-extreme contamination and extremely high Hg enrichment. However, the Hg levels in surface water did not influence the distribution or abundance of crabs. In contrast, Makaure et al. [86] compared Hg contamination and trophic dynamics of fishes between the protected Chivero and ASGM-impacted Mazowe reservoirs in Zimbabwe and revealed significant THg bioaccumulation in Mazowe with trophic magnification slopes (TMS) of 0.28. About 75% of the fish exceeded the 0.5 μg/g THg UNEP recommended threshold, potentially exposing humans and higher trophic fauna to health risks from fish consumption.

3.2.3. Central Africa

Mercury pollution in the lotic systems of Central Africa poses a growing environmental concern, primarily driven by ASGM activities [87]. Artisanal gold miners in Central Africa employ rudimentary processing techniques, often involving Hg [88,89]. The improper handling and disposal of this toxic element has led to substantial contamination of aquatic systems, endangering aquatic life and impacting the livelihoods of local communities [90]. According to UNEP [87], mercury pollution from ASGM in Central Africa is widespread across the region's water bodies, with far-reaching ecological and health consequences. In the south Kivu and some Ituri Basins in Butuzi, eastern DRC, atmospheric Hg emissions account for 29% of THg release from ASGM. In comparison, 71% is lost in mine wastes, of which nearly 50% is transported to streams via surface runoff, exerting adverse impacts on aquatic biota [87].

In the Kienke and Tchangue Basins of the Togo Gold District in Southern Cameroon, Hg levels of 0.106 mg/Kg in river sediments exceeded the WHO limits [91]. However, there was low sediment contamination (CF = 0.17) and low potential aquatic ecological risk (ERF = 4.24) from exposure to Hg in both basins. Additionally, Pascal et al. [92] investigated the influence of gold panning in nine rivers in the Fizi catchment of south Kivu, DRC. Hg levels ranged from 7.8 to 10.7 ng/L in rivers Kimuti, Mandje, and Kuwa, 12.9–18.10 ng/L in Kacumvi, Lubichako, Eto, and Makungu, and 20.3–41.3 ng/L in Misisi and Kambi, which were most polluted [92]. Similarly, Pascal et al. [93] noted Hg concentrations (mg/Kg) in stream sediments of nine rivers in the order Kimbi (89.8) > Misisi (41.1) > Makungu (36.9) > Eto (33.7) > Lubichako (30.1) > Kacumvi (25.3) > Kuwa (20) > Mandje (17.1) > Kimuti (13.6). Furthermore, the higher enrichment of stream sediment with Hg (ER = 24.5) implied a high ecological risk to aquatic biota (PERI = 539.5) in the Fizi catchment [93]. Ngueyep et al. [94] evaluated the effect of AGSM on the surface water quality of the Kadey River system in Batori, east Cameroon, and noted higher Hg concentrations up to 0.02 mg/L, exceeding the 0.001 mg/L WHO limits. Ngounouno et al. [95] evaluated the impact of gold mining on the Lom River of Wakaso, Adamawa, and Cameroon and the efficiency of drumstick tree (*Moringa oleifera*) seeds in bioremediating metal-polluted wastewater. Among other metallic contaminants, Hg concentrations for water (0.01–1.83 mg/L) and sediment (2–25 mg/kg) were both above the WHO limit (0.001 mg/L) and (1 mg/kg), respectively, in all the sampling points. This could be attributed to gold processing involving mercury amalgamation and burning [96]. Bella Atangana et al. [97] investigated the seasonal variability of Hg contamination of stream water in the Lom gold basin from ASGM, where the dry and wet season mean concentrations were 0.007 and 0.008 mg/L, respectively. In contrast, Ayiwouo et al. [98] observed a higher dry season mean value (0.005 mg/L) and a lower wet season mean value (0.004 mg/L) for Hg concentration in the Lom River, Gankombol region.

3.2.4. West Africa

Ghana

Rajaee et al. [99] evaluated the impacts of human and ecological Hg exposure and spatial distribution in the ASGM in Kejetia, Gorogo, and Bolgatanga, Upper East Ghana. The findings indicated elevated hair, urinary, and household soil Hg concentrations of 0.967 μg/g, 5.18 μg/L, and 3.77 μg/g, respectively, but low concentrations in sediment (0.005–0.248 μg/g; mean 0.036 μg/g) and fish (0.024–0.22 μg/g; mean 0.07 μg/g). Donkor et al. [36] reported seasonal variabilities of Hg in sediments from Ghana's Pra River Basin at 0.018–2.92 mg/kg in the rainy season and 0.01–0.043 mg/kg in the dry season. The high sediment enrichment with THg indicated severe contamination of streams in the Pra Basin. Further studies by Donkor et al. [37] in the basin revealed elevated seasonal Hg sediment concentrations between 0.265 mg/kg (rainy) and 0.019 mg/kg THg (dry), and 0.021 mg/kg (rainy) and 0.001 mg/kg MeHg (dry) from AGSM.

Recently, Affum et al. [100] also correlated the 0.18–0.56 μg/L Hg in water and 1.13–1.21 mg/kg Hg in sediments of the Bonsa River to gold mining and processing. The above Ghana EPS/WHO guideline value (0.001 mg/L) and Hg levels (0.045–0.061 mg/L) in River Bonsa water were also recently confirmed by Obiri-Yeboah et al. [101] and attributed to increased ASGM in the basin. However, Tulasi et al. [102] also attributed Hg methylation in Ghanaian river systems to cyanide (CN)-based Au mining activities. The authors correlated the increase in solubility and sediment transport of MeHg in the Apepe (4.58–14.82 ng/g) and Ankora (0.24–1.21 ng/g) rivers in southwest Ghana with elevated CN concentrations. A health risk assessment by Kortei et al. [40] on Hg toxicity in Ankobrah and Pra Rivers indicated 0.006–0.0093 mg/l in water and 0.4–0.6 mg/kg in fish tissue, with a corresponding target hazard quotient (THQ) > 1, demonstrating potentially harmful health effects.

Senegal

AGSM in Senegal is predominant in the southeastern Kedougou region, along the Mali-Guinea border, which hosts the 230 km^2 Sabodala Belt, the largest gold deposit (~400 tonnes) in West Africa [103]. Gold mining in the Gambia River, draining the Kedougou basin and characterised by THg levels above the sediment quality and probable concentration thresholds, has negatively impacted the basin's aquatic ecosystems. For instance, results from THg analysis of nine freshwater fish species indicated concentrations of between 0.03 and 0.5 mg/kg THg and between 0.5 and 1.05 mg/kg THg in two species of shellfish in the Gambia River, which were above the WHO guideline value of 0.5 mg/kg [22]. In the same basin, higher concentrations of up to 1.5 mg/kg, 0.4 mg/kg, and 0.32 mg/kg THg were detected in human hair in Bantako, Tinkoto, and Samekouta towns, respectively, of the Kedougou region [22]. Niane et al. [104] recently indicated a decrease in THg contamination in the Gambia River stream sediments from 9.9 mg/kg in 2014 to 1.16 mg/kg, with potential toxicity to aquatic biota. Another study by Gerson et al. [105] evidenced that ASGM in Kedougou contributed 4.2 ng/g and 22 ng/L THg in sediment and water, respectively.

Nigeria

In Nigeria, ASGM-related Hg contamination in the Manyera River in Niger State ranged between 0.014 and 0.025 mg/L in water and up to 0.021 mg/kg in sediment. Additionally, fish (*Heterotis niloticus*) from the river bioaccumulated 0.008 mg/kg [106].

In the northwest Anka ASGM region, 2.12 and 1.25 μg/g Hg were recorded from stream sediments and tailings, respectively [107]. Sani et al. [108] recently assessed ASGM pollution of three water sources around the Igade mining area in Niger State, where the Hg levels for the well (0.01 mg/L), borehole (0.005 mg/L), and river water (0.012 mg/L) were above the WHO and Nigeria recommended standards for potable water.

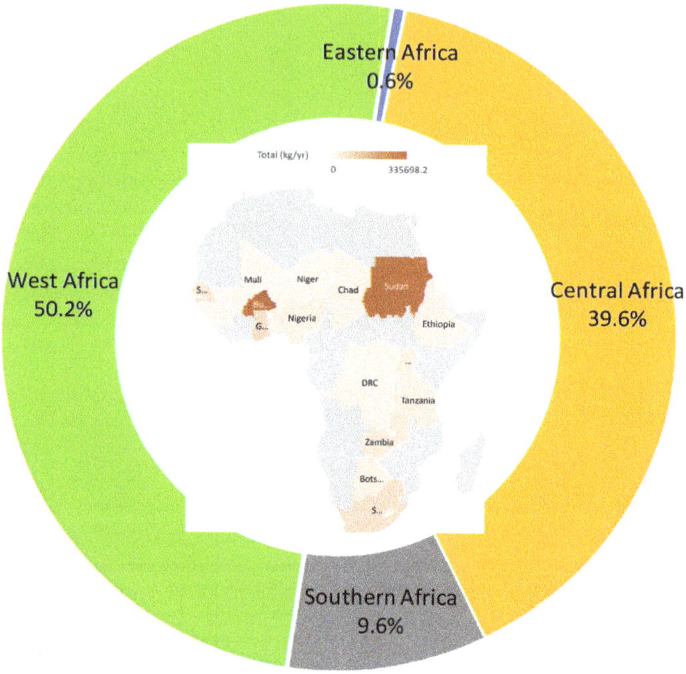

Figure 8. Contribution of THg in Eastern, Central, West, and Southern Africa showing the proportion of THg emissions in SSA from production of primary metals, other metals and minerals, and the extraction of fuel and other energy sources (data source: MIA, 2017–2023 reports).

Table 1 summarises the environmental occurrence and concentration of the different categories of Hg in aquatic ecosystems of the Eastern, Southern, Central, and Western sub-regions of SSA with evidence of ASGM-related impacts.

Table 1. Aquatic pollution in water, sediment, and biota in the Eastern, Southern, Central, and Western sub-regions of SSA from selected studies.

Region and Country	Aquatic Matrix	Hg Category and Value	References
1. Eastern			
Uganda	Okame River Nankuke River Nahweo River	Hg: 0.019 mg/L (Water) Hg: 0.0163 mg/L (Water) Hg: 0.0158 mg/L (Water)	[55]
	Namukombe stream	THg: 0.11 μg/g (Fish, *Oreochromis niloticus*); 1.21 mg/L (Water); 0.14 μg/g (Sediment)	[56,57]
	Rwizi River	Hg: 0.01–0.1 ug/g (Sediment); 0.01–0.3 μg/L (Water); 0.04 μg/g (Fish, *Barbus altianalis*); 0.09 μg/g (Fish, *Brycinus sadleri*)	[58]
	Lake Victoria	THg: 1.7–5.8 ng/L; MeHg: 0.2–1.0 ng/L (Water)	

Table 1. *Cont.*

Region and Country	Aquatic Matrix	Hg Category and Value	References
Tanzania	Lake Victoria Goldfield streams Mugusu mine streams L. Victoria, Nungwe Bay Rwamagasa Gold mine region streams	Hg: 7.0 μg/kg (Fish, *Tilapia* sp.); ug/g (Sediment) Hg: 0.35–6.0 mg/kg (Sediment) Hg: 2–35 μg/kg (Fish, *Tilapia* sp.; *Lates niloticus; Protopterus* sp.) Hg: 0.01–1.6 mg/kg (Sediment); 0.006–3.5 mg/kg (Fish, *Lates* sp., *Oreochromis* spp., *Clarias* sp.)	[58–61,63]
Kenya	Migori Goldbelt, Lake Victoria Basin	Hg: 430 μg/kg–149.9 mg/kg (Sediment) Hg: 0.36 μg/L–0.66 mg/L (Water); 0.26–355 mg/kg (Fish, *Rastrineobola argentea*)	[64–68]
Sudan	River Nile State Albenda region streams	Hg: 0.001–0.005 mg/L (Water); 0.017–0.094 mg/kg (Fish: *Oreochromis niloticus*); 0.085–0.172 mg/kg (Fish: *Lates niloticus*)	[72,74]
2. Southern South Africa	Berg River Olifants River Ga-Selati River Randfontein, West Johannesburg stream	THg: 50 ng/g (Sediment); MeHg: 1.3 ng/g (Sediment) MeHg: 68 ng/g (Sediment); THg: 0.1–6.1 μg/g (Fish, *Clarias* sp.; *Hydrocynus* sp., *Labeo* sp., *Schilbe* sp., *Labeobarbus* sp., *Oreochromis* sp.); 0.001–0.078 mg/g (Sediment) MeHg: 0.036 μg/L (Water) Hg: 10 μg/g (Sediment); 2 ng/L (Surface water); 223 ng/L (Ground water)	[76,79–81,84]
Zimbabwe	Farvic Gold mine area streams, Bulawayo. Chimanimani, East Zimbabwe Mazowe and Chivaro Reservoirs	Hg: 6–154 mg/kg (Sediment); 0.06–0.4 mg/L (Water) Hg: 0.1–0.3 mg/L (Water) THg: Above 0.5 μg/g (Fish)	[86–88]
3. Central Cameroon	Togo Gold District, Kianke, and Tchangue Basins Kadey River, East Cameroon Lom River, Adamawa Lom Basin Lom River, Gankombol	Hg: 0.106 mg/kg Hg: 0.02 mg/L (Water) Hg: 0.01–1.83 mg/L (Water) Hg: 0.007–0.008 mg/L (Water) Hg: 0.004–0.005 mg/L (Water)	[93] [96] [97] [98] [99,100]
Democratic Republic of Congo	Fizi Basin River systems, south Kivu	Hg: 7.8–41.3 ng/L (Water); 17.1–89.8 mg/kg (Sediment)	[94–96]
4. West Ghana	Kejetia Gorogo and Bolgatanga ASGM area streams, Upper East Ghana Pra River Basin Bonsa River	Hg: 0.05–0.248 μg/g (Sediment); 0.024–0.22 μg/g (Fish; species unidentified) Hg: 0.01–2.92 mg/kg (Sediment) THg: 0.019–0.265 mg/kg (Sediment) MeHg: 0.001 mg/kg (Sediment) Hg: 0.18 μg/L–0.061 mg/L (Water); 1.13–1.21 mg/kg (Sediment)	[101] [36] [37] [102,103]

Table 1. *Cont.*

Region and Country	Aquatic Matrix	Hg Category and Value	References
4. West			
	River Apepe and Ankora	MeHg: 0.24–14.82 ng/g (Sediment)	[104]
	Ankobrah and Pra Rivers	Hg: 0.006–0.0093 mg/L (Water); 0.04–0.6 mg/kg (Fish, *Oreochromis niloticus; Clarias angularis*)	[40]
Senegal	Gambia River	THg: 0.03–0.5 mg/kg (Fish); 0.5–1.05 mg/kg (Shellfish); 4.2 ng/g–9.9 mg/kg (Sediment); 22 ng/L (Water)	[106,107]
Nigeria	Manyera River, Niger State	Hg: 0.014–0.025 mg/L (Water); 0.021 mg/kg (Sediment); 0.008 mg/kg (Fish—*Heterotis niloticus*)	[108]
	Northwest Anka ASGM Region	Hg: 2.12 mg/g (Sediment)	[109]
	Igade mining area, Niger State	Hg: 0.01–0.012 mg/L (Water)	[110]

3.3. Regional Summary and Intercontinental Hg Pollution from ASGM

Table 2 reflects the overall trend in countries in sub-regions of sub-Saharan Africa, South America, and Asia where evidence of ASGM contributing to aquatic Hg pollution has been documented. Hg concentration in Eastern Africa's aquatic systems ranged from 5.8 ng/L in the Napoleon Gulf, L. Victoria, to 1.21 mg/L in the Namukombe stream, Uganda, and from 0.36 µg/L in the L. Victoria Basin rivers to 0.66 mg/L in the Lolgorian region in Kenya [54,64,66]. In the Alebedia area in River Nile State, Sudan, Ahmed et al. [70] reported 0.001 to 0.005 mg/L Hg in surface water. Sediments in Uganda's Namukombe stream contained Hg levels from 0.14 to 0.4 µg/L [54]. Campbell et al. [51] reported 629 ng/g Hg in sediments of L. Victoria, while Ikingura et al. [59] documented maximum Hg concentrations of 6 mg/kg in stream sediments around the Mugusu mine of the L. Victoria Basin of Tanzania. In Kenya, Tampushi et al. [66] investigated Hg release into stream sediments at the Lolgorian ASGM site, where maximum Hg concentrations reached 24.6 mg/kg. In Sudan's Dar Mali gold mining region, stream sediment Hg concentrations ranged between 0.001 and 0.005 mg/kg [71]. Hg levels of environmental concern have also been reported in Eastern Africa. For instance, evidence of fish contamination with Hg was reported in Uganda at 0.11 µg/g [54] and 2.0 µg/g to 3.5 mg/kg in Tanzania [61].

In South Africa, in the Ga-Selati and Berg Rivers (in Limpopo basin) and Phongolo, Olifants, and Upper Vaal Rivers, concentrations of 0.036 µg/L in surface water, 50 ng/g to 1.3 µg/g in sediments, and 0.001 to 6.1 ng/g Hg in fish tissue were reported for several studies [50,74,77,78,82]. Elevated Hg concentrations ranging between 6 and 154 mg/kg were reported in stream sediments of the Farvic Gold mining area in Bulawayo, Zimbabwe [84], and between 0.006 and 154 mg/kg in the tropical reserve area in Zimbabwe [85]. In central Africa's DRC, in the Fizi catchment of South Kivu, Hg release to water from gold production ranged from 7.8 to 41.4 ng/L and 13.6–89.8 mg/kg in sediments [93], while in the Kienke and Tchangue basins of Cameroon, sediment Hg levels reached 0.106 mg/kg [91]. In West Africa, Hg levels in the range 0.18 ug/L–0.0093 mg/L were reported by Affum et al. [100] and Kortei et al. [40] in the surface waters of Bonsa, Anrobrah, and Pra Rivers, respectively, in Ghana. Fish accumulated up to 0.6 mg/kg Hg in the Ankobrah-Pra River systems [40]. In Senegal, freshwater fish accumulated THg concentrations ranging between 0.03 and 1.05 mg/kg in the Gambia River [22]. In Nigeria's Manyera River, ASGM-produced Hg was detected in tissues of fish (*Heterotis niloticus*) up to 0.008 mg/kg.

Similarly, Hg pollution trends have been documented in Latin American and Asian ASGM environments. In Asia, Hg levels in surface water ranged from 0.002 µg/L in Myanmar [109] to 8 µg/L in the Philippines [110]. Murphy et al. [111] recorded contamination of sediments up to 64 ng/g in Cambodia; bioaccumulation of Hg in fish was between 90 ng/g and 50 µg/g. Abraham et al. [112] reported that bottom sediment contained 2.7 µug/g–0.02 mg/kg Hg in the Zamora River, Ecuador, while tailings from ASGM released 89–1535 µg/g Hg as mining effluents. In the Madeira River, Brazil, in Latin America, the THg accumulated up to 1242 µg/kg in fish [113]. Hg contamination was higher (0.12–1.92 mg/kg) in fish communities of the ASGM in the Cuyuni River Basin of Venezuela [114]. Hacon et al. [115] assessed the potential ecotoxicological risks of THg to trophic guilds of fish in 18 sites in five regions of the Amapá State territory in the Brazilian Northern Amazon region. Carnivorous fish had a mean concentration of 0.4 µg/g; omnivorous fish (*Pimelodus ornatus*) accumulated 1.8 µg/g Hg; and herbivorous fish, e.g., *Mylesinus paraschombourgkii* ("flaviano"), had unexpectedly high Hg levels (1.0 µg/g). At the same time, another herbivore, *Myloplus ternetzi* ("pacu"), bioaccumulated up to 0.85 µg/g Hg. The highest Hg levels were detected in carnivorous fish species (2.1 µg/g), including *Boulengerella cuvieri* ("pirapucu"), followed by *Cichla monoculus* ("tucunaré") and *Hoplias aimara* ("traírão"). The high Hg levels present a potential risk from local fish consumption in the region. A similar study in the western Amazon Basin Triple Frontier monitored the impact of THg on commercial fish [116]. In that study, carnivorous fish species had the highest concentrations of THg (e.g., *Hoplias malabaricus* with a THg of 0.205–2.818 µg/g; *Hoplerythrinus unitaeniatus* with a THg from 0.97 to 1.873 µg/g THg) above the maximum allowed by the Brazilian guidelines (1.0 µg/g), while the THg in the herbivores (e.g., *Pterygoplichthys pardalis*, with 0.092–0.279 µg/g THg) was below the acceptable limit. In the southeastern ASGM region of the Peruvian Amazon, Hg levels in lake and river sediments ranged between 64 and 86 ng/g and 20–53 ng/g, respectively [117]. Additionally, 5% of the piscivorous fish bioaccumulated Hg concentrations above the 500 ng/g limit recommended for human consumption (e.g., locally known as "chambira", "palomata", and "huasaco", bioaccumulated up to 1215, 80, and 500 ng/g, respectively). Some ASGM-impacted rivers in the ASGM regions, such as Quebrado Yarinal and Middle-Malinowski, had no fish during that study.

Table 2. Contribution to aquatic Hg from ASGM in SSA countries from 2000 to 2023 and selected studies from Asia and South America.

Sub-Region	Country	Environmental Compartment			References
		Water	Sediment/Tailings	Biota	
Eastern	Uganda	5.8 ng/L–1.21 mg/L	0.14–0.4 µg/g	Fish: 0.11 µg/g	[53–55]
	Tanzania		629 ng/g–6 mg/kg	Fish: 2 µg/kg–3.5 mg/kg Shrimp: 8µg/L Snail: 68 µg/L	[51,59–61]
	Kenya	0.36 ug/L–0.66 mg/L	430 ug/kg–1920 mg/kg (tailings) 0.001–24.6 mg/kg (sediment)	Fish: 0.26–355 mg/kg	[62,63,65,66]
	Sudan	0.001–0.005 mg/L	0.001–0.005 mg/kg (sediment)	Fish: 0.017–0.172 mg/kg	[70,72,118]
Southern	South Africa	0.036 µg/L	50 ng/g–1.3 mg/g	Fish: 0.001–6.1 ng/g	[50,74,77,78,82]
	Zimbabwe		0.06–154 mg/kg	Fish: 17–32% with Hg > 0.2 µg/g	[84–86]
Central	DRC	7.8–41.3 ng/L	13.6–88.8 mg/kg		[92]
	Cameroon	7.8 ng/L–0.83 mg/L	2–25 mg/kg		[91,93,94,97,98]
West	Ghana	0.18 µg/L–0.0093 mg/L	0.005 µg/g–2.92 mg/kg	Fish: 0.024 µg/g–0.6 mg/kg	[36,37,40,100,102,119]
	Senegal	22 ng/L	4.2 ng/g–9.9 mg/kg		[105]
	Nigeria	0.01–0.05 mg/L	2.12 ug/g–0.021 mg/kg	Fish: 0.008 mg/kg	[106–108]

Table 2. *Cont.*

Sub-Region	Country	Environmental Compartment			References
		Water	Sediment/Tailings	Biota	
Asia	Indonesia	0.7–9.9 µg/L			[34]
	Myanmar	0.002–0.008 µg/L			[109]
	Philippines	0.009–80 µg/L			[110]
	Thailand	0.7–6 µg/L			[120]
	Cambodia		64 ng/g	Fish: 90 ng/g–50 µg/g	[111]
Latin America	Ecuador		Sediment: 2.7 µg/g–0.02 mg/kg Tailings: 89–1535 µg/g		[112,121]
	Brazil		100 ng/g–1207 mg/kg	Fish: 100 ng/g–1242 µg/kg	[113,122,123]
	Venezuela	4.60 µg/L		0.12–1.92 mg/kg	[114]
	Brazil			Fish (Carnivorous 0.4 µg/g; Omnivore-*Pimelodus ornatus*, 1.8 µg/g; Herbivore-*Mylesinus paraschombourgkii*, 1 µg/g; *M. ternetzi*, 0.85 ug/g Hg)	[115]
	Brazil, Colombia, Peru			Fish (Carnivore, *Hoplias malabaricus*, 0.21–282 µg/g; Herbivore-*Pterygoplichthys pardalis*, 0.09–0.28 µg/g THg)	[116]
	Peru		Lake sediment: 64–86 ng/g River sediment: 20–53 ng/g	Fish("chambira", 1215 ng/g: "palomata", 80 ng/g; "Huasaco", 500 ng/g)	[117]

4. Socioeconomic, Environmental, and Human Health Impacts of ASGM in SSA

According to Hilson [124,125], ASGM is a significant economic driver in rural SSA, providing direct employment to more than 10 million miners in rural communities. For instance, in Ghana and Tanzania, more than 1 million people are directly employed in the ASGM sector, and at least 0.2 million of the ASGM community in Mali are women [124]. Furthermore, ASGM is a source of capital development beyond direct employment by diversifying livelihoods and micro-economies in rural SSA mining communities, thereby providing indirect employment to over 4 million people [126,127]. Therefore, across the sub-regions of SSA, the socioeconomic contribution of the ASGM sector to the rural transformation of the socioeconomic status and livelihoods of millions of people cannot be underestimated [128].

However, the largely informal organisation of the ASGM sector in SSA competes against the socioeconomic gains among mining communities across the region. Hilson et al. [129] noted that the formalisation of ASGM is a difficult undertaking regardless of the locality of the mining landscape and described the situation as a created and perpetual informality in the ASGM mining space. The biggest drawback to this is the imbalanced working conditions, such as the intensive labour demands involved with gold processing, higher prostitution levels, and abuse of drugs such as narcotics [124]. Competing interests between rural communities and ASGM entrepreneurs have frequently led to gang-related violence, thefts, and related vices. For instance, Mkodzongi [130] reported a dramatic increase in gang-related violence in the Shurugwi mining area of Zimbabwe's Midlands Province, with cases of robbery of cash, gold, and ore from miners and businesses. In addition, most ASGM entities operate under hazardous conditions, impacting the health and well-being of the predominantly poor rural communities [127,131].

ASGM miners are continuously exposed to gaseous Hg during the processing of gold by Au-Hg amalgam, which involves heating to vaporise the Hg. Hg is also released into aquatic systems during amalgamation processes and washed off during rain and flood events, becoming an important route for human exposure, primarily to MeHg (Figure 9). MeHg can bioaccumulate up the food chain, resulting in human exposure risks. Hence, communities within the ASGM regions that use Hg for gold processing may have a higher exposure through fish consumption, intoxication through inhalation, and physical contact, causing increased disease burden from elevated MeHg burden. For instance, in 2016, Steckling et al. [132] reported health impacts on 1.22–2.39 million miners globally from exposure to IHg. A critical factor determining the level of human exposure to MeHg is the interaction with the aquatic medium and the occurrence of IHg, which is usually associated with the aquatic environments synonymous with ASGM in SSA [133,134].

Figure 9. Unstructured ASGM and landscape degradation in the Democratic Republic of Congo, Sierra Leone, and Tanzania. Miners are highly exposed to Hg, MeHg, and other toxic elements associated with gold mining. The use of protective gear and proper handling or disposal of Hg in the ASGM industry in sub-Saharan Africa is largely unregulated (photo credits: N. Nyirabihogo, A. R. Thomas, and H. Nachilongo).

ASGM also exacerbates deforestation and soil erosion in many SSA countries [135–137]. The direct impact of ASGM is more focused on tropical countries in the global south, particularly SSA. According to Fisher et al. [138], several factors may increase the potential of exposure to Hg from ASGM, including (1) high concentrations of Hg in soils, tailings, and stream sediments; (2) a lack of employment or alternative livelihoods in many rural areas, which promotes low-level employment occupations such as the ASGM; (3) the availability and affordability of Hg required for gold processing and a lack of government control on Hg importation and unregulated use in the ASGM sector; and (4) the high relative market price of gold, which pushes the demand for Hg as a "quick" extraction method.

Additionally, reservoirs constructed for electricity generation in most river basins of SSA are potential convertors of Hg into toxic MeHg in the aquatic environment. Hg cycling in reservoirs typically increases MeHg production due to the ecological, biogeochemical, and hydrological changes in aquatic environments [139]. For instance, Hall et al. [140] observed elevated MeHg output and bioaccumulation from the deposition of organic matter during flooding that increased microbial decomposition and net MeHg retention in reservoirs in northwestern Ontario, Canada. In the boreal Canadian reservoirs, MeHg levels increased 3 to 6 times post-impoundment flooding and remained high several decades later [141]. In the tropics, reservoirs have been observed to increase MeHg production from Hg-laden surface sediments, which bioaccumulate to harmful levels in organisms [142]. In that study, THg accumulation ranged from 50 to 200 ng/g in reservoirs with a surface area between 80 and 400 km^2.

5. Considerations for Interventions and Remediation

5.1. Mitigation and Remediation Strategies

The Minamata Convention, enforced in 2017, aims to "protect the human health and the environment from anthropogenic emissions and releases of mercury and mercury compounds" [28,43,143]. Article 8 of the Convention focuses on controlling and/or reducing THg emissions to the atmosphere. Article 9 addresses efforts to prevent and/or reduce THg releases to land and water from point sources. To minimise exposure to Hg in the ASGM environment, capacity building, adoption of alternative methods, and technology transfer are addressed under Article 14 of the convention [43]. At the country level, Article 7 guides actions to "...reduce, and where feasible eliminate, the use of mercury and mercury compounds, and the emissions and releases to the environment of mercury from sources such as mining and processing" [43].

5.2. Minamata Convention Parties' Commitment in SSA

The Minamata Convention, initially signed by 76% of SSA countries, was established in line with Article 23 of the convention and is governed by the Conference of the Parties (COP) of the UNEP. The COP supports the implementation of the Convention through continuous evaluation and development. In the SSA, 4% of the parties ratified the convention in the 2014–2015 period, 39% in the next 4 years following the initial ratification, and an additional 15% had the convention officially binding in the 2020 to 2023 period. Presently, 58% of sub-Saharan countries have ratified the convention (Table 3). This is expected to positively impact the achievement of Article 19 through the promotion and adoption of Hg-free technologies in the ASGM sector.

Table 3. Participation of the SSA countries in accession and ratification of the Minamata Convention (data source: Minamata Convention Parties 2023).

No.	Action	Year	Party	% of SSA
1	Signature	2013–2014	Benin, Burkina Faso, CAR, Côte d'Ivoire, Djibouti, Gambia, Kenya, Malawi, Mali, Niger, Nigeria, South Africa, Togo, Uganda, Tanzania, Zambia, Ethiopia, Mozambique, Angola, Burundi, Cameroon, Chad, Congo, Gabon, Ghana, Guinea, Guinea-Bissau, Liberia, Senegal, Sierra Leone, Sudan, Zimbabwe	76
2	Accession	2014–2017	Botswana, Equatorial Guinea, Eritrea, Eswatini, Lesotho, Namibia, Rwanda	15
3	Ratification	2014–2015	Djibouti, Chad	4
		2016–2019	Benin, Burkina Faso, Côte d'Ivoire, Gambia, Mali, Niger, Nigeria, South Africa, Togo, Uganda, Zambia, the DRC, Ghana, Guinea-Bissau, Senegal, Sierra Leone	39
		2020–2023	CAR, Kenya, Malawi, Tanzania, Burundi, Cameroon, Zimbabwe	15

5.3. Towards Mercury-Free ASGM Technologies: Selected Case Studies

Article 19 (g) of the Minamata Convention commits all parties to actively and corporately collect relevant information and research on the "technical and economic availability of Hg-free products and processes and on best available techniques and best environmental practices to reduce and monitor emissions and releases of Hg and Hg compounds." Hg-free gold metallurgy technologies represent a crucial shift in the ASGM sector, aiming to mitigate the severe health and environmental impacts of using Hg [144]. The technologies are categorised into physical metallurgy (gravity, magnetic, and flotation), chemical/hydrometallurgy (chemical-based cyanidation, halide leaching, and use of alternative lixiviants), and pyrometallurgy (smelting, roasting, and calcination) [145].

Earlier interventions towards the sustainable development of Hg-free ASGM technologies in SSA were supported by the Canadian-based Artisanal Gold Council (AGC) in collaboration with the African Mining Development Council [146]. In Burkina Faso, the government supported the development, production, and establishment of chemical-free ASGM technology. The 1998–2004 "PRECAGEME" Project installed the equipment in 8 pilot-scale mining sites, each with 75–85% efficiency of extracting gold from rocks and 90–95% efficiency of free gold extraction [146]. In the Manica Province of Mozambique, a chemical-free small-scale gold mining technology using centrifugation and magnetic separation of Au was adopted in 2012 at the Clean Tech Mine [147]. Briefly, basins were used to collect mineral material from centrifuge drains, ball mill pipes, and centrifuges, which were water-panned, leaving a mixture of Au, Fe-minerals, and Fe-shavings. A magnet was repeatedly passed over the basins to trap magnetic material until 89–93% pure Au was retained.

The use of sodium borax in the Namayingo and Mubende districts, Uganda, is another alternative to Hg-free ASGM [148]. Sodium borax was used to remove impurities from gold concentrate. The method displayed was more efficient after gravitation and yielded higher gold quantities. The gold concentrate was placed in a plastic container. Equal amounts of borax and a few drops of water were then added to a clay bowl, which was then heated with a gasoline burner or acetylene flame to melt the mixture, after which the molten gold was deposited in the bottom of the bowl for collection. Gold has a melting point of 1063 °C. Borax, when heated, melts and lowers the melting point of the mixture, which then enables all the minerals to melt down and separate. As the process continues, the borax causes further oxidation and breakdown of the other minerals, except Au. Au is unaffected by this reaction and sinks to the bottom of the mixture intact for retrieval.

Other miners used the charcoal and blower method, which took about 30 min. The concentrate and borax mixture were placed in a plastic bag and into a clay bowl filled with charcoal. The charcoal was ignited and kept at high temperatures using a mechanical or hand-operated blower to melt the Au concentrate. The melted Au coalesced at the bottom of the bowl and was then retrieved. Separately, a two-stage gravity separation with a direct smelting method was successfully implemented in Buhere, Namayingo District, Uganda, to increase Au recovery in a Hg-free work environment [144].

In Ghana, a UNIDO-led Hg-free gold processing initiative in 2000 promoted the use of glass retorts for the extraction of Au from the concentrate [149]. However, this technology was not widespread due to the relatively longer processing time (~2 h), higher energy requirements, and the fragile nature of the retorts in the rugged mining environment. Styles et al. [150] identified direct smelting as the most preferred Hg-free Au processing method for ASGM in Ghana for processing small batches of Au concentrate. The technique involved smelting using crucibles containing concentrate and flux mixtures (e.g., potassium or sodium carbonate, silica sand, and borax) at 1200 °C on charcoal, gas, or palm-kernel shell fuel stoves. Smelting was advantageous over amalgamation since no Hg was used, thereby safer, cost-effective and non-technical.

The studies above show that the potential of achieving a Hg-free mining environment is apparent, provided efforts are tailored towards non-technical, cost-effective, time-efficient, and robust technologies applicable in the rugged SSA ASGM terrain [144,150].

6. Conclusions and Future Perspectives

Aquatic Hg pollution is rising due to the rapid expansion and intensification of ASGM in SSA. In general, West Africa reported the highest contribution to aquatic Hg pollution (50.2%), followed by Central Africa (39.6%) and Southern Africa (9.6%), while Eastern Africa contributed below 1%. Contamination of freshwater ecosystems was evident in the ASGM regions, from surface and groundwater, stream sediments, bioaccumulation in aquatic biota, and riparian vegetation. The subsequent environmental health risks to humans from fish consumption, water use, and exploitation of other water resources from Hg-contaminated aquatic ecosystems were also significant.

From the MIA report synthesis, the following gaps need urgent action to enhance the effective management of aquatic Hg pollution from ASGM in SSA: (1) limited documentation and information of mercury availability, quantification, use, and safe disposal; (2) inadequate capacity and resources to monitor and regulate Hg-related ASGM; (3) unregulated and illegal Hg use in the gold mining sector; and (4) poor implementation and low uptake of Hg-free alternative ASGM technologies.

We recommend the following options as a way forward for the management and control of Hg pollution in aquatic ecosystems in SSA: (1) the regulation and reduction of Hg use in ASGM through the development of alternative Hg-free gold mining technologies, including the physical (gravity, magnetic, and flotation), hydrometallurgical (borax, potassium carbonate, silica sand), and pyrometallurgical methods (smelting and roasting); (2) the identification, development, and strengthening of environmental policies and interventions to phase out the importation and use of Hg in the ASGM sector; for instance, the closure of ASGM that use Hg should be accompanied by offering alternative Hg-free livelihoods with an equal or better economic income; and (3) the implementation of health educational programs within the existing health care structures to inform communities within the ASGM regions of the negative health impacts of Hg use.

Implementing and enforcing regulations on the use of mercury in ASGM is essential. Environmental health and public policies should be implemented to balance the socioeconomic benefits and environmental impacts of ASGM in SSA. Governments and international organisations are increasingly recognising the need for stricter regulations. For example, the Minamata Convention aims to regulate and reduce Hg use, particularly in ASGM [43]. Encouraging the adoption of cleaner and more efficient gold extraction technologies is crucial. For example, gravity concentration, cyanide leaching, or other methods that minimise or eliminate the need for Hg can significantly reduce environmental and health impacts [144].

Furthermore, providing training and education to artisanal miners on the hazards of mercury and alternative practices is essential. Educational training would empower ASGM communities to make informed choices about sustainable and environmentally friendly mining practices. There is a need to focus efforts on regulatory measures, community engagement, and the promotion of cleaner technologies to mitigate the adverse effects associated with Hg use in ASGM.

Author Contributions: Conceptualization, M.M., C.M., K.O.O. and S.S.; methodology, M.M. and K.O.O.; validation, M.M., C.M., K.O.O. and S.S.; formal analysis, M.M., K.O.O. and S.S.; investigation, M.M. and K.O.O.; literature and data curation, M.M. and K.O.O.; writing—original draft, M.M., C.M., K.O.O. and S.S.; writing—review and editing, M.M., C.M., K.O.O. and S.S.; funding acquisition, S.S; supervision, C.M. and S.S. All authors have read and agreed to the published version of the manuscript.

Funding: This review was funded by the Oliver R Tambo Africa Research Chair Initiative (ORTARChI)—Environment and Development, of the Copperbelt University. ORTARChI is an initiative of the Canada's International Development Research Centre (IDRC); South Africa's National Research Foundation (NRF); and the Department of Science and Innovation (DSI), in partnership with the Oliver and Adelaide Tambo Foundation (OATF) and National Science and Technology Council (NSTC) of Zambia.

Data Availability Statement: Publicly available datasets were analysed in this review. The main datasets can be found here: [https://minamataconvention.org/en/parties/minamata-initial-assessments; https://minamataconvention.org; https://wedocs.unep.org/handle/20.500.11822/29830). More supporting data sources are available in the literature cited in this review.

Acknowledgments: We acknowledge the Oliver R. Tambo Africa Research Chair Initiative (ORTARChI) of the Copperbelt University for the administrative and logistical support during the preparation of the review. The findings and conclusions in this publication are those of the authors and should not be construed to represent any official position of the organizations that funded or supported the study.

Conflicts of Interest: The authors declare no conflict of interest.

References

1. Driscoll, C.T.; Mason, R.P.; Chan, H.M.; Jacob, D.J.; Pirrone, N. Mercury as a Global Pollutant: Sources, Pathways, and Effects. *Environ. Sci. Technol.* **2013**, *47*, 4967–4983. [CrossRef] [PubMed]
2. Mitra, S.; Chakraborty, A.J.; Tareq, A.M.; Emran, T.B.; Nainu, F.; Khusro, A.; Idris, A.M.; Khandaker, M.U.; Osman, H.; Alhumaydhi, F.A.; et al. Impact of Heavy Metals on the Environment and Human Health: Novel Therapeutic Insights to Counter the Toxicity. *J. King Saud Univ. Sci.* **2022**, *34*, 101865. [CrossRef]
3. Teng, H.; Altaf, A.R. Journal of Hazardous Materials Advances Elemental Mercury (Hg 0) Emission, Hazards, and Control: A Brief Review. *J. Hazard. Mater. Adv.* **2022**, *5*, 100049. [CrossRef]
4. Park, J.; Zheng, W. Human Exposure and Health Effects of Inorganic and Elemental Mercury. *J. Prev. Med. Public Health* **2012**, *45*, 344–352. [CrossRef] [PubMed]
5. Streets, D.G.; Horowitz, H.M.; Jacob, D.J.; Lu, Z.; Levin, L.; Ter Schure, A.F.H.; Sunderland, E.M. Total Mercury Released to the Environment by Human Activities. *Environ. Sci. Technol.* **2017**, *51*, 5969–5977. [CrossRef] [PubMed]
6. Marrugo-Negrete, J.; Verbel, J.O.; Ceballos, E.L.; Benitez, L.N. Total Mercury and Methylmercury Concentrations in Fish from the Mojana Region of Colombia. *Environ. Geochem. Health* **2008**, *30*, 21–30. [CrossRef]
7. Harika, V.K.; Penki, T.R.; Loukya, B.; Samanta, A.; Xu, G.; Sun, C.; Grinberg, I.; Deepak, F.L.; Amine, K.; Aurbach, D.; et al. Sustainable Existence of Solid Mercury Nanoparticles at Room Temperature and Their Applications. *Chem. Sci.* **2021**, *12*, 3226–3238. [CrossRef]
8. Sawicka, B.; Umachandran, K.; Fawzy, M.; Mahmoud, A.E.D. 27—Impacts of Inorganic/Organic Pollutants on Agroecosystems and Eco-Friendly Solutions. In *Egbuna the Military and Health*; Mtewa, A.G., Egbuna, C., Eds.; Elsevier: Amsterdam, The Netherlands, 2021; pp. 523–552. ISBN 978-0-12-821556-2.
9. Zhang, W.; Zhang, X.; Tian, Y.; Zhu, Y.; Tong, Y.; Li, Y.; Wang, X. Risk Assessment of Total Mercury and Methylmercury in Aquatic Products from Offshore Farms in China. *J. Hazard. Mater.* **2018**, *354*, 198–205. [CrossRef]
10. Fleck, J.A.; Marvin-DiPasquale, M.; Eagles-Smith, C.A.; Ackerman, J.T.; Lutz, M.A.; Tate, M.; Alpers, C.N.; Hall, B.D.; Krabbenhoft, D.P.; Eckley, C.S. Mercury and Methylmercury in Aquatic Sediment across Western North America. *Sci. Total Environ.* **2016**, *568*, 727–738. [CrossRef]
11. Kalisińska, E.; Łanocha-Arendarczyk, N.; Kosik-Bogacka, D.I. Mercury, Hg. In *Mammals and Birds as Bioindicators of Trace Element Contaminations in Terrestrial Environments*; Kalisińska, E., Ed.; Springer International Publishing: Cham, Switzerland, 2019; pp. 593–653. ISBN 978-3-030-00121-6.
12. de Almeida Rodrigues, P.; Ferrari, R.G.; dos Santos, L.N.; Conte Junior, C.A. Mercury in Aquatic Fauna Contamination: A Systematic Review on Its Dynamics and Potential Health Risks. *J. Environ. Sci.* **2019**, *84*, 205–218. [CrossRef]
13. Celo, V.; Lean, D.R.S.; Scott, S.L. Abiotic Methylation of Mercury in the Aquatic Environment. *Sci. Total Environ.* **2006**, *368*, 126–137. [CrossRef] [PubMed]
14. La Colla, N.S.; Botté, S.E.; Marcovecchio, J.E. Mercury Cycling and Bioaccumulation in a Changing Coastal System: From Water to Aquatic Organisms. *Mar. Pollut. Bull.* **2019**, *140*, 40–50. [CrossRef] [PubMed]
15. Regnell, O.; Watras, C.J. Microbial Mercury Methylation in Aquatic Environments—A Critical Review of Published Field and Laboratory Studies Published Field and Laboratory Studies. *Environ. Sci. Technol.* **2018**, *53*, 4–19. [CrossRef] [PubMed]
16. UNEP. *Global Mercury Assessment*; UNEP: Nairobi, Kenya, 2018.
17. Amoatey, P.; Baawain, M.S. Effects of Pollution on Freshwater Aquatic Organisms. *Water Environ. Res.* **2019**, *91*, 1272–1287. [CrossRef] [PubMed]
18. Amundsen, P.; Henriksson, M.; Poste, A.; Prati, S.; Power, M. Ecological Drivers of Mercury Bioaccumulation in Fish of a Subarctic Watercourse. *Environ. Toxicol. Chem.* **2023**, *42*, 873–887. [CrossRef] [PubMed]
19. Cuellar-Valencia, O.M.; Murillo-García, O.E.; Rodriguez-Salazar, G.A.; Bolívar-García, W. Bioaccumulation of Mercury in Direct-Developing Frogs: The Aftermath of Illegal Gold Mining in a National Park. *Herpetol. J.* **2023**, *33*, 6–13. [CrossRef]
20. Kraus, J.M.; Wanty, R.B.; Schmidt, T.S.; Walters, D.M.; Wolf, R.E. Variation in Metal Concentrations across a Large Contamination Gradient Is Reflected in Stream but Not Linked Riparian Food Webs. *Sci. Total Environ.* **2021**, *769*, 144714. [CrossRef]
21. Lino, A.S.; Kasper, D.; Guida, Y.S.; Thomaz, J.R.; Malm, O. Total and Methyl Mercury Distribution in Water, Sediment, Plankton and Fish along the Tapajós River Basin in the Brazilian Amazon. *Chemosphere* **2019**, *235*, 690–700. [CrossRef]
22. Niane, B.; Moritz, R.; Guédron, S.; Ngom, P.M.; Pfeifer, H.R.; Mall, I.; Poté, J. Effect of Recent Artisanal Small-Scale Gold Mining on the Contamination of Surface River Sediment: Case of Gambia River, Kedougou Region, Southeastern Senegal. *J. Geochem. Explor.* **2014**, *144*, 517–527. [CrossRef]
23. Niane, B.; Devarajan, N.; Poté, J.; Moritz, R. Quantification and Characterization of Mercury Resistant Bacteria in Sediments Contaminated by Artisanal Small-Scale Gold Mining Activities, Kedougou Region, Senegal. *J. Geochem. Explor.* **2019**, *205*, 106353. [CrossRef]
24. Streets, D.G.; Horowitz, H.M.; Lu, Z.; Levin, L.; Thackray, C.P.; Sunderland, E.M. Five Hundred Years of Anthropogenic Mercury: Spatial and Temporal Release Profiles. *Environ. Res. Lett.* **2019**, *14*, 084004. [CrossRef]
25. Streets, D.G.; Horowitz, H.M.; Lu, Z.; Levin, L.; Thackray, C.P.; Sunderland, E.M. Global and Regional Trends in Mercury Emissions and Concentrations, 2010–2015. *Atmos. Environ.* **2019**, *201*, 417–427. [CrossRef]
26. Seccatore, J.; Veiga, M.; Origliasso, C.; Marin, T.; De Tomi, G. An Estimation of the Artisanal Small-Scale Production of Gold in the World. *Sci. Total Environ.* **2014**, *496*, 662–667. [CrossRef] [PubMed]

27. Hilson, G. Farming, Small-Scale Mining and Rural Livelihoods in Sub-Saharan Africa: A Critical Overview. *Extr. Ind. Soc.* **2016**, *3*, 547–563. [CrossRef]
28. Al-Sulaiti, M.M.; Soubra, L.; Al-Ghouti, M.A. The Causes and Effects of Mercury and Methylmercury Contamination in the Marine Environment: A Review. *Curr. Pollut. Rep.* **2022**, *8*, 249–272. [CrossRef]
29. Wang, Y.; Xie, Q.; Xu, Q.; Xue, J.; Zhang, C.; Wang, D. Mercury Bioaccumulation in Fish in an Artificial Lake Used to Carry out Cage Culture. *J. Environ. Sci.* **2019**, *78*, 352–359. [CrossRef] [PubMed]
30. Calao-Ramos, C.; Bravo, A.G.; Paternina-Uribe, R.; Marrugo-Negrete, J.; Díez, S. Occupational Human Exposure to Mercury in Artisanal Small-Scale Gold Mining Communities of Colombia. *Environ. Int.* **2021**, *146*, 106216. [CrossRef] [PubMed]
31. Taux, K.; Kraus, T.; Kaifie, A. Mercury Exposure and Its Health Effects in Workers in the Artisanal and Small-Scale Gold Mining (ASGM) Sector—A Systematic Review. *Int. J. Environ. Res. Public Health* **2022**, *19*, 81. [CrossRef]
32. Mallongi, A.; Parkpian, P.; Pataranawat, P.; Chinwetkitvanich, S. Mercury Distribution and Its Potential Environmental and Health Risks in Aquatic Habitat at Artisanal Buladu Gold Mine in Gorontalo Province, Indonesia. *Pak. J. Nutr.* **2015**, *14*, 1010–1025. [CrossRef]
33. Meutia, A.A.; Bachriadi, D.; Gafur, N.A. Environment Degradation, Health Threats, and Legality at the Artisanal Small-Scale Gold Mining Sites in Indonesia. *Int. J. Environ. Res. Public Health* **2023**, *20*, 6774. [CrossRef]
34. Barkdull, N.M.; Carling, G.T.; Rey, K.; Yudiantoro, D.F. Comparison of Mercury Contamination in Four Indonesian Watersheds Affected by Artisanal and Small-Scale Gold Mining of Varying Scale. *Water. Air. Soil Pollut.* **2019**, *230*, 214. [CrossRef]
35. Hilson, G. The Environmental Impact of Small-Scale Gold Mining in Ghana. *Geogr. J.* **2002**, *168*, 57. [CrossRef]
36. Donkor, A.K.; Bonzongo, J.-C.J.; Nartey, V.K.; Adotey, D.K. Heavy Metals in Sediments of the Gold Mining Impacted Pra River Basin, Ghana, West Africa. *Soil Sediment Contam. Int. J.* **2005**, *14*, 479–503. [CrossRef]
37. Donkor, A.K.; Bonzongo, J.C.; Nartey, V.K.; Adotey, D.K. Mercury in Different Environmental Compartments of the Pra River Basin, Ghana. *Sci. Total Environ.* **2006**, *368*, 164–176. [CrossRef] [PubMed]
38. Telmer, K.H.; Veiga, M.M. World Emissions of Mercury from Artisanal and Small—Scale Gold Mining. In *Mercury Fate and Transport in the Global Environment: Emissions, Measurements and Models*; Mason, R., Pirrone, N., Eds.; Springer US: Boston, MA, USA, 2009; pp. 131–172. [CrossRef]
39. Yoshimura, A.; Suemasu, K.; Veiga, M.M. Estimation of Mercury Losses and Gold Production by Artisanal and Small-Scale Gold Mining (ASGM). *J. Sustain. Metall.* **2021**, *7*, 1045–1059. [CrossRef]
40. Kortei, N.K.; Heymann, M.E.; Essuman, E.K.; Kpodo, F.M.; Akonor, P.T.; Lokpo, S.Y.; Boadi, N.O.; Ayim-Akonor, M.; Tettey, C. Health Risk Assessment and Levels of Toxic Metals in Fishes (*Oreochromis noliticus* and *Clarias anguillaris*) from Ankobrah and Pra Basins: Impact of Illegal Mining Activities on Food Safety. *Toxicol. Rep.* **2020**, *7*, 360–369. [CrossRef]
41. Basta, P.C.; de Vasconcellos, A.C.S.; Hallwass, G.; Yokota, D.; Pinto, D.d.O.d.R.; de Aguiar, D.S.; de Souza, C.C.; Oliveira-da-Costa, M. Risk Assessment of Mercury-Contaminated Fish Consumption in the Brazilian Amazon: An Ecological Study. *Toxics* **2023**, *11*, 800. [CrossRef]
42. Barbieri, F.L.; Gardon, J. Hair Mercury Levels in Amazonian Populations: Spatial Distribution and Trends. *Int. J. Health Geogr.* **2009**, *8*, 71. [CrossRef]
43. UNEP. *Minamata Convention on Mercury*; UNEP: Nairobi, Kenya, 2013.
44. Anan, T.; Toda, E. Analysis of National Priorities from Minamata Initial Assessments Minamata Convention on Mercury. In Proceedings of the 14th International Conference on Mercury as a Global Pollutant (ICMGP 2019), Krakow, Poland, 8–13 September 2019.
45. van Straaten, P. Mercury Contamination Associated with Small-Scale Gold Mining in Tanzania and Zimbabwe. *Sci. Total Environ.* **2000**, *259*, 105–113. [CrossRef]
46. Nuwagira, U.; Mubiru, D.; Yasin, I.; Nasasira, P. Impact of Artisanal Gold Mining on Wetland Health in Buhweju District, Southwestern Uganda. *East Afr. J. Environ. Nat. Resour.* **2023**, *6*, 297–310. [CrossRef]
47. Fayiga, A.O.; Ipinmoroti, M.O.; Chirenje, T. Environmental Pollution in Africa. *Environ. Dev. Sustain.* **2018**, *20*, 41–73. [CrossRef]
48. Abdoul, K.A.M.; Alassane, Y.A.K.; Alphonse, S.A.; Emmanuel, A.; Olivier, D.; Daouda, M.; Dominique, C.K.S. Mercury Pollution in Africa: A Review. *J. Environ. Chem. Ecotoxicol.* **2022**, *14*, 32–49. [CrossRef]
49. Yabe, J.; Ishizuka, M.; Umemura, T. Current Levels of Heavy Metal Pollution in Africa. *J. Vet. Med. Sci.* **2010**, *72*, 1257–1263. [CrossRef] [PubMed]
50. Walters, C.R.; Somerset, V.S.; Leaner, J.J.; Nel, J.M. A Review of Mercury Pollution in South Africa: Current Status. *J. Environ. Sci. Health Part A* **2011**, *46*, 1129–1137. [CrossRef] [PubMed]
51. Campbell, L.M.; Dixon, D.G.; Hecky, R.E. A Review of Mercury in Lake Victoria, East Africa: Implications for Human and Ecosystem Health. *J. Toxicol. Environ. Health Part B* **2003**, *6*, 325–356. [CrossRef] [PubMed]
52. Issah Musah-Surugu, J.; Ahenkan, A.; Bawole, J.N.; Yeboah-Assiamah, E. Rural Poverty and Artisanal Mining in Sub-Saharan Africa. *Int. J. Rural Manag.* **2017**, *13*, 162–181. [CrossRef]
53. Barasa, B.; Kakembo, V.; Karl, T. Characterization of Artisanal Gold Mining Activities in the Tropics and Their Impact on Sediment Loading and Stream Flow in the Okame River Catchment, Eastern Uganda. *Environ. Earth Sci.* **2016**, *75*, 1076. [CrossRef]
54. Omara, T.; Karungi, S.; Kalukusu, R.; Nakabuye, B.V.; Kagoya, S.; Musau, B. Mercuric Pollution of Surface Water, Superficial Sediments, *Nile tilapia* (*Oreochromis nilotica* Linnaeus 1758 [Cichlidae]) and Yams (*Dioscorea alata*) in Auriferous Areas of Namukombe Stream, Syanyonja, Busia, Uganda. *PeerJ* **2019**, *7*, e7919. [CrossRef]

55. Basooma, A.; Teunen, L.; Semwanga, N.; Bervoets, L. Trace Metal Concentrations in the Abiotic and Biotic Components of River Rwizi Ecosystem in Western Uganda, and the Risks to Human Health. *Heliyon* **2021**, *7*, e08327. [CrossRef]
56. Campbell, L.M.; Hecky, R.E.; Muggide, R.; Dixon, D.G.; Ramlal, P.S. Variation and Distribution of Total Mercury in Water, Sediment and Soil from Northern Lake Victoria, East Africa. *Biogeochemistry* **2003**, *65*, 195–211. [CrossRef]
57. Ikingura, J.R.; Akagi, H. Monitoring of Fish and Human Exposure to Mercury Due to Gold Mining in the Lake Victoria Goldfields, Tanzania. *Sci. Total Environ.* **1996**, *191*, 59–68. [CrossRef] [PubMed]
58. Ikingura, J.R.; Mutakyahwa, M.K.D.; Kahatano, J.M.J. Mercury and Mining in Africa with Special Reference to Tanzania. *Water Air Soil Pollut.* **1997**, *97*, 223–232. [CrossRef]
59. Ikingura, J.R.; Akagi, H.; Mujumba, J.; Messo, C. Environmental Assessment of Mercury Dispersion, Transformation and Bioavailability in the Lake Victoria Goldfields, Tanzania. *J. Environ. Manag.* **2006**, *81*, 167–173. [CrossRef] [PubMed]
60. Chibunda, R.T. Comparative Sensitivity of *Caridina nilotica, Haplochromis nubilus, Bulinus africanus* and *Bulinus forskalii* from Lake Victoria, Tanzania to Mercury Chloride. *Chem. Ecol.* **2008**, *24*, 207–212. [CrossRef]
61. Taylor, H.; Appleton, J.D.; Lister, R.; Smith, B.; Chitamweba, D.; Mkumbo, O.; Machiwa, J.F.; Tesha, A.L.; Beinhoff, C. Environmental Assessment of Mercury Contamination from the Rwamagasa Artisanal Gold Mining Centre, Geita District, Tanzania. *Sci. Total Environ.* **2005**, *343*, 111–133. [CrossRef]
62. Ogola, J.S.; Mitullah, W.V.; Omulo, M.A. Impact of Gold mining on the Environment and Human Health: A Case Study in the Migori Gold Belt, Kenya. *Environ. Geochem. Health* **2002**, *24*, 141–157. [CrossRef]
63. Odumo, O.B.; Mustapha, A.O.; Patel, J.P.; Angeyo, H.K. Multielemental Analysis of Migori (Southwest, Kenya) Artisanal Gold Mine Ores and Sediments by EDX-ray Fluorescence Technique: Implications of Occupational Exposure and Environmental Impact. *Bull. Environ. Contam. Toxicol.* **2011**, *86*, 484–489. [CrossRef]
64. Ngure, V.; Davies, T.; Kinuthia, G.; Sitati, N.; Shisia, S.; Oyoo-Okoth, E. Concentration Levels of Potentially Harmful Elements from Gold Mining in Lake Victoria Region, Kenya: Environmental and Health Implications. *J. Geochem. Explor.* **2014**, *144*, 511–516. [CrossRef]
65. Odumo, B.O.; Carbonell, G.; Angeyo, H.K.; Patel, J.P.; Torrijos, M.; Rodríguez Martín, J.A. Impact of Gold Mining Associated with Mercury Contamination in Soil, Biota Sediments and Tailings in Kenya. *Environ. Sci. Pollut. Res.* **2014**, *21*, 12426–12435. [CrossRef]
66. Tampushi, L.L.; Onyari, J.M.; Muthama, N.J. Environmental Distribution and Risk of Exposure of Heavy Metal Pollutants from Lolgorian Artisanal Gold Mining in Kenya. *Bull. Environ. Contam. Toxicol.* **2022**, *109*, 310–316. [CrossRef]
67. Ondayo, M.A.; Watts, M.J.; Hamilton, E.M.; Mitchell, C.; Mankelow, J.; Osano, O. Artisanal Gold Mining in Kakamega and Vihiga Counties, Kenya: Potential Human Exposure and Health Risk. *Environ. Geochem. Health* **2023**, *45*, 6543–6565. [CrossRef] [PubMed]
68. Abdelrahman, M.S. *How Mercury Is Poisoning a Nation and Gross Mismanagement Is Aggravating the Problem*; Sudan's Gold Curse Briefing Paper No. 1; Sudan Transparency Tracker: Khartoum, Sudan, 2022.
69. Yan, J.; Li, R.; Ali, M.U.; Wang, C.; Wang, B.; Jin, X.; Shao, M.; Li, P.; Zhang, L.; Feng, X. Mercury Migration to Surface Water from Remediated Mine Waste and Impacts of Rainfall in a Karst Area—Evidence from Hg Isotopes. *Water Res.* **2023**, *230*, 119592. [CrossRef] [PubMed]
70. Ahmed, A.; Purwanto, P.; Sunoko, H. Consequences of Mercury Used by Artisanal and Small-Scale Gold Mining Processes—A Case of River Nile State Sudan. *J. Ecol. Eng.* **2019**, *20*, 106–115. [CrossRef]
71. Ali, M.; Elhagwa, A.; Elfaki, J. An Investigation of Mercury Distribution in the Soils around Gold Mining Area at Dar-Mali Locality, River Nile State, Sudan. *Eurasian J. Soil Sci.* **2018**, *7*, 365–372. [CrossRef]
72. Mubarak, E.A.T.; Ali, A.A. Determination of Some Heavy Metals Content in the Body of Two Popular Fish Species *O. niloticus* and *L. niloticus*, in Lake Nubia, Wadi Halfa, Sudan. *J. Aquac. Mar. Biol.* **2020**, *9*, 170–175. [CrossRef]
73. Oosthuizen, M.A.; John, J.; Somerset, V. Mercury Exposure in a Low-Income Community in South Africa. *S. Afr. Med. J.* **2010**, *100*, 366–371. [CrossRef]
74. van Rooyen, D.; Erasmus, J.H.H.; Gerber, R.; Nachev, M.; Sures, B.; Wepener, V.; Smit, N.J.J. Bioaccumulation and Trophic Transfer of Total Mercury through the Aquatic Food Webs of an African Sub-Tropical Wetland System. *Sci. Total Environ.* **2023**, *889*, 164210. [CrossRef]
75. Dabrowski, J.M.; Ashton, P.J.; Murray, K.; Leaner, J.J.; Mason, R.P. Anthropogenic Mercury Emissions in South Africa: Coal Combustion in Power Plants. *Atmos. Environ.* **2008**, *42*, 6620–6626. [CrossRef]
76. Pacyna, E.G.; Pacyna, J.M.; Steenhuisen, F.; Wilson, S. Global Anthropogenic Mercury Emission Inventory for 2000. *Atmos. Environ.* **2006**, *40*, 4048–4063. [CrossRef]
77. Kading, T.J.; Mason, R.P.; Leaner, J.J. Mercury Contamination History of an Estuarine Floodplain Reconstructed from a 210Pb-Dated Sediment Core (Berg River, South Africa). *Mar. Pollut. Bull.* **2009**, *59*, 116–122. [CrossRef]
78. Govaerts, A.; Verhaert, V.; Covaci, A.; Jaspers, V.L.B.; Berg, O.K.; Addo-Bediako, A.; Jooste, A.; Bervoets, L. Distribution and Bioaccumulation of POPs and Mercury in the Ga-Selati River (South Africa) and the Rivers Gudbrandsdalslågen and Rena (Norway). *Environ. Int.* **2018**, *121*, 1319–1330. [CrossRef] [PubMed]
79. Lusilao-Makiese, J.G.; Cukrowska, E.M.; Tessier, E.; Amouroux, D.; Weiersbye, I. The Impact of Post Gold Mining on Mercury Pollution in the West Rand Region, Gauteng, South Africa. *J. Geochem. Explor.* **2013**, *134*, 111–119. [CrossRef]
80. Lusilao-Makiese, J.G.; Tessier, E.; Amouroux, D.; Tutu, H.; Chimuka, L.; Weiersbye, I.; Cukrowska, E.M. Seasonal Distribution and Speciation of Mercury in a Gold Mining Area, North-West Province, South Africa. *Toxicol. Environ. Chem.* **2014**, *96*, 387–402. [CrossRef]

81. Lusilao-Makiese, J.G.; Tessier, E.; Amouroux, D.; Tutu, H.; Chimuka, L.; Weiersbye, I.; Cukrowska, E.M. Mercury Speciation and Dispersion from an Active Gold Mine at the West Wits Area, South Africa. *Environ. Monit. Assess.* **2016**, *188*, 47. [CrossRef] [PubMed]
82. Verhaert, V.; Teuchies, J.; Vlok, W.; Wepener, V.; Addo-Bediako, A.; Jooste, A.; Blust, R.; Bervoets, L. Bioaccumulation and Trophic Transfer of Total Mercury in the Subtropical Olifants River Basin, South Africa. *Chemosphere* **2019**, *216*, 832–843. [CrossRef] [PubMed]
83. Spiegel, S.J.; Veiga, M.M. Building Capacity in Small-Scale Mining Communities: Health, Ecosystem Sustainability, and the Global Mercury Project. *Ecohealth* **2005**, *2*, 361–369. [CrossRef]
84. Green, C.S.; Lewis, P.J.; Wozniak, J.R.; Drevnick, P.E.; Thies, M.L. A Comparison of Factors Affecting the Small-Scale Distribution of Mercury from Artisanal Small-Scale Gold Mining in a Zimbabwean Stream System. *Sci. Total Environ.* **2019**, *647*, 400–410. [CrossRef]
85. Dalu, T.; Dube, T.; Dondofema, F.; Cuthbert, R.N. Illegal Mining Impacts on Freshwater Potamonautid Crab in a Subtropical Austral Highland Biosphere Reserve. *Sci. Total Environ.* **2023**, *896*, 165251. [CrossRef]
86. Makaure, J.; Dube, T.; Stewart, D.; Razavi, N.R. Mercury Exposure in Two Fish Trophic Guilds from Protected and ASGM-Impacted Reservoirs in Zimbabwe and Possible Risks to Human Health. *Arch. Environ. Contam. Toxicol.* **2023**, *84*, 199–213. [CrossRef]
87. UNEP. *Environmental Assessment of Mercury Pollution in Two Artisanal Gold Mining Sites in Eastern Democratic Republic of the Congo*; UNEP: Nairobi, Kenya, 2016.
88. Shandro, J.A.; Veiga, M.M.; Chouinard, R. Reducing Mercury Pollution from Artisanal Gold Mining in Munhena, Mozambique. *J. Clean. Prod.* **2009**, *17*, 525–532. [CrossRef]
89. Esdaile, L.J.; Chalker, J.M. The Mercury Problem in Artisanal and Small-Scale Gold Mining. *Chem. A Eur. J.* **2018**, *24*, 6905–6916. [CrossRef] [PubMed]
90. Bashir, I.; Lone, F.A.; Bhat, R.A.; Mir, S.A.; Dar, Z.A.; Dar, S.A. Concerns and Threats of Contamination on Aquatic Ecosystems. In *Bioremediation and Biotechnology*; Springer International Publishing: Cham, Switzerland, 2020; pp. 1–26. ISBN 9783030356910.
91. Mandeng, E.P.B.; Bidjeck, L.M.B.; Bessa, A.Z.E.; Ntomb, Y.D.; Wadjou, J.W.; Doumo, E.P.E.; Dieudonné, L.B. Contamination and Risk Assessment of Heavy Metals, and Uranium of Sediments in Two Watersheds in Abiete-Toko Gold District, Southern Cameroon. *Heliyon* **2019**, *5*, e02591. [CrossRef] [PubMed]
92. Pascal, N.M.; Kiteba, S.; Dieudonne, M.; Jean-Noel, M. Physicochemical Characterization of the Waters of Nine Rivers Draining the FIZI Gold Panning Areas in South Kivu: Environmental Impact Study. *Int. J. Eng. Appl. Sci.* **2020**, *7*, 29. [CrossRef]
93. Pascal, N.M.; Dieudonné, M.E.; Jean-Noël, M.K. Evaluation of the Level of Mercury Pollution in the Sediments of the Rivers Draining the Gold Panning Sites in the Territory of Fizi, Eastern Democratic Republic of Congo. *J. Geosci. Environ. Prot.* **2020**, *08*, 97–111. [CrossRef]
94. Ngueyep, M.L.L.; Kingni, K.S.; Ngounouno, N.M.; Ndi, A.A. The Impact of Gold Mining Exploitation on the Physicochemical Quality of Water: Case of Batouri (Cameroon). *Int. J. Energy Water Resour.* **2021**, *5*, 159–173. [CrossRef]
95. Ngounouno, M.A.; Ngueyep, L.L.M.; Kingni, S.T.; Nforsoh, S.N.; Ngounouno, I. Evaluation of the Impact of Gold Mining Activities on the Waters and Sediments of Lom River, Wakaso, Cameroon and the Restorative Effect of Moringa Oleifera Seeds. *Appl. Water Sci.* **2021**, *11*, 113. [CrossRef]
96. Veiga, M.M.; Angeloci-Santos, G.; Meech, J.A. Review of Barriers to Reduce Mercury Use in Artisanal Gold Mining. *Extr. Ind. Soc.* **2014**, *1*, 351–361. [CrossRef]
97. Bella Atangana, M.; Ndam Ngoupayou, J.; Deliege, J.-F. Hydrogeochemistry and Mercury Contamination of Surface Water in the Lom Gold Basin (East Cameroon): Water Quality Index, Multivariate Statistical Analysis and Spatial Interpolation. *Water* **2023**, *15*, 2502. [CrossRef]
98. Ayiwouo, M.N.; Yamgouot, F.N.; Ngueyep Mambou, L.L.; Kingni, S.T.; Ngounouno, I. Impact of Gold Mining on the Water Quality of the Lom River, Gankombol, Cameroon. *Heliyon* **2022**, *8*, e12452. [CrossRef]
99. Rajaee, M.; Long, R.; Renne, E.; Basu, N. Mercury Exposure Assessment and Spatial Distribution in A Ghanaian Small-Scale Gold Mining Community. *Int. J. Environ. Res. Public Health* **2015**, *12*, 10755–10782. [CrossRef]
100. Affum, A.O.; Dede, S.O.; Nyarko, B.J.B.; Acquaah, S.O.; Kwaansa-Ansah, E.E.; Darko, G.; Dickson, A.; Affum, E.A.; Fianko, J.R. Influence of Small-Scale Gold Mining and Toxic Element Concentrations in Bonsa River, Ghana: A Potential Risk to Water Quality and Public Health. *Environ. Earth Sci.* **2016**, *75*, 178. [CrossRef]
101. Obiri-Yeboah, A.; Nyantakyi, E.K.; Mohammed, A.R.; Yeboah, S.I.I.K.; Domfeh, M.K.; Abokyi, E. Assessing Potential Health Effect of Lead and Mercury and the Impact of Illegal Mining Activities in the Bonsa River, Tarkwa Nsuaem, Ghana. *Sci. Afr.* **2021**, *13*, e00876. [CrossRef]
102. Tulasi, D.; Fajon, V.; Kotnik, J.; Shlyapnikov, Y.; Adotey, D.K.; Serfor-Armah, Y.; Horvat, M. Mercury Methylation in Cyanide Influenced River Sediments: A Comparative Study in Southwestern Ghana. *Environ. Monit. Assess.* **2021**, *193*, 180. [CrossRef] [PubMed]
103. Persaud, A.W.; Telmer, K.H.; Costa, M.; Moore, M.L. Artisanal and Small-Scale Gold Mining in Senegal: Livelihoods, Customary Authority, and Formalization. *Soc. Nat. Resour.* **2017**, *30*, 980–993. [CrossRef]

104. Niane, B.; Guédron, S.; Feder, F.; Legros, S.; Ngom, P.M.; Moritz, R. Impact of Recent Artisanal Small-Scale Gold Mining in Senegal: Mercury and Methylmercury Contamination of Terrestrial and Aquatic Ecosystems. *Sci. Total Environ.* **2019**, *669*, 185–193. [CrossRef] [PubMed]
105. Gerson, J.R.; Driscoll, C.T.; Hsu-Kim, H.; Bernhardt, E.S. Senegalese Artisanal Gold Mining Leads to Elevated Total Mercury and Methylmercury Concentrations in Soils, Sediments, and Rivers. *Elementa* **2018**, *6*, 11. [CrossRef]
106. Idowu, O.S.; Muruf, A.K.; Osaguona, P.; Ajayi, J. Mercury Contamination in Artisanal Gold Mining Area of Manyera River, Niger State Nigeria. *E3 J. Environ. Res. Manag.* **2013**, *4*, 326–0333.
107. Adewumi, A.J.; Laniyan, T.A. Contamination, Sources and Risk Assessments of Metals in Media from Anka Artisanal Gold Mining Area, Northwest Nigeria. *Sci. Total Environ.* **2020**, *718*, 137235. [CrossRef]
108. Sani, A.H.; Amanabo, M.; Achimugu, M.D. Assessment of Heavy Metal Pollution of Drinking Water Sources and Staple Food Cultivars around Artisanal Mining Site in Igade-Mashegu, Niger State, Nigeria. *World J. Biol. Pharm. Health Sci.* **2023**, *14*, 306–319. [CrossRef]
109. Kawakami, T. Diffusin of Mercury from Artisanal Small-Scale Gold Mining (ASGM) Sites in Myanmar. *Int. J. GEOMATE* **2019**, *17*, 228–235. [CrossRef]
110. Akagi, H.; Castillo, E.S.; Cortes-Maramba, N.; Francisco-Rivera, A.T.; Timbang, T.D. Health Assessment for Mercury Exposure among Schoolchildren Residing near a Gold Processing and Refining Plant in Apokon, Tagum, Davao Del Norte, Philippines. *Sci. Total Environ.* **2000**, *259*, 31–43. [CrossRef] [PubMed]
111. Murphy, T.P.; Irvine, K.N.; Sampson, M.; Guo, J.; Parr, T. Mercury Contamination along the Mekong River, Cambodia. *Asian J. Water Environ. Pollut.* **2008**, *6*, 1–9. [CrossRef]
112. Abrahan, M.; Diana, J.F.; Jürgen, M. Levels of MN, ZN, PB and HG in Sediments of the Zamora River, Ecuador. *Rev. Int. Contam. Ambient.* **2018**, *34*, 245–249.
113. Bastos, W.R.; Dórea, J.G.; Bernardi, J.V.; Lauthartte, L.C.; Mussy, M.H.; Lacerda, L.D.; Malm, O. Mercury in fish of the Madeira river (temporal and spatial assessment), Brazilian Amazon. *Environ. Res.* **2015**, *140*, 191–197. [CrossRef] [PubMed]
114. Garcia-Sanchez, A.; Contreras, F.; Adams, M.; Santos, F. Mercury Contamination of Surface Water and Fish in a Gold Mining Region (Cuyuní River Basin, Venezuela). *Int. J. Environ. Pollut.* **2008**, *33*, 260–274. [CrossRef]
115. Hacon, S.d.S.; Oliveira-da-Costa, M.; Gama, C.d.S.; Ferreira, R.; Basta, P.C.; Schramm, A.; Yokota, D. Mercury Exposure through Fish Consumption in Traditional Communities in the Brazilian Northern Amazon. *Int. J. Environ. Res. Public Health* **2020**, *17*, 5269. [CrossRef]
116. da Silva, S.F.; Oliveira, D.C.; Pereira, J.P.G.; Castro, S.P.; Costa, B.N.S.; de Oliveira Lima, M. Seasonal Variation of Mercury in Commercial Fishes of the Amazon Triple Frontier, Western Amazon Basin. *Ecol. Indic.* **2019**, *106*, 105549. [CrossRef]
117. Moreno-Brush, M.; Rydberg, J.; Gamboa, N.; Storch, I.; Biester, H. Is Mercury from Small-Scale Gold Mining Prevalent in the Southeastern Peruvian Amazon? *Environ. Pollut.* **2016**, *218*, 150–159. [CrossRef]
118. Ali, M.; Hery, S.; Putri, S.T. Mercury Toxicity Potential from Artisanal and Small—Scale Gold Mines in Lebong Regency, Bengkulu Province. *E3S Web Conf.* **2018**, *73*, 06002. [CrossRef]
119. Rajaee, M.; Obiri, S.; Green, A.; Long, R.; Cobbina, S.J.; Nartey, V.; Buck, D.; Antwi, E.; Basu, N. Integrated Assessment of Artisanal and Small-Scale Gold Mining in Ghana—Part 2: Natural Sciences Review. *Int. J. Environ. Res. Public Health* **2015**, *12*, 8971–9011. [CrossRef]
120. Pataranawat, P.; Parkpian, P.; Polprasert, C.; Delaune, R.D.; Jugsujinda, A. Mercury Emission and Distribution: Potential Environmental Risks at a Small-Scale Gold Mining Operation, Phichit Province, Thailand. *J. Environ. Sci. Health—Part A Toxic/Hazard. Subst. Environ. Eng.* **2007**, *42*, 1081–1093. [CrossRef] [PubMed]
121. Ramírez Requelme, M.E.; Ramos, J.F.F.; Angélica, R.S.; Brabo, E.S. Assessment of Hg-Contamination in Soils and Stream Sediments in the Mineral District of Nambija, Ecuadorian Amazon (Example of an Impacted Area Affected by Artisanal Gold Mining). *Appl. Geochem.* **2003**, *18*, 371–381. [CrossRef]
122. da Penha Rhodes, V.; de Lena, J.C.; Santolin, C.V.A.; da Silva Pinto, T.; Mendes, L.A.; Windmöller, C.C. Speciation and Quantification of Hg in Sediments Contaminated by Artisanal Gold Mining in the Gualaxo Do Norte River, Minas Gerais, SE, Brazil. *Environ. Monit. Assess.* **2018**, *190*, 49. [CrossRef] [PubMed]
123. de Almeida Ferreira, C.M.; Egler, S.G.; Yallouz, A.V.; Ignácio, Á.R.A. Semiquantitative Determination of Total Mercury in Pygocentrus Nattereri Kner, 1858 and Sediment at the Plateau of Upper Paraguai River, Brazil. *Chemosphere* **2017**, *174*, 604–612. [CrossRef] [PubMed]
124. Hilson, G. "Creating" Rural Informality: The Case of Artisanal Gold Mining in Sub-Saharan Africa. *SAIS Rev. Int. Aff.* **2013**, *33*, 51–64. [CrossRef]
125. Hilson, G. Small-Scale Mining, Poverty and Economic Development in Sub-Saharan Africa: An Overview. *Resour. Policy* **2009**, *34*, 1–5. [CrossRef]
126. Hilson, G. Artisanal Mining, Small Holder Farming and Livelihood Diversificatio in Rural Sub-Saharan Africa: An Introduction. *J. Int. Dev.* **2011**, *23*, 1031–1041. [CrossRef]
127. Schwartz, F.W. A Review of the Scope of Artisanal and Small-Scale Mining Worldwide, Poverty, and the Associated Health Impacts. *GeoHealth* **2021**, *5*, e2020GH000325. [CrossRef]
128. Mkodzongi, G. Artisanal and Small-Scale Gold Mining and Rural Transformation in Post-Land Reform Zimbabwe: A Broad Overview. *J. Rural Stud.* **2023**, *100*, 103027. [CrossRef]

129. Hilson, G.; Goumandakoye, H.; Diallo, P. Land Use Policy Formalizing Artisanal Mining 'Spaces' in Rural Sub-Saharan Africa: The Case of Niger. *Land Use Policy* **2019**, *80*, 259–268. [CrossRef]
130. Mkodzongi, G. The Extractive Industries and Society The Rise of 'Mashurugwi' Machete Gangs and Violent Conflicts in Zimbabwe's Artisanal and Small-Scale Gold Mining Sector. *Extr. Ind. Soc.* **2020**, *7*, 1480–1489. [CrossRef]
131. WHO. *Environmental and Occupational Health Hazards Associated with Artisanal and Small-Scale Gold Mining*; WHO: Geneva, Switzerland, 2016.
132. Steckling, N.; Tobollik, M.; Plass, D.; Hornberg, C.; Ericson, B.; Fuller, R.; Bose-O'Reilly, S. Global Burden of Disease of Mercury Used in Artisanal Small-Scale Gold Mining. *Ann. Glob. Health* **2017**, *83*, 234–247. [CrossRef] [PubMed]
133. Shoko, D.S.M. Small-Scale Mining and Alluvial Gold Panning within the Zambezi Basin: An Ecological Time Bomb and Tinderbox for Future Conflicts among Riparian States. *Bloom. Businessweek* **2012**, *10*, 12–14.
134. Hilson, G. *Artisanal and Small-Scale Mining and Agriculture Exploring Their Links in Rural Sub-Saharan Africa*; International Institute for Environment and Development, JSTOR: London, UK, 2016.
135. Ncube-Phiri, S.; Ncube, A.; Mucherera, B.; Ncube, M. Artisanal Small-Scale Mining: Potential Ecological Disaster in Mzingwane District, Zimbabwe. *Jàmbá J. Disaster Risk Stud.* **2015**, *7*, 158. [CrossRef] [PubMed]
136. Mhangara, P.; Tsoeleng, L.T.; Mapurisa, W. Monitoring the Development of Artisanal Mines in South Africa. *J. S. Afr. Inst. Min. Metall.* **2020**, *120*, 299–306. [CrossRef] [PubMed]
137. Girard, V.; Molina-Millán, T.; Vic, G. *Artisanal Mining in Africa*; NOVAFRICA Working Paper; NOVAFRICA: Carcavelos, Portugal, 2022.
138. Fisher, J.A.; Schneider, L.; Fostier, A.-H.; Guerrero, S.; Guimarães, J.R.D.; Labuschagne, C.; Leaner, J.J.; Martin, L.G.; Mason, R.P.; Somerset, V.; et al. A Synthesis of Mercury Research in the Southern Hemisphere, Part 2: Anthropogenic Perturbations. *Ambio* **2023**, *52*, 918–937. [CrossRef] [PubMed]
139. Hsu-Kim, H.; Eckley, C.S.; Achá, D.; Feng, X.; Gilmour, C.C.; Jonsson, S.; Mitchell, C.P.J. Challenges and Opportunities for Managing Aquatic Mercury Pollution in Altered Landscapes. *Ambio* **2018**, *47*, 141–169. [CrossRef] [PubMed]
140. Hall, S.; Bradley, T.; Moore, J.; Kuykindall, T.; Minella, L. Acute and Chronic Toxicity of Nano-Scale TiO_2 Particles to Freshwater Fish, Cladocerans, and Green Algae, and Effects of Organic and Inorganic Substrate on TiO_2 Toxicity. *Nanotoxicology* **2009**, *3*, 91–97. [CrossRef]
141. Bodaly, R.A.; Jansen, W.A.; Majewski, A.R.; Fudge, R.J.P.; Strange, N.E.; Derksen, A.J.; Green, D.J. Post-impoundment Time Course of Increased Mercury Concentrations in Fish in Hydroelectric Reservoirs of Northern Manitoba, Canada. *Arch. Environ. Contam. Toxicol.* **2007**, *53*, 379–389. [CrossRef]
142. Baptista-Salazar, C.; Quadra, G.R.; Sobek, A.; Jonsson, S. Insights into the Factors Influencing Mercury Concentrations in Tropical Reservoir Sediments. *Environ. Sci. Process. Impacts* **2021**, *23*, 1542–1553. [CrossRef]
143. Selin, H. Global Environmental Law and Treaty-Making on Hazardous Substances: The Minamata Convention and Mercury Abatement. *Glob. Environ. Polit.* **2014**, *14*, 1–19. [CrossRef]
144. Keane, S.; Bernaudat, L.; Davis, K.J.; Stylo, M.; Mutemeri, N.; Singo, P.; Twala, P.; Mutemeri, I.; Nakafeero, A.; Etui, I.D. Mercury and Artisanal and Small-Scale Gold Mining: Review of Global Use Estimates and Considerations for Promoting Mercury-Free Alternatives. *Ambio* **2023**, *52*, 833–852. [CrossRef] [PubMed]
145. Poloko, N. Physical Separation Methods, Part 1: A Review. *IOP Conf. Ser. Mater. Sci. Eng.* **2019**, *641*, 012023. [CrossRef]
146. AMDC. *Artisanal and Small–Scale Mining in Africa: Selected Countries Policy Profile Review on ASM*; Africa Mining Development Centre/African Union: Addis Ababa, Ethiopia, 2017.
147. Drace, K.; Kiefer, A.M.; Veiga, M.M.; Williams, M.K.; Ascari, B.; Knapper, K.A.; Logan, K.M.; Breslin, V.M.; Skidmore, A.; Bolt, D.A.; et al. Mercury-Free, Small-Scale Artisanal Gold Mining in Mozambique: Utilization of Magnets to Isolate Gold at Clean Tech Mine. *J. Clean. Prod.* **2012**, *32*, 88–95. [CrossRef]
148. NEMA. *The National Action Plan for Artisanal and Small–Scale Gold Mining in Uganda*; National Environmental Management Authority NEMA: Kampala, Uganda, 2019.
149. Babut, M.; Sekyi, R.; Rambaud, A.; Potin-Gautier, M.; Tellier, S.; Bannerman, W.; Beinhoff, C. Improving the Environment Management of Small-Scale Gold Mining in Ghana: A Case Study of Dumasi. *J. Clean. Prod.* **2003**, *11*, 215–221. [CrossRef]
150. Styles, M.T.; Amankwah, R.K.; Al-Hassan, S.; Nartey, R.S. The Identification and Testing of a Method for Mercury-Free Gold Processing for Artisanal and Small-Scale Gold Miners in Ghana. *Int. J. Environ. Pollut.* **2010**, *41*, 289. [CrossRef]

water

MDPI

Article

Study of the Spatiotemporal Variations, Source Determination, and Potential Ecological Risk of Organophosphate Esters in Typical Coastal Tourist Resorts in China

Xuezhi Cao [1], Bingbing Wang [2], Xinxin Liu [2], Jichun Cheng [1] and Shiliang Wang [2,*]

1 School of History and Culture, Qufu Normal University, Qufu 273165, China; hjxyjz@126.com (X.C.); chengjcqf@163.cn (J.C.)
2 School of Life Science, Qufu Normal University, Qufu 273165, China
* Correspondence: wangshiliang@tsinghua.org.cn

Abstract: Investigation of the environmental occurrence and behavior of organophosphate esters (OPEs) is very important and is becoming a hot topic in the academic community. In this study, 12 targeted OPEs in the water and sediment from 19 coastal tourist resorts in the Shandong Peninsula in China were analyzed to show their concentrations, spatial variations, and potential ecological risks. The results showed that the total concentrations of OPEs (ΣOPEs) were in the range of 18.52–3069.43 ng/L in the water and 3.20–568.76 ng/g in the sediment. The dominant OPEs in the water and sediment were tris (2-chloroisopropyl) phosphate (TCIPP), tris (2-chloroethyl) phosphate (TCEP), and triethyl phosphate (TEP). The OPE concentrations in the water were ranked as dry season > normal season > wet season. The sewage treatment plants near tourist resorts were important contributors to the level of OPEs in the water. Triphenyl phosphate (TPHP), tri-n-butyl phosphate (TNBP), and resorcinol-bis(diphenyl)phosphate (RDP) had a relatively higher ecological risk than other OPEs in the water samples. Industrial emissions might be the main source of OPEs in the coastal tourist resorts of Shandong Peninsula in China. The results of this study verified that OPEs occur in the water and sediment of coastal tourist resorts, and more attention should be given to the existence of OPEs and the safety of aquatic environments near coastal tourist resorts.

Keywords: organophosphate esters; spatial distribution; ecological risk assessment; coastal tourist resorts

Citation: Cao, X.; Wang, B.; Liu, X.; Cheng, J.; Wang, S. Study of the Spatiotemporal Variations, Source Determination, and Potential Ecological Risk of Organophosphate Esters in Typical Coastal Tourist Resorts in China. *Water* **2023**, *15*, 3976. https://doi.org/10.3390/w15223976

Academic Editors: Weiying Feng, Fang Yang and Jing Liu

Received: 4 October 2023
Revised: 9 November 2023
Accepted: 14 November 2023
Published: 16 November 2023

1. Introduction

There are many types of organophosphate esters (OPEs), such as chlorinated (Cl)-OPEs, alkyl-OPEs, aryl-OPEs, and brominated (Br)-OPEs [1,2]. They are used for a wide variety of consumer goods, including plastics, foams, textiles, electronics, and construction materials. Chlorinated OPEs are mostly utilized as flame retardants, whereas non-chlorinated OPEs are predominantly used as plasticizers [3,4]. OPEs were initially employed due to the restricted use of brominated flame retardants (BFRs), as they are more environmentally friendly and cost-effective [5]. Furthermore, OPEs have also been extensively utilized as additives [6]; they are often added to materials physically rather than combined with chemical substances and can be released into the environment by volatilization, leaching, and other similar methods [4]. OPEs are ubiquitous and widely distributed worldwide in various environmental matrices, including in the air, dust [3], water [7], sediment [8], soil [9], and biota [10]. In recent years, the focus of research has also shifted from investigating their levels in various environmental media to their effects on human health [11,12]. Numerous studies have indicated that exposure to OPEs may have negative effects on human health, such as reproductive inhibition, neurotoxicity, carcinogenicity, and endocrine disruption [13]. More thorough studies of their source and fate in the environment are required.

OPEs have been found in surface water all around the world. Additional research from the USA, Australia, and Europe, including Germany, Italy, and Spain, and Asia, including South Korea and Japan, revealed that the mean value of OPEs was in the range of 76–2230 ng/L, and Cl-OPEs were predominant [4]. OPEs have also been detected in Chinese waters, primarily in Taihu Lake and the Pearl River Delta [14,15]. The water in the Pearl River Delta contained in the range of 15–1790 ng/L, the water in Taihu Lake had concentrations in the range of 166–1530 ng/L, and the sediment in the lake had 2.8–47.5 ng/g [14,15]. Eight OPEs have been found in European river sediments, ranging from 2.5181 ng/g [16]. Fourteen OPEs were found in sediment from the North American Great Lakes, ranging from 0.44 to 47.8 ng/g [17].

The warm temperature and curved coastline of the Shandong Peninsula have proven ideal for the development of coastal tourism. Shandong Peninsula has the largest number of national and provincial coastal tourist resorts in China. However, many environmental issues have also arisen as a result of the rapid development of coastal tourist resorts. It is essential to protect the environmental quality of water bodies during the further development of coastal tourist resorts. Numerous man-made substances have been released into the Bohai Sea as a result of extensive anthropogenic activity along the shore, including urban and industrial refuse [18,19], which may have an immediate effect on human health. Thus, a more in-depth investigation of the contamination status and bioaccumulation of OPEs in the environment needs to be carried out.

The main purpose of this study was to (i) evaluate the presence and spatial distribution of 12 OPEs in the surface water and sediment of coastal tourist resorts, (ii) quantify the partitioning of OPEs between the water and sediment, (iii) analyze the seasonal correlations, the correlation of OPE levels in surface water and sediment, and the correlation between the physical and chemical characteristics of OPEs and their concentration, and (iv) identify potential sources of target OPEs in sampling sites; and (v) evaluate the ecological risk posed by OPEs around typical coastal tourist resorts in China.

2. Materials and Methods

2.1. Sample Collection

The water samples (S1–S19) were taken in August 2018 (wet season), December 2019 (dry season), and April 2019 (normal season) from coastal tourist resorts of the Shandong Peninsula. The sediment sampling locations were the same as the water sampling locations and were taken in December 2019. The sampling stations are shown in Figure 1 and summarized in Tables S1 and S2.

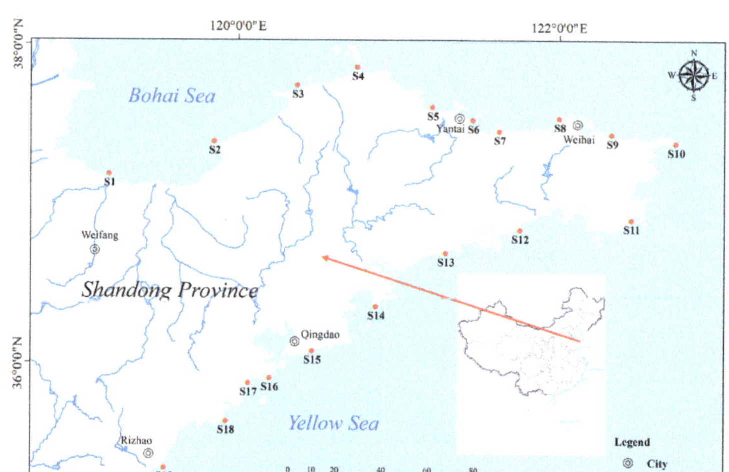

Figure 1. Sampling location map in the coastal tourist resorts of Shandong Peninsula.

A stainless-steel water sampler was used to simultaneously capture two 1 L water samples at each location. Before laboratory analysis, all samples were stored in brown glass jars that had been cleaned and then were refrigerated immediately after collection. To obtain more representative samples, at each monitoring site, surface water samples were taken five times and held in stainless-steel drums. Then, 1 L of the combined water sample was obtained, transferred into PP bottles, and kept frozen for no longer than 12 h. Milli-Q water in PP bottles was utilized as water sample blanks and was preserved identically to the samples. Stainless-steel grab samplers with a 0.2 m^2 surface area and 30 cm depth were used to gather sediment samples. Utilizing a previously cleaned stainless-steel scoop, sediment samples were collected and put into aluminum storage containers. All samples were kept at a temperature of −20 °C until further analysis.

2.2. Chemicals and Materials

Twelve OPEs were purchased from the Sigma-Aldrich (St. Louis, MO, USA), for which detailed information such as the CAS number, formula, molecular weight, Log K_{ow}, Log K_{oc}, Log P_{ow}, boiling point, water solubility, and application are listed in Tables S3 and S4. Before instrumental analysis, isotopically labeled internal standards were purchased from Toronto Research Chemicals Inc. in Canada and Sigma-Aldrich in the UK and were employed as stand-ins, as listed in Table S5. The target OPEs in this research were classified into three categories, tris (2-chloroethyl) phosphate (TCEP), and tris (2-chloroisopropyl) phosphate (TCIPP), tris (1,3-dichloro-2-propyl) phosphate (TDCIPP), and tetrakis (2-chlorethyl) dichloroisopentyl-diphosphate (V6) were classified as Cl-OPEs, bisphenol A diphenylphosphate (BDP), resorcinol-bis (diphenyl) phosphate (RDP), triphenyl phosphate (TPHP), and tris (methylphenyl) phosphate (TMPP) were classified as aryl-OPEs, and tris (2-chloroethyl) phosphate (TCEP), tripropyl phosphate (TPP), tris (2-butoxyethyl) phosphate (TBOPE), and tri-n-butyl phosphate (TNBP) were classified as alkyl-OPEs. The purity of the standard materials exceeded 98%.

The following chemicals were purchased from J.T. Baker (Darmstadt, Germany): HPLC-grade acetonitrile, methanol, n-hexane, dichloromethane, and ethyl acetate. Formic acid was purchased from CNW Technologies in Düsseldorf, Germany, with a purity of 99%. Oasis HLB cartridges were purchased (500 g, 6 mL) from Waters Company in the Milford, MA, USA and were used to perform solid phase extraction. Glass fiber filters were purchased (0.22 μm) from Whatman Company in the Maidstone, UK. Ultrapure water (18.2 MΩ cm) was generated by the Milli-Q Advantage A10 system from the Millipore Company in the Billerica, MA, USA.

2.3. Pretreatment and Analytical Procedure

The methods of water sample pretreatment were consistent with previous investigations [20–22]. The method was as follows: 1 L of water was first filtered through a 0.22 μm glass fiber filter, after which 20 ng of standard substitution was added to the filtered sample to remove solids such as particles. Next, using a vacuum pump, the sample was passed slowly, at a speed of 3 mL/min, through an Oasis HLB cartridge, which was pretreated with 6 mL of ultrapure water and 6 mL of methanol. Vacuum drying was carried out for 1 h after the extraction process to load the water sample completely. This was followed by three additions of 2 mL of dichloromethane and ethyl acetate (1:1, *v*/*v*). Subsequently, the eluate was concentrated to a dry state by passing through a gentle stream of nitrogen to a fixed volume of 1 mL with methanol/ultrapure water (1:3). Finally, the extract was transferred to a chromatography vial for instrumental analysis.

The technique outlined by [23] was used to extract OPEs from sediment. After being freeze-dried, sediment samples were run through a 60-mesh filter (0.25 mm) before 20 ng of the deuterated surrogate standard was added to 10 g of homogenous sediment and cultured overnight. The stainless-steel extractor was filled with the mixture, which had been prepared with 2 g of copper powder and 5 g of silica gel for purification. We used n-hexane: acetone (1:1, *v*/*v*) rapid solvent extraction to extract the samples. According to [24], extraction conditions were as follows: the pressure extraction of 1500 psi, rinse volume of 60%, continuous nitrogen purge time of 60 s, temperature extraction of 100 °C, and two extraction cycles. The extracted substances were dried, and the remaining material was restructured with 1 mL of methanol/ultrapure water (1:3) during analysis.

2.4. Instrumental Analysis

The OPE concentrations of water and sediment were determined by an ultra-performance liquid chromatography-tandem mass spectrometry (UPLC-MS/MS) system (an Agilent 1290 Infinity LC with an ABSCIEX QTRAP 5500 triple quadruple mass spectrometry LC-MS/MS system) (Agilent Technologies, Inc., Santa Clara, CA, USA). Comprehensive information on the UPLC-MS/MS analytical parameters is given in Table S5. A Waters Xbridge BEH-C18 XP column (4.6 mm × 100 mm, 2.5 μm) from Waters Corporation (Milford, MA, USA) preserved by a Waters Xbridge BEH-C18 XP VanGuard pre-column (2.1 mm × 5 mm, 2.5 μm) from Waters Corporation (Milford, MA, USA) was used to analyze samples. Until they were injected into the instrument, the samples were kept at 10 °C. A partial loop was used to inject 10 μL of the sample once every 5 min at 45 °C. In this study, ultrapure water acidified with 0.1% formic acid was used as mobile phase A, and acetonitrile was used as mobile phase B for chromatographic separation. The flow rate was set at 0.2 mL/min. The gradient progressed as follows: 40% B for 0.5 min; 40–80% B for 3.5 min; 80% B for 2.0 min; 80–100% B for 2.0 min; 100% B for 4.5 min, back to 40% B for 1.0 min, and held for 2.5 min. A positive ion pattern electrospray ionization (ESI +/−) source in multiple reaction monitoring (MRM) patterns was employed for mass spectrometry analysis. In addition, the pH value and salinity of the water and total organic content (TOC) were also measured, and the results are shown in Table S6.

2.5. Quality Assurance and Control

Blanks of field and laboratory procedures were conducted for each batch of analyses. To reduce the probability of contamination, plastic and rubber materials were not used for sampling, sample storage, or sample transportation. The glass was cleaned in methanol and pure water before being baked for four hours at 450 °C. For the water samples, in addition to the samples themselves, field blanks and laboratory procedure blanks were calculated, and for the sediment samples, only laboratory procedure blanks were used.

Strict control measures were required during sample collection and analysis for quality assurance. For any batch of analysis, field blank and laboratory procedure blank were both required to be performed. All glassware used in the laboratory needs to be baked overnight at 450 °C again to reduce contamination during the experiment. Vessels were cleaned three times with ultrapure water, acetone, and ethyl acetate prior to the sample-collection procedure. The findings showed that there was background contamination. TEP, TNBP, TCIPP, TCEP, and TDCIPP were the primary OPEs with background concentrations in the field blanks for the water samples, as indicated in Table S7. Additionally, TBOPE, TEP, TDCIPP, and TCIPP were found in lab-related blanks. Background contamination of sediment samples was rare. A combination of internal and external standard methods was used in the analysis for quality control. The internal standards are given in Table S5 and were added before instrumental analysis. The Marking Line, encompassing 6 concentrations with correlation coefficients higher than 0.993, was obtained to quantify the target compounds. The concentrations corresponding to signal/noise ratios of 3 and 10, respectively, were designated as the limit of detection (LOD) and the limit of quantification (LOQ) [20]. The average blank plus the triple standard deviation of the procedure blanks was used to define the LOD of a chemical [25]. The LOQ was the analyte concentration that corresponded to the sample blank value plus 10 standard deviations [25]. The LODs of target OPEs in sediment and water were in the range of 0.01–0.32 ng/g and 0.01–0.35 ng/L, respectively. The LOQs of target OPEs in water and sediment were between 0.02–0.51 ng/L and 0.01–0.67 ng/g, respectively. To evaluate the extraction rates and procedures, known concentrations of the 12 OPEs standards and surrogate standards were generated concurrently with the genuine samples and were introduced into the blank samples prior to extraction. Before conducting instrumental analysis using the recovery determination method, the inter-standard chemicals specified in Table S8 were added to the samples for quantization. For water and sediment, the recovery of the 12 OPEs was 70–112% and 61–112%, respectively.

2.6. Datum Analysis

The pollutant distribution between water and sediment, particularly through secondary releases of organic contaminants from sediments, was assumed to have a substantial effect on the quality of the aquatic environment [26,27]. Therefore, the distribution coefficient (K_d) of OPEs between water and sediment needed to be calculated. A significant factor that modifies the affinity of OPE molecules for sorbents is the amount of organic carbon in the substance [28,29]. Therefore, the normalized organic carbon/water partition coefficient (K_{oc}) of the OPEs needed to be calculated.

$$K_d = (C_s \times 1000)/C_w$$

$$K_{oc} = (K_d \times 100)/\mathrm{TOC}$$

K_d was utilized to examine how OPEs transformed between the water and sediment in the tourist vacation districts of the Shandong Peninsula. K_{oc} was calculated from the values of K_d and TOC. TOC was the percentage of organic carbon in the sediment (%).

In the correlation analysis, SPSS 16.0 (SPSS Inc., Chicago, IL, USA) and Origin2021 (OriginLab Inc., Northampton, MA, USA) were used to perform the correlation analysis and principal component analysis (PCA). The correlation analysis included the correlation between the scientific properties of the 12 OPEs and their concentrations, the correlation between seasons, and the correlation of the OPE concentrations in surface water and sediment. In the PCA study, the eigenvalue had to be bigger than 1 to meet the PC extraction criteria.

2.7. Ecological Risk Assessment

According to some previous studies [7,24], the risk evaluation was based on the risk quotient (*RQ*), which was calculated by the measured environmental concentration (*MEC*). The MEC for individual OPE derived from the data of measurement was the average concentration. In the current investigation, the median effective concentration (EC50) and median lethal concentration (LC50) obtained from [30] were used to define the predicted no-effect concentration (*PNEC*). Three trophic levels (fish, algae, and aquatic invertebrates) were taken into consideration for the examination of the *RQ* values of the OPEs. The following equations were used to determine the *RQ* values of the OPEs [31]:

$$\mathrm{RQ} = \mathrm{MEC}/\mathrm{PNEC}$$

$$\mathrm{PNEC} = \mathrm{EC(L)50}/\mathrm{AF}$$

where *AF* represents an assessment factor of 1000 when short-term toxicity data are used to derive *PNEC* [30].

The OPEs represented no ecological risk when the *RQ* was less than 0.01. OPEs exhibited a low risk when $0.01 \leq RQ < 0.10$. OPEs exhibited a moderate risk when $0.10 \leq RQ < 1$. OPEs exhibited a high risk when $RQ \geq 1$ [32].

Since it was uncommon for a sample to contain a single pollutant, the ecotoxicological impact of OPE mixes was evaluated in accordance with the concentration-added effects [33]. The following equations were used to determine the *RQ* values of the mixtures:

$$RQ_{mix,algae} = \sum_{i=1}^{n} \frac{MEC_i}{EC(L)50_{i,\ algae}} \times AF$$

$$RQ_{mix,aquatic\ invertebrate} = \sum_{i=1}^{n} \frac{MEC_i}{EC(L)50_{i,aquatic\ invertebrate}} \times AF$$

$$RQ_{mix,fish} = \sum_{i=1}^{n} \frac{MEC_i}{EC(L)50_{i,fish}} \times AF$$

The *RQ* values of mixes in water samples for fish, algae, and aquatic invertebrates were represented by the $RQ_{mix,fish}$, $RQ_{mix,algae}$, and $RQ_{mix,aquatic\ invertebrates}$, respectively. The compound concentration in the mixed water samples was called the MEC. The median effective (deadly) concentrations of the ingredients in mixed samples of fish, algae, and aquatic invertebrates were $EC(L)50_{i,fish}$, $EC(L)50_{i,algae}$, and $EC(L)50_{i,aquatic\ invertebrates}$, respectively.

Additionally, based on sediment (marine water) *PNEC* and marine water *PNEC* from the previous study [30], an ecological risk assessment for the 12 OPEs was also carried out.

2.8. Data Analysis

Statistical analyses of this study were conducted with IBM SPSS Statistics v20. Students' *t*-test was used to assess the variance of the OPE concentrations and partition coefficients at different sampling stations. Principal component analysis (PCA) was used with the software SIMCA 13.0 to investigate possible sources of OPEs.

3. Results and Discussion

3.1. Concentration and Composition of OPEs in the Surface Water

The OPEs detected in surface water from coastal tourist resorts in the Shandong Peninsula are shown in Figure 2, and the concentration information is summarized in Table S9. All target OPEs were detected in surface water, demonstrating their prevalence in an aquatic environment. The aggregate concentrations of the 12 OPEs ranged from 18.52 to 3069.43 ng/L, with an average value of 561.39 ng/L. The majority of the OPEs had a high detection frequency (DF); their DFs were in the range of 26–100%. In all water samples, higher concentrations of Cl-OPEs and alkyl-OPEs were found in comparison to aryl-OPEs. However, the concentration of TPP was much lower. The ∑OPE concentrations in surface water were 3744.68 ng/L (the maximum value was 361.42 ng/L, the minimum value was 48.37 ng/L, and the average value was 197.09 ng/L), 2028.63 ng/L (the maximum value was 210.07 ng/L, the minimum value was 26.38 ng/L, and the average value was 106.77 ng/L), and 693.32 ng/L (the maximum value was 105.34 ng/L, the minimum value was 14.02 ng/L, and the average value was 50.70 ng/L) during the dry, normal, and wet season, respectively.

The concentration range of Cl-OPEs, alkyl-OPEs, and aryl-OPEs during the dry season was 0.94–204.53, 0.13–77.47, and 0.28–10.51 ng/L, respectively. The concentration range of Cl-OPEs, alkyl-OPEs, and aryl-OPEs during the normal season was 0.34–98.66, 0.33–98.66, and 0.10–4.66 ng/L, respectively. The concentration range of Cl-OPEs, alkyl-OPEs, and aryl-OPEs during the wet season was 0.11–66.24, 0.18–26.20, and 0.17–2.16 ng/L, respectively. These results revealed that the levels of all OPEs exhibited obvious seasonal variations, showing the highest level in the dry season and the lowest level in the wet season. Seasonal variations of OPEs in the present study were different from the results observed by a previous study that found organic contaminants in the wet season were generally higher than in other seasons because of rainfall erosion [34]. In this study, the high river flow and the heavy rainfall in the wet season might decrease the concentrations of OPEs in water [21]. The correlation analysis of OPE concentrations in the three seasons is shown in Figure S1. The different compositions and temporal fluctuations of the target OPEs can be seen. However, some compounds, such as TPP and all aryl-OPEs (TPHP, TMPP, BDP, and RDP), were not detectable in the wet season, which might be led by the water dilution or the solar irradiation degradation in the wet season [35].

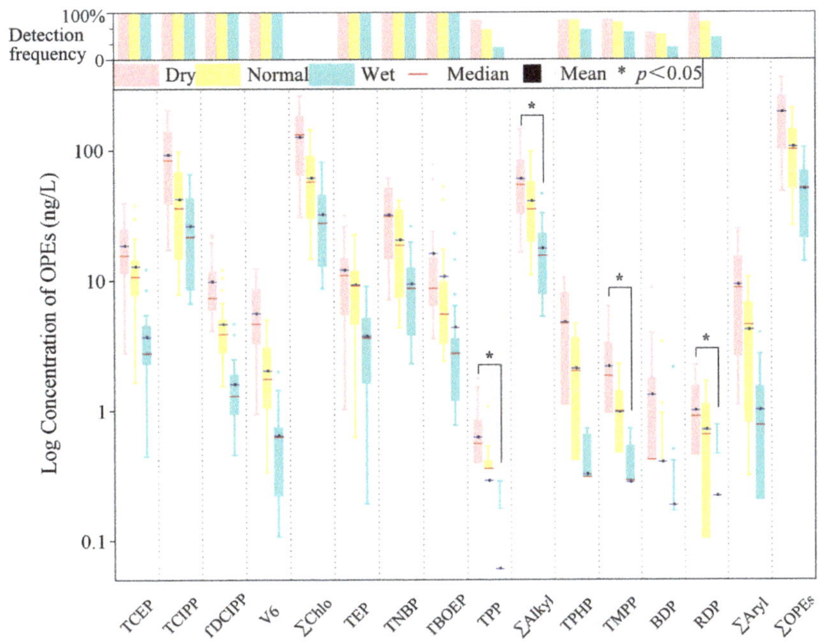

Figure 2. The concentrations of OPEs in surface water from coastal tourist resorts.

The percentage of OPEs varied from season to season. The percentage of target OPEs in surface water for each of the three seasons is shown in Figure 3. Cl-OPEs were the dominant compound in the water samples, while the content of alkyl-OPEs was lower, and the aryl-OPE occurred the least. Cl-OPEs accounted for 64.5%, 57.9%, and 63.4% of the total concentration during the dry, normal and wet seasons, respectively. Likewise, the alkyl-OPEs were accounting for 30.8%, 38.2%, and 34.6%, respectively. Aryl-OPE accounted for 4.7%, 3.9%, and 2.0%, respectively. Therefore, the percentages of the three types of OPEs were ranked as Cl-OPEs > alkyl-OPEs > aryl-OPEs across seasons. These results may have been related to the fact that Cl-OPEs are persistent, with a half-life of 8.6–21.3, with higher water solubility (C_{wsat}) and lower octanol-water partitioning coefficients (K_{ow}). In addition, in the surface water of the Shandong Peninsula, substantial amounts of TCIPP are typically found. This might be related to the fact that TCIPP is widely used in the production of polyurethane foam and polyvinyl chloride (PVC) plastic as an important additive. Moreover, the high water solubility of TCEP (7000 mg/L) and TCIPP (1600 mg/L) might facilitate their migration with runoff and diffusion into surface water.

The average levels of OPEs observed in this study were compared with other regions in the world, such as Asia, Europe, and North America. The OPE concentration in this study was often lower than at the coast near Dalian [36] and the Yellow Sea [37], especially the Northern Yellow Sea [38], and was below the value of the East China Sea [38], Northern South China Sea [38], Pearl River Estuary [39], Greater Bay area [40], Hong Kong [14], Tokyo Bay [14], Thermaikos Gulf [41], Marseille Bay [42], and San Francisco Bay [43]. These results implied that the water in the coastal tourist resorts of the Shandong Peninsula in China contained low levels of OPEs. However, high levels of OPEs or some of their homologs were observed in the water of some resorts. Because coastal tourism resorts have higher requirements for seawater quality, the quality monitoring of the water coming from the surrounding rivers should be strengthened.

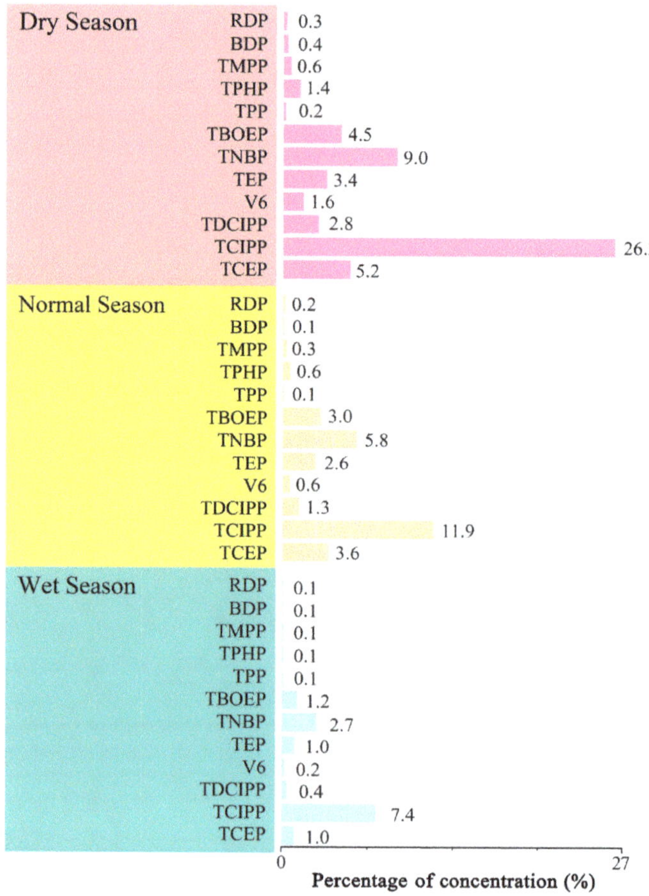

Figure 3. The percentage of OPE concentrations in the surface water during three seasons.

In conclusion, apparent temporal changes in OPEs were found in the seawater of coastal tourist resorts in the Shandong Peninsula. The average concentrations of OPEs were in the order of dry season > normal season > wet season. The average levels of the three types of OPEs were ordered as Cl-OPEs > alkyl-OPEs > aryl-OPEs. TCIPP had the highest level, and TPP exhibited the lowest level.

3.2. *Concentrations and Composition of OPEs in Sediment*

The concentration and DF of OPEs in sediment from coastal tourist resorts in the Shandong Peninsula are shown in Figure 4, and the concentration information is summarized in Table S10. All the target OPEs, with DF values ranging from 53% to 100%, were discovered in sediments. Individual OPEs for TCEP, TCIPP, TDCIPP, TEP, TNBP, TBOEP, TPP, and RDP had the highest DF (75–100%), followed by TPHP, TMPP, and BDP (60–75%). The DF of V6 was under 55%. In comparison to aryl-OPEs, Cl-OPEs, and alkyl-OPEs were generally found more often. Cl-OPEs accounted for the highest level of OPEs in sediment, at 62.7% of all OPEs, followed by alkyl-OPEs (31.7%) and aryl-OPEs (5.6%). The most prevalent substances in sediment were TCIPP and TEP; together, they accounted for nearly 71% of the ∑OPE concentrations. Compared to alkyl-OPEs and aryl-OPEs, Cl-OPEs were generally more commonly found due to increased usage or in relation to the TOC content and salinity in the sediments of the coastal areas. The Pearson correlation analysis revealed that TCEP, TEP, TPHP, and TMPP were correlated with TOC ($p \leq 0.05$); V6, TNBP, and TPP were negatively correlated with TOC ($p \leq 0.05$); and the others were not correlated with TOC ($p \geq 0.05$). In addition, we found that TPHP, TMPP, and TDCIPP were significantly correlated with salinity ($p \leq 0.05$), BDP was correlated with salinity but not significantly ($p \leq 0.05$), TCIPP and RDP were negatively correlated with salinity, and the others were not correlated with salinity ($p \geq 0.05$). This suggested that the content of OPEs in the sediment was not significantly influenced by TOC and salinity but mainly by the emission source and their transport in water. The correlation of OPE concentrations in sediment with salinity and TOC is displayed in Figure S2.

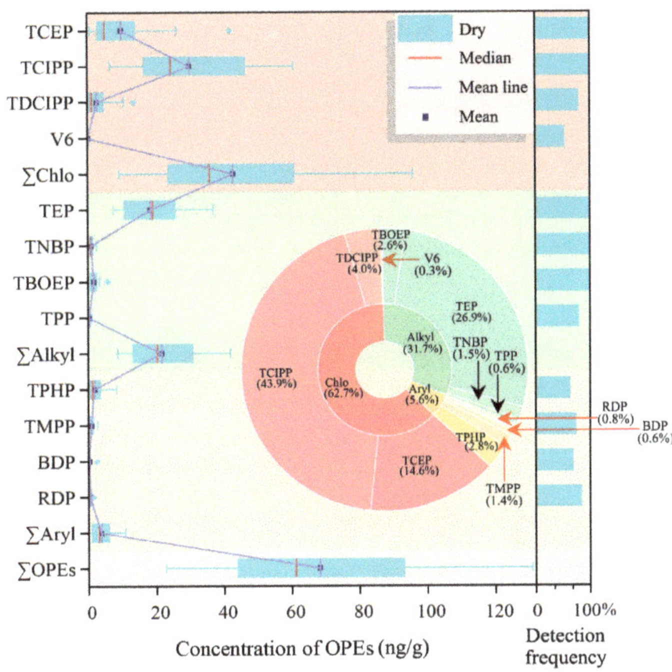

Figure 4. The concentrations of OPEs in sediment from the coastal tourist resorts.

During the dry season, the range of ∑OPE concentration values in sediments from the coastal tourist resorts of the Shandong Peninsula was 3.20–568.76 ng/g, with a median and average of 26.89 ng/g and 107.95 ng/g. TCIPP and TEP had higher average concentrations at 29.93 and 18.36 ng/g, respectively. Other OPEs had mean values less than 10 ng/g. Cl-OPEs levels were the highest among the three types of OPEs, with a mean of 42.78 ng/g. Alkyl-OPE and aryl-OPE were the next highest, with a mean of 21.6 and 3.8 ng/g, respectively. The content of OPEs in the sediment was significantly less than in the water.

Compared with other coastal sediments of Asia, the OPE content of the Shandong Peninsula was lower than that of the Liao River Estuary [44], Yellow River Estuary [45], Jiaozhou Bay [29], Pearl River Delta [46], and Korean coast [47]. The OPE content of this study was higher than the Yellow Sea, except for V6, which at 0.17 ng/g was lower than in the Yellow Sea [48]. On the contrary, the OPE content of this study was lower than the East China Sea, except for TCIPP at 29.93 ng/g [49]. Compared with the coastal sediments of Europe, the OPE content of this study was lower than the Gulf of Lion [50]. The OPE content of this study was lower than in Marseille Bay, except TCIPP at 29.93 ng/g [42]. Compared with the coastal sediments of North America, the OPE content of this study was lower than the Palos Verdes Shelf (PVS), US [51].

In conclusion, the order of the different types of OPE concentrations in the sediment of the Shandong Peninsula coastal tourist resorts from high to low was Cl-OPEs, alkyl-OPEs, and aryl-OPEs, which was the same as the OPE concentrations in surface water. TCIPP levels were the highest, and V6 levels were the lowest.

3.3. Spatial Distribution of OPEs in the Shandong Coastal Tourist Resorts

For OPEs in the water of various vacation resorts across the three seasons, apparent spatial differences were also noted, as shown in Figure 5. For instance, the OPE levels of sites S13, S16, and S17 ranged from more than 325 ng/L during the dry season to around 200 ng/L during the normal season and less than 106 ng/L during the wet season. The highest OPE concentration was recorded at the S16 site in Qingdao, at 361.42, 210.07, and 105.34 ng/L in the dry, normal, and wet seasons, respectively. The major OPE at site S16 was TCIPP, which had concentrations of 204.53, 98.66, and 66.24 ng/L in the dry, normal, and wet seasons, respectively. Site S16 and S17 were located at the entrance to Jiaozhou Bay, near the Yellow Sea. Qingdao, in Shandong Province, is a highly industrialized and urbanized city with a thriving metallurgical industry, shipbuilding industry, print industry, automotive, electroplate industry, battery, and machine manufacturing industry. Jiaozhou Bay is affected by pollution discharges from Qingdao and its surrounding areas, resulting in the high OPE concentrations at these two locations. Therefore, industrial emissions might be the main factor leading to the high level of OPEs at site S16. During the three seasons, Site S10 had the lowest concentration of OPEs. This location is surrounded by natural areas and has a healthy ecological setting. It is far from urban centers and rarely suffers direct damage from industrial pollutants. On the other hand, TCIPP was the predominant OPE at site S10, with concentrations of 17.30, 7.88, and 6.70 ng/L during the dry, normal, and wet seasons, respectively. The concentrations at this site were obviously lower than at other locations. In summary, the levels of OPEs in different water samples of tourism resorts were significantly different. Tourist resorts that were more affected by nearby industrial activities had higher concentrations of OPEs, while those that were farther away from industrial areas had lower concentrations of OPEs. Therefore, residential activities and industrial pollutant emissions from the cities surrounding tourist resorts were the key factors determining OPE concentrations and high pollution levels.

During the dry season, apparent spatial variations for OPEs were also seen in the sediment of several tourist resorts, as shown in Figure 5. For instance, the values of sites S1 and S14–S16 were higher than 100 ng/g during the dry season. Sites S5–S9, S12, and S17 were between 55 ng/g and 100 ng/g, and S2–S4, S10–S11, S13, and S18–S19 were below 55 ng/g. Laizhou Bay is an important bay in Bohai Bay, and its water has poor self-purification ability, resulting in the easy accumulation of OPEs in the coastal sediments. Therefore, the ΣOPE value was high in sediments at site S1. Sites S15 and S16 are in Qingdao, with concentrations of 120.29 and 130.96 ng/g, respectively. Sites S15 and S16 are located on both sides of the entrance to Jiaozhou Bay, which serves as a discharge area for pollutants from Qingdao. Therefore, the highly urbanized and industrialized setting in Qingdao was the major cause of the high values of ΣOPEs in the sediment of S15 and S16. The OPE levels of water at sites S15 and S16 were also higher than other sites. Therefore, the migration of OPEs from water to sediment may also be an important reason for the high concentration of OPEs in sediment. Site S10 is in Weihai, where the concentration was 22.76 ng/g, which was the only value below 30 ng/g. The site is surrounded by natural areas and is ecologically sound. It is removed from urban centers and is hardly ever directly affected by industrial pollutants. The data showed that the concentration in the sediment was basically the same as that in the water. This phenomenon might have been related to the high OPE content in the surface water of the tourist resorts, indicating that the OPE contents in the surface water have a direct impact on the sediment content. The correlation of the OPE concentration in the water and sediment during the dry season is shown in Figure S2.

In conclusion, the highest OPEs in surface water and sediment were detected at S16 in Qingdao. The high levels of OPEs in the coastal water and sediment of Qingdao were related to the degree of urbanization and industrialization. The OPE concentrations of the tourist resorts that were most affected by nearby industry were higher, while the OPE concentrations of the tourist resorts that were far away from the industrial areas were lower. Therefore, the emission of industrial and domestic pollutants from nearby cities was the main factor affecting the content of OPEs near tourist resorts. These findings showed that urban activities have a significant influence on OPE spatial distribution.

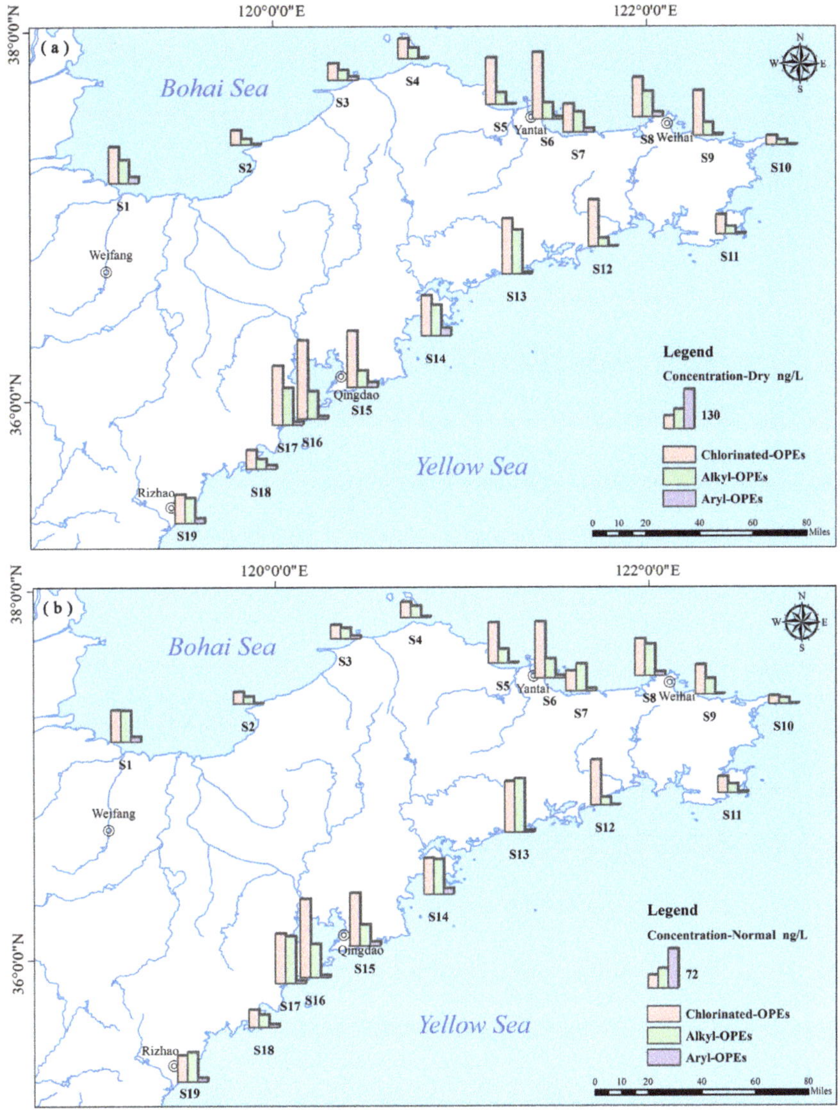

Figure 5. *Cont.*

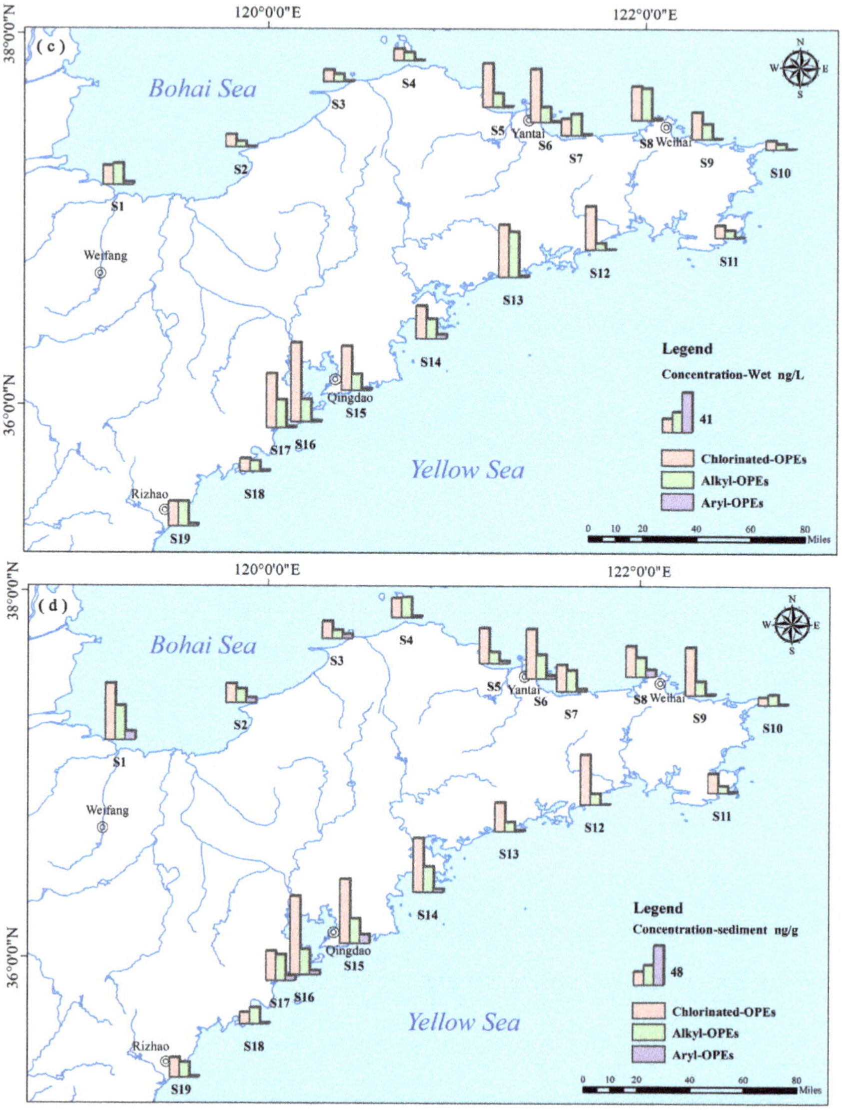

Figure 5. Spatial distribution of the three OPEs in the surface water during the dry season (**a**), normal season (**b**), wet season (**c**), and in the sediment (**d**).

TCIPP was the primary substance in the sediment of Shandong Peninsula coastal tourist resorts, accounting for 59.37%, 43.90%, and 49.97% of the total concentration of OPEs at sites S14–S16, respectively. TEP was the major substance in the sediment of site S1, accounting for up to 30.31%. The content of TCIPP in the sediments at many points was higher than TEP and was the highest content compound.

In addition, TCIPP and TNBP were predominant in water, accounting for 47.2% and 16.2% in the dry season, 39.6% and 29.8% in the normal season, and 51.7% and 18.6% in the wet season, respectively. There was a high TNBP content in the water samples. This may have been due to the impact of intensive mining activities in coastal cities. TNBP is widely used as an additive in hydraulic oils, lubricants for machinery and equipment, etc. [1].

TCIPP and TEP were predominant in sediment, accounting for 43.91% and 26.93%. TCIPP accounted for up to 59.37%, 43.90%, and 49.97% of the total concentration of OPEs at sites S14–S16, respectively. TEP accounted for up to 30.31% at site S1. A similar trend of regional variation in the water samples revealed that one of the major sources of sediment OPEs was OPE deposition in coastal water.

3.4. Water-Sediment Partitioning of OPEs

The Log K_d was calculated to assess the behavior of OPEs in partitioning between water and sediment, and the Log K_{oc} was calculated to evaluate the absorption of OPEs in sediment. The results are displayed in Figure 6. OPEs absorbed by sediments may be resuspended in coastal waters because of environmental disturbance. Therefore, in aquatic environments, OPEs can either be derived from or converge to the sediment. The K_d and K_{oc} of OPEs in the solid/liquid phase were important parameters in terms of their transformation and migration in the environment. In soils with various structures and levels of organic carbon [52], comparing the K_{oc} of various OPEs showed that the adsorption of OPEs can be affected by the amount of soil organic carbon. The correlation of other physical and chemical properties, such as soil organic carbon, with the concentration of OPEs, is shown in Figure S3.

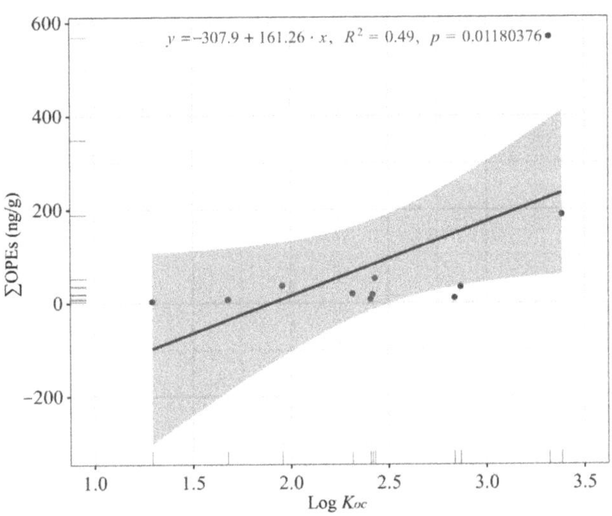

Figure 6. Correlations between ΣOPEs and Log K_{oc} in the sediment.

The mean value of the Log K_d of OPEs ranged from 1.53 to 3.25, and the OPE with the highest Log K_d value was TEP. Except for V6 (1.67) and TBOEP (1.53), the Log K_d values of the OPEs were all higher than 2.0, indicating that the sediment generally had a strong sorption capacity for these compounds. The strongest sorption capacity was for TEP, and the weakest sorption capacity was for V6 and TBOEP. TEP was absorbed on sediment more readily than other OPEs, and V6 and TBOEP were difficult to absorb on sediment. The mean value of the Log K_{oc} of OPEs ranged from 1.29 to 4.04. The OPE with the highest Log K_{oc} value was TEP, and the lowest Log K_{oc} value was V6. K_{oc} was shown to be inversely correlated with the overall amount of soil organic carbon, indicating that high levels of organic carbon may specifically increase the mobility of certain OPEs in soil [27]. In other words, the K_{oc} value increased with a decreasing TOC value, and the Log K_{oc} value decreased with an increasing K_{oc} value. For the same OPEs, there was a higher migration rate of sediment OPEs or a lower content in sediment and higher content in water. Thus, TEP was the only type of OPE that was more abundant in sediment than water.

3.5. Source Identification

The PCA of 12 OPEs was determined to indicate potential sources of OPEs in Shandong Peninsula coastal tourist resorts. Two principal components (PCs) were identified, accounting for 71.1% of the overall variance, as displayed in Figure 7a, with PC1 and PC2 obtaining respective weights of 53.8% and 17.3%. Except for TBOEP and TCIPP, all OPEs were close to the *x*-axis between the first and fourth quadrants, which indicated that TBOEP and TCIPP were not similar to the other OPE sources. Large loadings of V6, RDP, TEP, TDCIPP, TPP, TPHP, and BDP were seen in PC1. Previous research showed that TDCIPP and V6 in the environment typically come from discarded furniture and appliances, including building supplies, cotton drapes, and hardwood furniture [45]. Thermoplastics commonly contain TMPP, TPP, and TPHP [53], and electronic recycling factories are the main environmental sources [54]. Considering the above, waste recycling facilities were identified as OPE source 1.

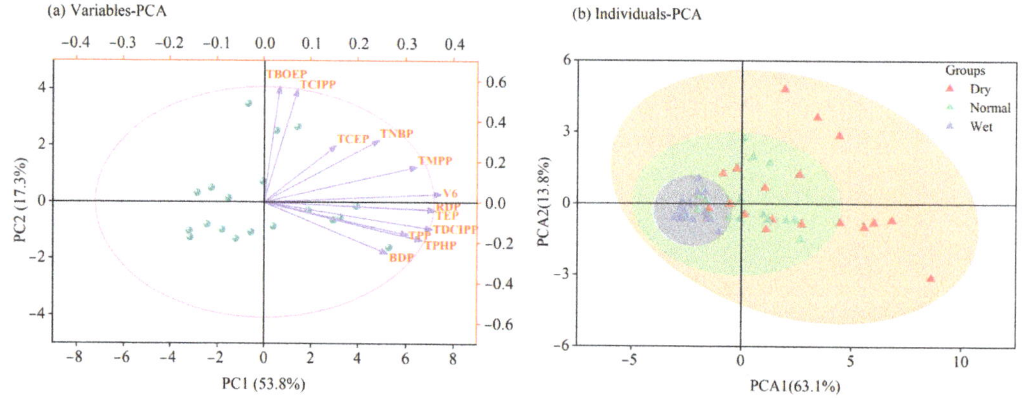

Figure 7. The variables-PCA and individuals-PCA of OPEs during three seasons.

For Cl-OPEs in the water of coastal tourist resorts, source 2 was heavily weighted. TCIPP and TBOEP were of similar origin, and the TCIPP concentration was the highest. The prime source of Cl-OPEs in the aquatic environment was wastewater discharge from sewage treatment plants (STP) due to the lower removal rate of Cl-OPEs [4,6,55]. TBOEP and TNBP, two non-Cl-OPEs, were frequently found in STP effluents [56,57]. Because of the sewage emissions and surface water corrosion, PC1 was identified as a municipal source.

RDP and TEP showed a high association, suggesting the same source. The high TEP levels in this sampling region were in heavy industrial cities like Qingdao. Therefore, it was important to note the release of OPEs from industry in Qingdao. TEP is a common industrial catalyst that contributes significantly to environmental pollution when released from chemical facilities [53]. Numerous coastal tourist resorts are in proximity to industrial cities heavily influenced by the chemical industry. Therefore, source 3 was the chemical industry, and PC2 served as a hub for extensive industries. OPEs in surface water could have originated from a variety of sources, including STP wastewater, waste recycling plants, large-scale chemical industries, textile manufacturing, and plastic processing, after taking all the aforementioned factors into consideration.

As seen in Figure 7b, PC1 and PC2 identified the locations of sampling that had adequate seasonal variation. Two major components were identified, accounting for 79.6% of the total variance, with 63.1% and 13.8% for PC1 and PC2, respectively. The PC1 and PC2 results for the samples gathered during the wet season scored negatively. Most samples collected in the dry and normal seasons scored positively on PC1.

3.6. Ecological Risk Assessment of OPEs

Three trophic-level creatures (fish, algae, and aquatic invertebrates) were chosen to investigate the RQ of mixed and individual OPEs. RQs for 12 OPEs were higher during the dry season than the normal and wet seasons, according to the cluster analysis. Throughout the three seasons, no OPEs presented a moderate or high risk (RQ $\geq$ 0.1) to fish, algae, or aquatic invertebrates. The highest RQs for the three trophic levels for individual OPEs were often less than 0.01, demonstrating that there was no ecological risk for individual OPEs. However, compared to other OPEs, TPHP (dry and normal season), TNBP (dry and normal season), and RDP (dry season) posed a comparatively greater danger to these trophic levels, even though the RQs were typically greater than 0.01 and the ecological risk was low. The EC(L)50 values of TPP and RDP for the three trophic levels, TCEP and TNBP for fish and TPHP for algae, were not found, so we did not analyze these items specifically here. In general, the ecological risk of the OPEs tended to be low.

Figure 8 displays the RQ value of OPEs at 19 sites. During the dry season, sites S2 and S10 had no ecological risk, sites S8 and S14 were moderate risk, and the others were low risk. During the normal season, sites S2, S4, S5, S9, S10, and S12 had no ecological risk, and the others were low risk. In the wet season, sites S1, S8, and S14 were low risk, and the others were no risk.

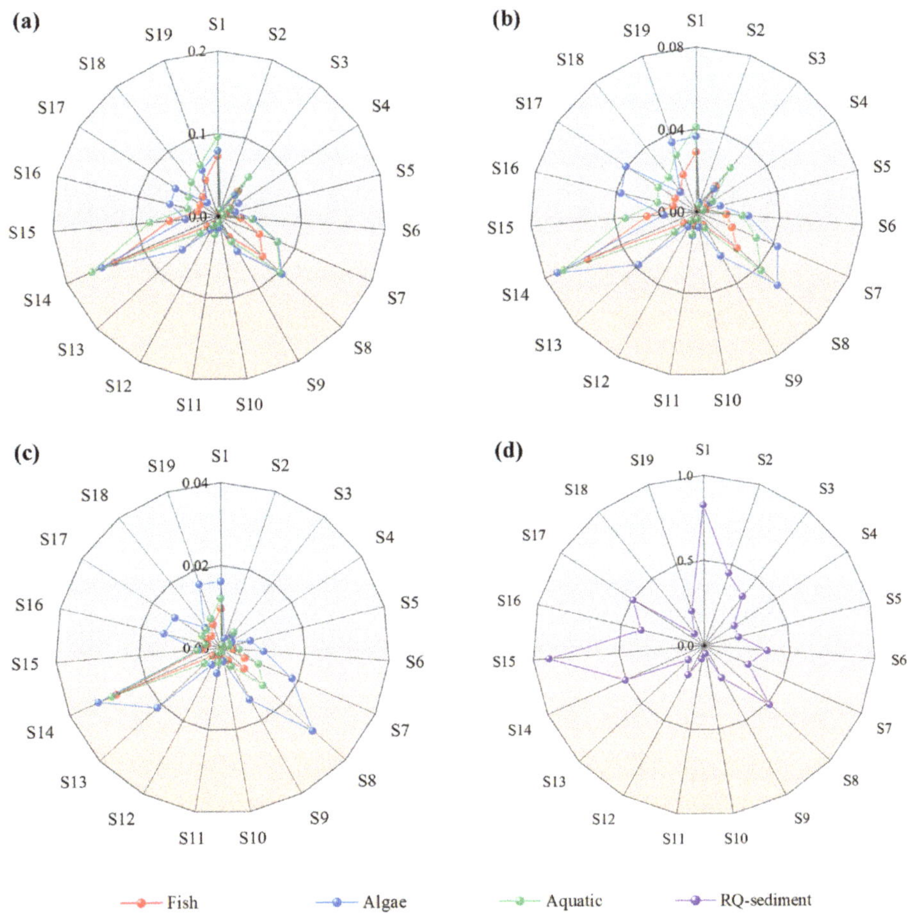

Figure 8. RQ value of OPEs in water at each sampling location during the dry season (**a**), normal season (**b**), wet season (**c**), and in sediment (**d**).

Another approach to ecological risk assessment was used in this study: direct calculation of RQ values based on the marine water PNEC [30]. The analysis revealed that all OPEs posed no ecological risk during the three seasons, except TDCIPP, TPHP, and BDP. TDCIPP had moderate ecological risk during the dry and normal seasons and low ecological risk during the wet season. TPHP had low ecological risk during the dry and normal seasons and no ecological risk during the wet season. BDP was a high ecological risk during the dry season and a medium risk during the normal and wet seasons. The values of marine water PNEC for TCEP, TPP, and RDP during the three seasons were not found, so we did not analyze these items here. In general, the target OPEs posed no ecological risk. The outcome of this analysis method was consistent with the aforementioned analytical results, based on EC(L)50 values for the three trophic levels (fish, algae, and aquatic invertebrates).

For sediment, TCIPP and TPHP showed moderate risk, and the rest of the substances were moderate or low risk. The NCEP of TCEP, TPP, and RDP were not found, so their ecological risk level in sediment was not analyzed here. In terms of sites, the RQ values were between 0.1 and 1, except for S10, S11, and S18, where the RQ values were between 0.01 and 0.1, indicating that only three sites were at low ecological risk, while the others were at moderate ecological risk.

4. Conclusions

This study is the first to investigate the spatiotemporal variations, partition characteristics, potential sources, and ecological risk of OPEs in the water and sediment of coastal tourist resorts in China. The results indicate that OPEs were widespread in the water and sediment of coastal tourist resorts. Cl-OPEs were the dominant type, and TCIPP was dominant in both the water and sediment. The water exhibited the highest concentrations of $\sum$OPEs during the dry season. Therefore, tourism activities in coastal tourist resorts should not be carried out during the dry season. Higher concentrations of $\sum$OPEs were observed in the study areas with denser populations and developed industry/agriculture, implying that human activities had influenced their spatial distribution. The LogK_{ow} of OPEs might have an important impact on the partition between water and sediment via the adsorption process. The PCA results verified that STP effluent may be an important source of OPEs in the water and sediment of coastal tourist resorts. The results show that OPEs presented a generally low degree of risk in the water of coastal tourist resorts of the Shandong Peninsula. Nevertheless, higher concentrations of OPEs in the water and sediment of some coastal tourist resorts were detected. Therefore, it is necessary to strengthen the environmental management of coastal tourist resorts in the future, as tourism demands high environmental quality. For example, a real-time monitoring system for water quality in tourist resorts should be established, and the water quality management of surrounding sewage plants and rivers entering the sea should be strengthened. Since the current sewage treatment technology of STPs pays little attention to the removal of organic pollutants, future studies should consider this.

Supplementary Materials: The following supporting information can be downloaded at: https://www.mdpi.com/article/10.3390/w15223976/s1, Table S1. List of the coastal tourist resorts in Shandong Province; Table S2. Sampling coordinates about surface water during the dry, normal, and wet seasons from coastal tourist resorts of Shandong Peninsula; Table S3. The name, abbreviation and properties of the most common in the present study; Table S4. The application of organophosphate esters (OPEs); Table S5. Optimized MS/MS parameters for the targeted OPEs and their surrogate standards in the MRM mode; Table S6. Some characteristics of the water and sediment; Table S7. The recoveries and concentrations of targeted OPEs in blanks; Table S8. Coefficient of association, recovery, method limit of detections (LOD) and method limit of quantifications (LOQ) of the target OPEs; Table S9. Concentration range and the detection frequencies of each OPE in the water during the dry, normal and wet season (ng/L); Table S10. Concentration range and the detection frequencies of each OPE in the sediment during the dry season (ng/g); Figure S1. Correlation of OPEs concentration in surface water between three seasons; Figure S2. Correlation of OPEs concentration in water and sediment during dry season and correlation of OPEs concentration in sediment with salinity and TOC; Figure S3. Correlation between the concentration of OPEs in water and sediment and their physical and chemical properties.

Author Contributions: X.C., conceptualization, methodology, software, and writing—original draft; B.W., methodology, software, and investigation; X.L., software and investigation; J.C., conceptualization, methodology, and investigation; S.W., methodology, writing—review and editing, and supervision. All authors have read and agreed to the published version of the manuscript.

Funding: This work is supported by the Project of the Natural Science Foundation of Shandong Province, China (No. ZR2023DM032) and the China Postdoctoral Science Foundation (No. 2013M540103).

Data Availability Statement: All data will be made available on request.

Acknowledgments: We are very grateful to Qufu Normal University for providing us with the necessary experimental conditions and environment.

Conflicts of Interest: The authors declare no conflict of interest.

References

1. Ye, L.J.; Li, J.H.; Gong, S.; Herczegh, S.M.; Zhang, Q.; Letcher, R.J.; Su, G.Y. Established and emerging organophosphate esters (OPEs) and the expansion of an environmental contamination issue: A review and future directions. *J. Hazard. Mater.* **2023**, *459*, 132095. [CrossRef] [PubMed]
2. Greaves, A.K.; Letcher, R.J. A Review of Organophosphate Esters in the Environment from Biological Effects to Distribution and Fate. *B Environ. Contam. Tox.* **2017**, *98*, 2–7. [CrossRef] [PubMed]
3. Hou, M.; Shi, Y.; Na, G.; Cai, Y. A review of organophosphate esters in indoor dust, air, hand wipes and silicone wristbands: Implications for human exposure. *Environ. Int.* **2021**, *146*, 106261. [CrossRef] [PubMed]
4. Wei, G.-L.; Li, D.-Q.; Zhuo, M.-N.; Liao, Y.-S.; Xie, Z.-Y.; Guo, T.-L.; Li, J.-J.; Zhang, S.-Y.; Liang, Z.-Q. Organophosphorus flame retardants and plasticizers: Sources, occurrence, toxicity and human exposure. *Environ. Pollut.* **2015**, *196*, 29–46. [CrossRef]
5. Li, J.; Xie, Z.; Mi, W.; Lai, S.; Tian, C.; Emeis, K.-C.; Ebinghaus, R. Organophosphate Esters in Air, Snow, and Seawater in the North Atlantic and the Arctic. *Environ. Sci. Technol.* **2017**, *51*, 6887–6896. [CrossRef]
6. Chen, M.-H.; Ma, W.-L. A review on the occurrence of organophosphate flame retardants in the aquatic environment in China and implications for risk assessment. *Sci. Total Environ.* **2021**, *783*, 147064. [CrossRef]
7. Qi, Y.; Yao, Z.; Ma, X.; Ding, X.; Shangguan, K.; Zhang, M.; Xu, N. Ecological risk assessment for organophosphate esters in the surface water from the Bohai Sea of China using multimodal species sensitivity distributions. *Sci. Total Environ.* **2022**, *820*, 153172. [CrossRef]
8. Liao, C.; Kim, U.-J.; Kannan, K. Occurrence and distribution of organophosphate esters in sediment from northern Chinese coastal waters. *Sci. Total Environ.* **2020**, *704*, 135328. [CrossRef]
9. Han, B.; Chen, L.; Li, Y.; Yu, L.; Zhang, J.; Tao, S.; Liu, W. Spatial distribution and risk assessment of 11 organophosphate flame retardants in soils from different regions of agricultural farmlands in mainland China. *Sci. Total Environ.* **2022**, *842*, 156806. [CrossRef]
10. Pantelaki, I.; Voutsa, D. Occurrence, analysis and risk assessment of organophosphate esters (OPEs) in biota: A review. *Mar. Pollut. Bull.* **2020**, *160*, 111547. [CrossRef]
11. Du, J.; Li, H.; Xu, S.; Zhou, Q.; Jin, M.; Tang, J. A review of organophosphorus flame retardants (OPFRs): Occurrence, bioaccumulation, toxicity, and organism exposure. *Environ. Sci. Pollut. Res.* **2019**, *26*, 22126–22136. [CrossRef] [PubMed]
12. Li, Z.; Zhu, Y.; Wang, D.; Zhang, X.; Jones, K.C.; Ma, J.; Wang, P.; Yang, R.; Li, Y.; Pei, Z.; et al. Modeling of Flame Retardants in Typical Urban Indoor Environments in China during 2010–2030: Influence of Policy and Decoration and Implications for Human Exposure. *Environ. Sci. Technol.* **2021**, *55*, 11745–11755. [CrossRef] [PubMed]
13. Yan, Z.F.; Feng, C.L.; Jin, X.W.; Qiao, Y.; Wang, J.D.; Huang, N.N.; Bai, Y.C. Unbalanced Pollution and Ecological Risk of Organophosphate Esters in Chinese Surface Water and Land Use Under Multiple Driving Factors. *Rev. Environ. Contam. T* **2023**, *261*, 18. [CrossRef]
14. Lai, N.L.S.; Kwok, K.Y.; Wang, X.; Yamashita, N.; Liu, G.; Leung, K.M.Y.; Lam, P.K.S.; Lam, J.C.W. Assessment of organophosphorus flame retardants and plasticizers in aquatic environments of China (Pearl River Delta, South China Sea, Yellow River Estuary) and Japan (Tokyo Bay). *J. Hazard. Mater.* **2019**, *371*, 288–294. [CrossRef] [PubMed]
15. Wang, X.; Zhu, L.; Zhong, W.; Yang, L. Partition and source identification of organophosphate esters in the water and sediment of Taihu Lake, China. *J. Hazard Mater.* **2018**, *360*, 43–50. [CrossRef] [PubMed]
16. Wolschke, H.; Sühring, R.; Massei, R.; Tang, J.R. Ebinghaus, Regional variations of organophosphorus flame retardants—Fingerprint of large river basin estuaries/deltas in Europe compared with China. *Environ. Pollut.* **2018**, *236*, 391–395. [CrossRef]
17. Cao, D.; Guo, J.; Wang, Y.; Li, Z.; Liang, K.; Corcoran, M.B.; Hosseini, S.; Bonina, S.M.C.; Rockne, K.J.; Sturchio, N.C.; et al. Organophosphate Esters in Sediment of the Great Lakes. *Environ. Sci. Technol.* **2017**, *51*, 1441–1449. [CrossRef]
18. Xing, R.G.; Zhang, P.; Zheng, N.; Ji, H.; Shi, B.; Ge, L.K.; Ma, H.R. Organophosphate esters in the seawater of the Bohai Sea: Environmental occurrence, sources and ecological risks. *Mar. Pollut. Bull.* **2023**, *191*, 114883. [CrossRef]

19. Shi, B.; Wang, T.; Yang, H.; Zhou, Y.; Yang, L.; Yoon, S.; Kim, T.; Khim, J. Perfluoroalkyl acids in rapidly developing coastal areas of China and South Korea: Spatiotemporal variation and source apportionment. *Sci. Total Environ.* **2021**, *761*, 143297. [CrossRef]
20. Li, W.; Yuan, Y.; Wang, S.; Liu, X. Occurrence, spatiotemporal variation, and ecological risks of organophosphate esters in the water and sediment of the middle and lower streams of the Yellow River and its important tributaries. *J. Hazard. Mater.* **2023**, *443*, 130153. [CrossRef]
21. Lian, M.; Lin, C.; Wu, T.; Xin, M.; Gu, X.; Lu, S.; Cao, Y.; Wang, B.; Ouyang, W.; Liu, X.; et al. Occurrence, spatiotemporal distribution, and ecological risks of organophosphate esters in the water of the Yellow River to the Laizhou Bay, Bohai Sea. *Sci. Total Environ.* **2021**, *787*, 147528. [CrossRef] [PubMed]
22. Qi, Y.; He, Z.; Yuan, J.; Ma, X.; Du, J.; Yao, Z.; Wang, W. Comprehensive evaluation of organophosphate ester contamination in surface water and sediment of the Bohai Sea, China. *Mar. Pollut. Bull.* **2021**, *163*, 112013. [CrossRef] [PubMed]
23. Luo, Q.; Wang, S.; Sun, L.; Wang, H. Simultaneous accelerated solvent extraction and purification for the determination of 13 organophosphate esters in soils by gas chromatography-tandem mass spectrometry. *Environ. Sci. Pollut. Res.* **2018**, *25*, 19546–19554. [CrossRef] [PubMed]
24. Luo, Q.; Gu, L.; Wu, Z.; Shan, Y.; Wang, H.; Sun, L. Distribution, source apportionment and ecological risks of organophosphate esters in surface sediments from the Liao River, Northeast China. *Chemosphere* **2020**, *250*, 126297. [CrossRef]
25. Shrivastava, A.; Gupta, V. Methods for the determination of limit of detection and limit of quantitation of the analytical methods. *Chron. Young Sci.* **2011**, *2*, 21. [CrossRef]
26. Zhao, Z.; Gong, X.; Zhang, L.; Jin, M.; Cai, Y.; Wang, X. Riverine transport and water-sediment exchange of polycyclic aromatic hydrocarbons (PAHs) along the middle-lower Yangtze River, China. *J. Hazard. Mater.* **2021**, *403*, 123973. [CrossRef]
27. Kim, S.; Carlson, K. Temporal and Spatial Trends in the Occurrence of Human and Veterinary Antibiotics in Aqueous and River Sediment Matrices. *Environ. Sci. Technol.* **2007**, *41*, 50–57. [CrossRef]
28. Chen, Z.; An, C.; Elektorowicz, M.; Tian, X. Sources, behaviors, transformations, and environmental risks of organophosphate esters in the coastal environment: A review. *Mar. Pollut. Bull.* **2022**, *180*, 113779. [CrossRef]
29. Wu, T.; Mao, L.; Liu, X.; Wang, B.; Lin, C.; Xin, M.; He, M.; Ouyang, W. Seasonal occurrence, allocation and ecological risk of organophosphate esters in a typical urbanized semi-closed bay. *Environ Pollut.* **2021**, *290*, 118074. [CrossRef]
30. Homepage—ECHA. Available online: https://echa.europa.eu (accessed on 25 March 2023).
31. Liu, Y.; Song, N.; Guo, R.; Xu, H.; Zhang, Q.; Han, Z.; Feng, M.; Li, D.; Zhang, S.; Chen, J. Occurrence and partitioning behavior of organophosphate esters in surface water and sediment of a shallow Chinese freshwater lake (Taihu Lake): Implication for eco-toxicity risk. *Chemosphere* **2018**, *202*, 255–263. [CrossRef]
32. Wang, X.; Zhu, Q.; Yan, X.; Wang, Y.; Liao, C.; Jiang, G. A review of organophosphate flame retardants and plasticizers in the environment: Analysis, occurrence and risk assessment. *Sci. Total Environ.* **2020**, *731*, 139071. [CrossRef] [PubMed]
33. Backhaus, T.; Faust, M. Predictive Environmental Risk Assessment of Chemical Mixtures: A Conceptual Framework. *Environ. Sci. Technol.* **2012**, *46*, 2564–2573. [CrossRef] [PubMed]
34. Struger, J.; Grabuski, J.; Cagampan, S.; Sverko, E.; McGoldrick, D.; Marvinm, C.H. Factors influencing the occurrence and distribution of neonicotinoid insecticides in surface waters of southern Ontario, Canada. *Chemosphere* **2017**, *169*, 516–523. [CrossRef] [PubMed]
35. Xu, L.; Zhang, B.; Hu, Q.P.; Liu, Y.; Shang, T.; Zeng, X.Y.; Yu, Z.Q. Occurrence and spatio-seasonal distribution of organophosphate tri- and di-esters in surface water from Dongting Lake and their potential biological risk. *Environ. Pollut.* **2021**, *282*, 1170313. [CrossRef]
36. Zhang, L.; Wang, Y.; Tan, F.; Yang, Y.; Wu, X.; Wang, W.; Liu, D. Tidal variability of polycyclic aromatic hydrocarbons and organophosphate esters in the coastal seawater of Dalian, China. *Sci. Total Environ.* **2020**, *708*, 134441. [CrossRef]
37. Zhong, M.; Tang, J.; Guo, X.; Guo, C.; Li, F.; Wu, H. Occurrence and spatial distribution of organophosphorus flame retardants and plasticizers in the Bohai, Yellow and East China seas. *Sci. Total Environ.* **2020**, *741*, 140434. [CrossRef]
38. Zheng, H.; Cai, M.; Yang, C.; Gao, Y.; Chen, Z.; Liu, Y. Terrigenous export and ocean currents' diffusion of organophosphorus flame retardants along China's adjacent seas. *Environ. Pollut.* **2022**, *299*, 118873. [CrossRef]
39. Shi, Y.; Zhang, Y.; Du, Y.; Kong, D.; Wu, Q.; Hong, Y.; Wang, Y.; Tam, N.F.Y.; Leung, J.Y.S. Occurrence, composition and biological risk of organophosphate esters (OPEs) in water of the Pearl River Estuary, South China. *Environ. Sci. Pollut. Res.* **2020**, *27*, 14852–14862. [CrossRef]
40. Gao, X.; Lin, Y.; Li, J.; Xu, Y.; Qian, Z.; Lin, W. Spatial pattern analysis reveals multiple sources of organophosphorus flame retardants in coastal waters. *J. Hazard. Mater.* **2021**, *417*, 125882. [CrossRef]
41. Pantelaki, I.; Voutsa, D. Organophosphate esters in inland and coastal waters in northern Greece. *Sci. Total Environ.* **2021**, *800*, 149544. [CrossRef]
42. Schmidt, N.; Castro-Jiménez, J.; Oursel, B.; Sempéré, R. Phthalates and organophosphate esters in surface water, sediments and zooplankton of the NW Mediterranean Sea: Exploring links with microplastic abundance and accumulation in the marine food web. *Environ. Pollut.* **2021**, *272*, 115970. [CrossRef]
43. Shimabuku, I.; Chen, D.; Wu, Y.; Miller, E.; Sun, J.; Sutton, R. Occurrence and risk assessment of organophosphate esters and bisphenols in San Francisco Bay, California, USA. *Sci. Total Environ.* **2022**, *813*, 152287. [CrossRef]

44. Luo, Q.; Wu, Z.; Wang, C.; Gu, L.; Li, Y.; Wang, H. Seasonal variation, source identification, and risk assessment of organophosphate ester flame retardants and plasticizers in surficial sediments from Liao River estuary wetland, China. *Mar. Pollut. Bull.* **2021**, *173*, 112947. [CrossRef] [PubMed]
45. Wang, Y.; Wu, X.; Zhang, Q.; Hou, M.; Zhao, H.; Xie, Q.; Du, J.; Chen, J. Organophosphate esters in sediment cores from coastal Laizhou Bay of the Bohai Sea, China. *Sci. Total Environ.* **2017**, *607–608*, 103–108. [CrossRef]
46. Tan, X.-X.; Luo, X.-J.; Zheng, X.-B.; Li, Z.-R.; Sun, R.-X.; Mai, B.-X. Distribution of organophosphorus flame retardants in sediments from the Pearl River Delta in South China. *Sci. Total Environ.* **2016**, *544*, 77–84. [CrossRef] [PubMed]
47. Choi, W.; Lee, S.; Lee, H.-K.; Moon, H.-B. Organophosphate flame retardants and plasticizers in sediment and bivalves along the Korean coast: Occurrence, geographical distribution, and a potential for bioaccumulation. *Mar. Pollut. Bull.* **2020**, *156*, 111275. [CrossRef] [PubMed]
48. Zhong, M.; Wu, H.; Mi, W.; Li, F.; Ji, C.; Ebinghaus, R.; Tang, J.; Xie, Z. Occurrences and distribution characteristics of organophosphate ester flame retardants and plasticizers in the sediments of the Bohai and Yellow Seas, China. *Sci. Total Environ.* **2018**, *615*, 1305–1311. [CrossRef]
49. Fan, Q.; Zou, X.; Gao, J.; Cheng, Y.; Wang, C.; Feng, Z.; Ding, Y.; Zhang, C. Assessing ecological risk of organophosphate esters released from sediment with both of total content and desorbable content. *Sci. Total Environ.* **2021**, *772*, 144907. [CrossRef]
50. Alkan, N.; Alkan, A.; Castro-Jiménez, J.; Royer, F.; Papillon, L.; Ourgaud, M.; Sempéré, R. Environmental occurrence of phthalate and organophosphate esters in sediments across the Gulf of Lion (NW Mediterranean Sea). *Sci. Total Environ.* **2021**, *760*, 143412. [CrossRef]
51. Kim, U.-J.; Oh, J.K.; Kannan, K. Occurrence, Removal, and Environmental Emission of Organophosphate Flame Retardants/Plasticizers in a Wastewater Treatment Plant in New York State. *Environ. Sci. Technol.* **2017**, *51*, 7872–7880. [CrossRef]
52. Cristale, J.; Dantas, R.F.; De Luca, A.; Sans, C.; Esplugas, S.; Lacorte, S. Role of oxygen and DOM in sunlight induced photodegradation of organophosphorous flame retardants in river water. *J. Hazard. Mater.* **2017**, *323*, 242–249. [CrossRef] [PubMed]
53. Verbruggen, E.M.J.; Rila, J.P.; Traas, T.P.; Posthuma-Doodeman, C.J.A.M.; Posthumus, R. Environmental Risk Limits for Several Phosphate Esters, with Possible Application as Flame Retardant. Rijksinstituut voor Volksgezondheid en Milieu RIVM. 2006. Available online: https://rivm.openrepository.com/handle/10029/255802 (accessed on 29 March 2023).
54. Matsukami, H.; Tue, N.M.; Suzuki, G.; Someya, M.; Tuyen, L.H.; Viet, P.H.; Takahashi, S.; Tanabe, S.; Takigami, H. Flame retardant emission from e-waste recycling operation in northern Vietnam: Environmental occurrence of emerging organophosphorus esters used as alternatives for PBDEs. *Sci. Total Environ.* **2015**, *514*, 492–499. [CrossRef] [PubMed]
55. Liang, K.; Liu, J. Understanding the distribution, degradation and fate of organophosphate esters in an advanced municipal sewage treatment plant based on mass flow and mass balance analysis. *Sci. Total Environ.* **2016**, *544*, 262–270. [CrossRef] [PubMed]
56. Cui, K.; Wen, J.; Zeng, F.; Li, S.; Zhou, X.; Zeng, Z. Occurrence and distribution of organophosphate esters in urban soils of the subtropical city, Guangzhou, China. *Chemosphere* **2017**, *175*, 514–520. [CrossRef]
57. Meyer, J.; Bester, K. Organophosphate flame retardants and plasticisers in wastewater treatment plants. *J. Environ. Monit.* **2004**, *6*, 599–605. [CrossRef]

water

MDPI

Review

Systematic Review of Contaminants of Emerging Concern (CECs): Distribution, Risks, and Implications for Water Quality and Health

Weiying Feng [1,*], Yuxin Deng [1], Fang Yang [2], Qingfeng Miao [3] and Su Kong Ngien [4]

1 School of Materials Science and Engineering, Beihang University, Beijing 100191, China; dengyx@buaa.edu.cn
2 State Key Laboratory of Environmental Criteria and Risk Assessment, Chinese Research Academy of Environmental Sciences, Beijing 100012, China; yang.fang@craes.org.cn
3 College of Water Conservancy and Civil Engineering, Inner Mongolia Agricultural University, Hohhot 010018, China; 15049121836@126.com
4 Faculty of Civil Engineering Technology, Universiti Malaysia Pahang, Lebuhraya Persiaran Tun Khalil Yaakob, Gambang 26300, Pahang, Malaysia; ftkaump@ump.edu.my
* Correspondence: fengweiying@buaa.edu.cn

Abstract: The introduction of contaminants of emerging concern (CECs) into the environment has raised concerns due to the significant risks they pose to both ecosystems and human health. In this sys-tematic review, we investigate research trends on CECs worldwide over the past 10 years, focus-ing on four critical aspects: (i) the identification and distribution of typical CECs across various media, (ii) the sources and environmental behavior of CECs, (iii) the implications of CECs expo-sure on human health, and (iv) risk assessment and control measures for CECs. The review re-veals a comprehensive understanding of the typical types and distribution of CECs in different environmental media, shedding light on their prevalence and potential impact on ecosystems. Furthermore, insights into the sources and behavior of CECs provide crucial information for de-vising effective strategies to mitigate their release into the environment. By examining the health effects of EC exposure, we highlight the importance of considering potential risks to human well-being. This aspect of the review emphasizes the significance of monitoring and managing CECs to safeguard public health. The review also synthesizes the advancements in risk assessment methodologies and control measures for CECs, which are essential for developing comprehensive regulations and guidelines to manage these contaminants effectively. Drawing from the findings, we identify future research directions for CECs in aquatic environments.

Keywords: contaminants of emerging concern; POPs; microplastics; endocrine disruptor; antibiotic

Citation: Feng, W.; Deng, Y.; Yang, F.; Miao, Q.; Ngien, S.K. Systematic Review of Contaminants of Emerging Concern (CECs): Distribution, Risks, and Implications for Water Quality and Health. *Water* **2023**, *15*, 3922. https://doi.org/10.3390/w15223922

Academic Editor: Christos S. Akratos

Received: 9 October 2023
Revised: 4 November 2023
Accepted: 6 November 2023
Published: 10 November 2023

1. Introduction

Contaminants of emerging concern (CECs) are a group of pollutants that have recently raised concerns due to their potential ecological and human health-related risks. Many of these contaminants have not yet been included in existing management protocols, and current efforts to prevent and mitigate their risks are insufficient. CECs often exhibit widespread distribution, unclear baseline levels, and host a wide array of unforeseen dangers [1,2]. In 2003, Jerald Schnoor, a member of the American Academy of Engineering, proposed the concept of "CECs" and stated that the environmental occurrence of any synthetic or naturally occurring chemicals and microorganisms could cause significant toxic effects and pose health hazards [3]. The number of chemicals registered by the American Chemical Society's global Chemical Abstracts Service exceeds 142 million [4]. According to a report developed by the United Nations Environment Program and the International Council of Chemical Associations, over 350,000 chemicals and mixtures are used commercially worldwide [5].

To control CECs, it is crucial to conduct investigations, monitoring, and environmental risk assessments for contaminants such as persistent organic pollutants (POPs) and endocrine disruptors. Such efforts are important for establishing and enhancing environmental risk management. This can strengthen our management of pollution sources, help create lists of CECs under control, and advance environmental risk management measures in the context of prohibition, restriction, and emission limitation. On 4 May 2022, China proposed an action plan for the control of CECs with the aim of completing environmental risk screenings for high-yield and high-usage chemical substances by 2025, along with conducting environmental risk assessments for a batch of chemical substances. The management of CECs is closely tied to social security, and without increased control efforts, it could hinder industrial upgrades and impede sustainable development. Consequently, the identification of CECs and the enhancement of product access conditions may have short-term impacts on certain industries. However, in the long run, it can drive and facilitate the entire industry's green transformation, thereby reducing the harmful effects of CECs on human health.

Research on CECs can be categorized into four aspects: types and distribution, sources and environmental behaviors, exposure and health effects, and risk assessment. At present, the CECs of global concern include POPs controlled by international conventions, endocrine disruptors, antibiotics, and microplastics. Analyzing the research hotspots and evolutionary trends of these CECs has shed light on their dynamics. Moreover, many studies have explored the relationships and distributions of these four typical CECs in different countries worldwide. These findings provide valuable theoretical insights for the screening, treatment, and control of CECs, allowing for more effective strategies to safeguard the environment and human health.

2. Types and Distribution of CECs

CECs are a heterogeneous set of pollutants that includes pharmaceuticals, personal care products, insecticides, flame retardants, industrial additives, surfactants, plasticizers, and nanomaterials [6–8] Based on the current global research trends on CECs, the CECs discussed in this study can be classified as follows: POPs (Persistent Organic Pollutants), antibiotics, EDCs (Endocrine-Disrupting Chemicals), and microplastics. With the wide application of chemicals in industry, agriculture, etc., the introduction of CECs into the environment is inevitable. Although we know some of the chemical structures of CECs, this is only the tip of the iceberg, and there is limited understanding of their structure and content and even less of their toxicity [9,10]. Compared to other nations, China's CECs pose more critical pollutant crises. China is the largest user of antibiotics in the world, with a total production of 248,000 tons, of which 52 percent is used in animals and 48 percent in humans, according to a 2013 survey [11]. The average concentration of antibiotics in China's rivers is 303 ng/L, which is three times that of the United States and 15 times that of Germany. And the density of antibiotic discharge in the eastern basin is more than six times that of the western basin [12–15]. Furthermore, the abundance of microplastics detected in freshwater is 2–3 times higher than that reported in other nations [16,17]. In the five major river basins in China, the ecological risks caused by antibiotics are ranked in descending order: Jianghan Plain > Yangtze River Delta > Chaohu River Basin > Yellow River Delta > Pearl River Delta, and the main types of antibiotic pollution are sulfamethoxazole and erythromycin [18].

In developing countries, a higher quantity of CECs is released into the environment due to inadequate post-treatment of pollutants. For instance, in Brazilian rivers, the concentration of acetylsalicylic acid was measured at 20,960 ng/L, caffeine at 14,955 ng/L, and the concentration of acetaminophen exceeded 30,000 ng/L [19]. India is one of the largest pharmaceutical suppliers globally, accounting for 20% of global exports [20]. However, due to the lack of CECs management strategies, the concentrations of certain antibiotics in Indian wastewater treatment plants were found to be 40 times higher than in European, Australian, and other North American countries [21]. CECs are typically detected in densely populated catchment areas, often associated with wastewater treatment plants [22,23].

However, CECs have also been found in rural areas far from known sources of pollution. In the water and sediments of 21 sites in eight US national parks located in the northern Colorado Plateau, CECs were detected, indicating that CECs can be transported through atmospheric deposition to relatively pristine high-altitude and high-latitude regions [24]. Masoner et al. conducted CECs testing in 19 landfill sites across the United States, where bisphenol A was the most detected CEC, and areas with higher levels of precipitation had higher CEC concentrations in leachate compared to lower-precipitation areas [25]. Gao et al. measured 15 types of organochlorine pesticides in Xiamen, China. Among them, HCHs (hexachlorocyclohexane and isomers) showed an increasing trend with depth, while the concentrations of DDT (dichlorodiphenyldichloroethane and its isomers) and OCPs (organochlorine pesticides) decreased with depth, potentially due to the properties of the contaminants [26]. In terms of horizontal distribution, heavier pollution was found in Tong'an District and Xiang'an District, where irrigation was the primary cultivation method, and higher pollution levels were observed in the northeastern part of Xiamen City, possibly due to atmospheric transport and deposition in the region [26].

Moreover, CECs like tetrabromodiphenol A and its derivatives are commonly found in drinking water, rivers, and lakes [27]. China's sludge contains 749 pollutants across 35 categories, and a significant portion of these are CECs [28,29]. Unfortunately, these chemicals not only affect the environment but also find their way into the human body through bioaccumulation in the food chain, even reaching the bloodstream [30].

Based on the search conducted in the core database of Web of Science with a limited retrieval time from 2013 to 2023, the following results were obtained for each search term: (1) "Antibiotic" (Figure 1): 1646 articles retrieved; (2) "Persistent Organic Pollutants" (Figure 2): 4902 articles retrieved; (3) "Microplastic" (Figure 3): 938 articles retrieved; and (4) "Endocrine Disruptor" (Figure 4): 2957 articles retrieved. These numbers represent the total count of articles related to each specific search term within the given time frame. Please note that these figures may change as new research is published or as the database is updated.

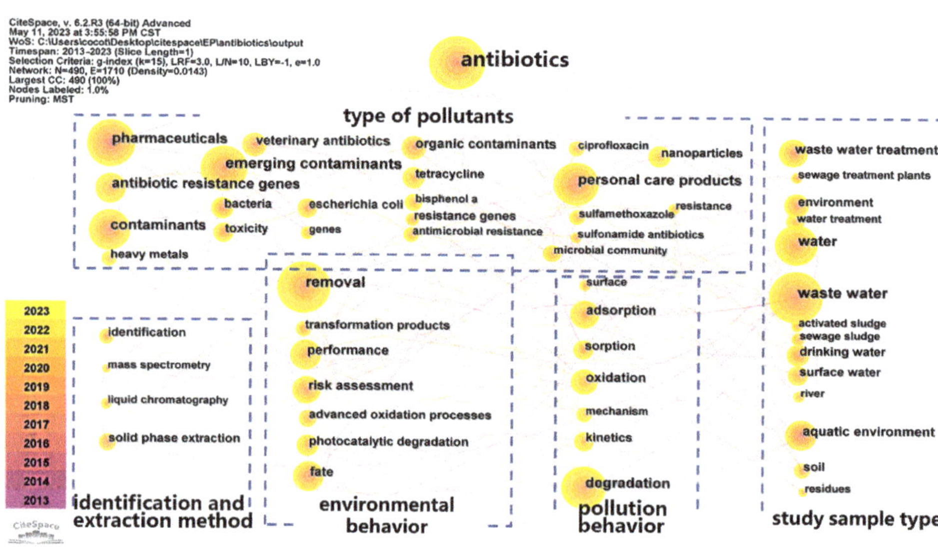

Figure 1. Co-occurrence map of keywords in studies on antibiotics environmental contamination from 2013 to 2023.

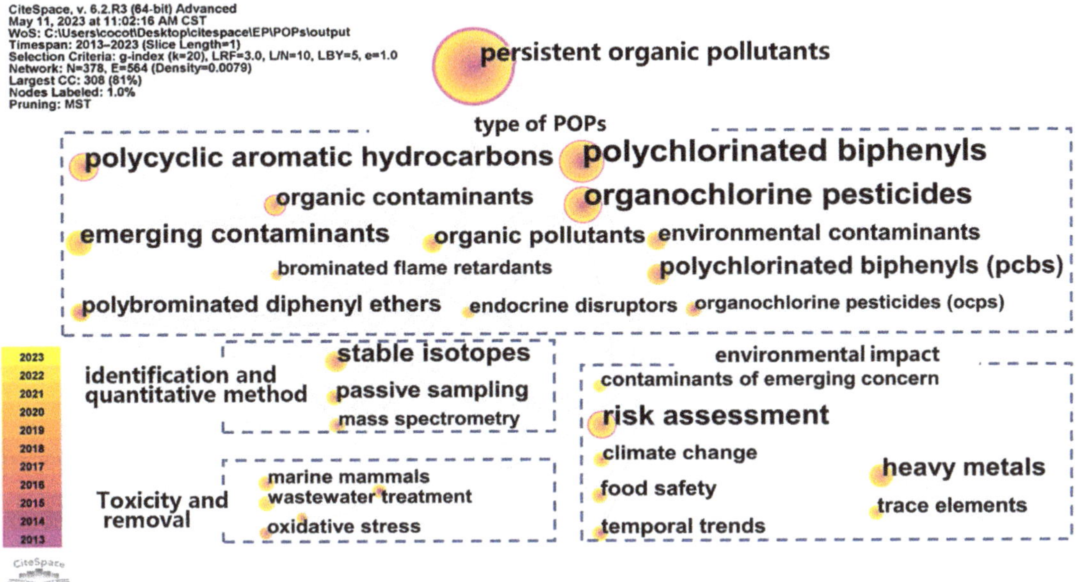

Figure 2. Keyword co-occurrence map of studies on POPs and environmental pollution from 2013 to 2023.

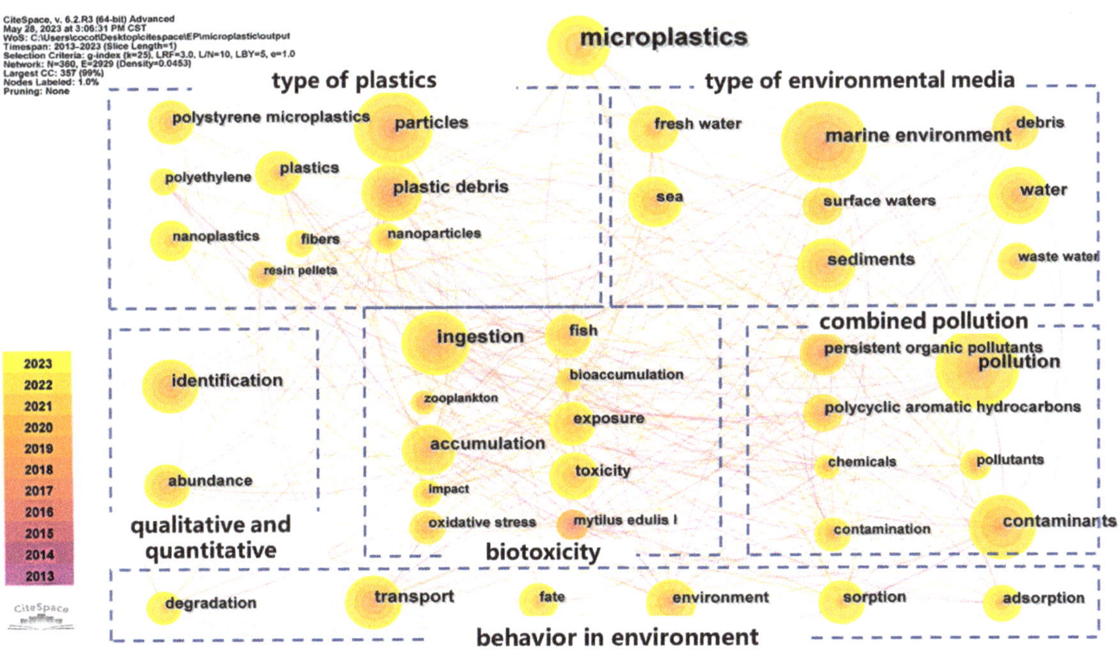

Figure 3. Keyword co-occurrence of research on microplastics and environmental pollution from 2013 to 2023.

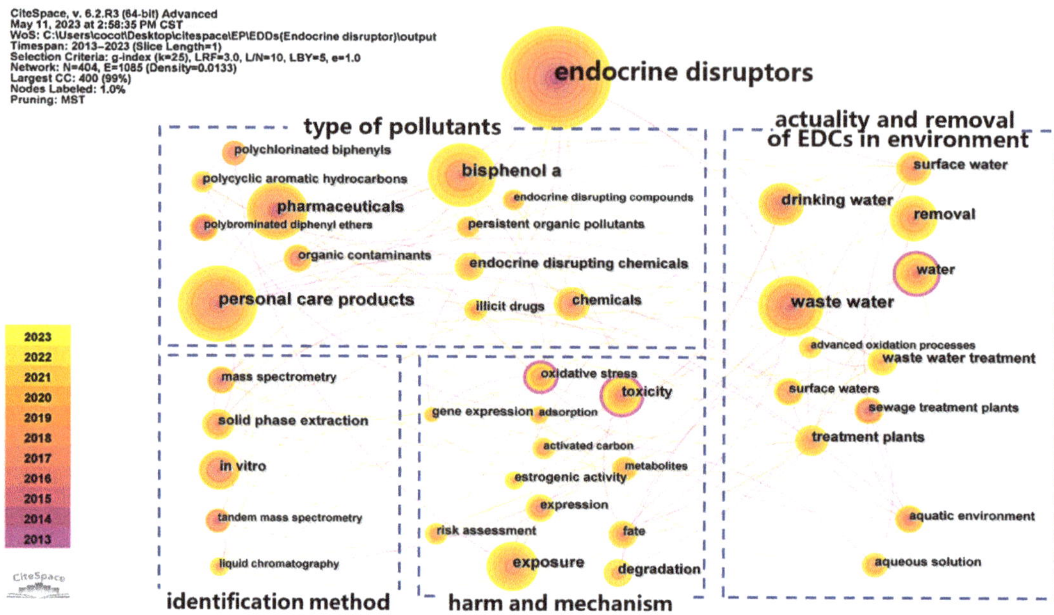

Figure 4. Co-occurrence of keywords in research on EDCs and environmental pollution from 2013 to 2023.

Over the past 10 years, research on antibiotics has mainly focused on sample types, identification methods, environmental behaviors, pollution behavior, and the interaction between antibiotic types and other related pollutants (such as those found in personal care products and pharmaceuticals) (Figure 1). Mass spectrometry, liquid chromatography, and solid-phase extraction were used to identify the antibiotics. The study samples mainly included wastewater and aquatic environments (Figure 1). The research on antibiotics has shown a trend towards multidisciplinary and integrated studies. Researchers no longer focus solely on the drugs themselves, but also pay greater attention to the behavior of antibiotics in the environment and their relationship with other pollutants. They are increasingly interested in the interactions between different substances and are approaching actual environmental conditions. This means that future research on antibiotics may involve more interdisciplinary collaboration. Research on POPs has focused on their type, identification methods, biohazards, and environmental impacts (Figure 2). Polychlorinated biphenyls and organochlorine pesticides were the main POP types, accounting for more than 30% of the total. Passive sampling, mass spectrometry, and stable isotope analysis were used to identify the types and their environmental behaviors (Figure 2). In the case of MPs (Figure 3), the research primarily focuses on the types of plastics, environmental media, qualitative and quantitative characteristics, biological toxicity, composite pollution, and environmental behavior. Through studies on the generation and decomposition of different plastic types in the environment, the forms and migration behaviors of microplastics in environmental media such as water, soil, and sediment, particle size, morphology, concentration analysis of microplastics, toxic effects on organisms, composite pollution with other pollutants, and environmental behavior processes of microplastics, a comprehensive understanding of the sources, distribution characteristics, and environmental effects of microplastics can be achieved. This helps to evaluate the ecological risks and develop suitable management and control measures. The research on EDCs mainly focuses on the types, identification methods, hazards and mechanisms, and environmental media (Figure 4). Researchers identify and quantify endocrine-disrupting chemicals through various techniques, such as chemical analysis, biological detection, molecular sieves, and mass spectrometry. They investigate the harm and potential carcinogenic risk effects of endocrine disruptors on human reproductive, immune, and nervous systems. Currently, research mainly focuses on endocrine disruptors in wastewater, and further studies are needed to strengthen research on drinking water in order to better protect drinking water hygiene and food safety.

3. Sources and Environmental Behavior of CECs

CiteSpace 6.2 was used for keyword co-occurrence analysis of the obtained literature (Figure 5). In the field of microplastics, research clustering has mainly focused on the ecological risks of polyvinyl chloride and plastic pollution over the past 10 years (Figure 5a). In the past decade, many studies have focused on POPs contained in personal care products, polychlorinated biphenyls, and bisphenol A, as well as environmental disturbances and endocrine disorders caused by POPs (Figure 5b). Recently, endocrine disruptors and biochar-related processes have garnered a lot of attention in the context of research (Figure 5c). The main research on antibiotics focuses on the study of resistance genes, especially in the context of adsorption and photocatalysis (Figure 5d).

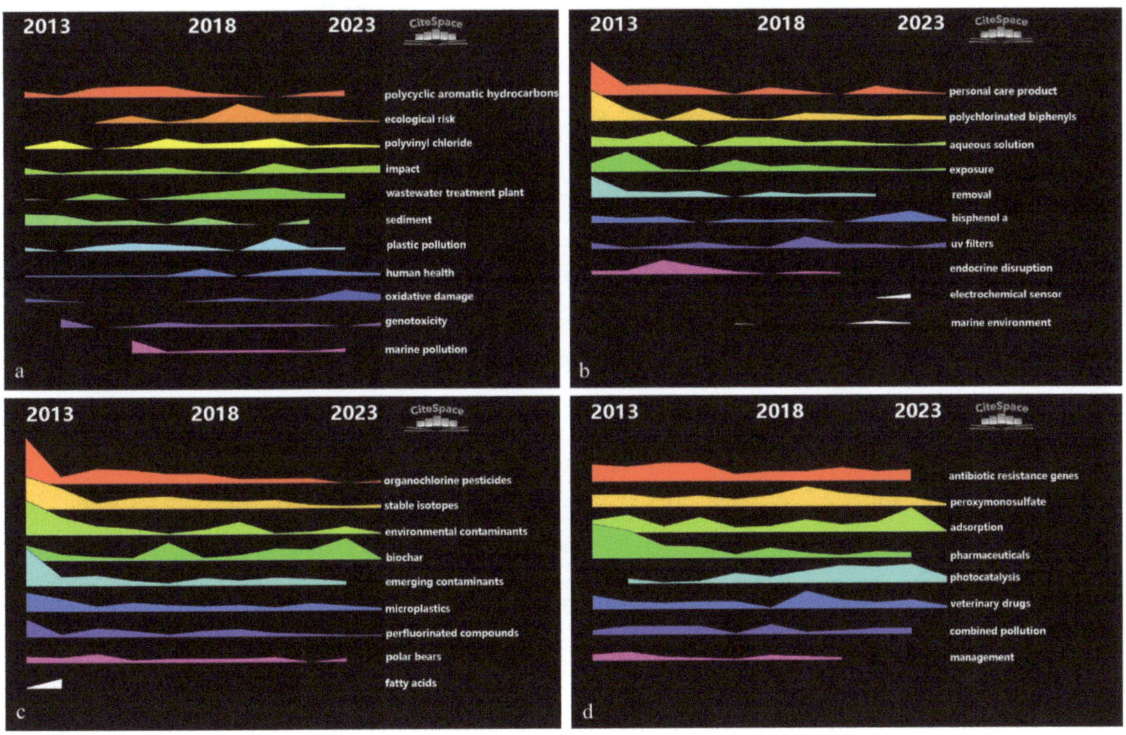

Figure 5. Landscape view of each keyword from 2013 to 2023 ((**a**): microplastics; (**b**): POPs; (**c**): EDCs; (**d**): antibiotics on the right side represents the cluster, and the smaller the antibiotics number, the larger the cluster. Landscape view can be used to intuitively show the research hotspots and evolutionary trajectory).

The production and use of toxic and harmful chemicals are the primary sources of CECs [31,32]. CECs may originate from industrial waste, municipal waste, agricultural waste, hospital, and laboratory wastewater [33]. They can enter the environment through various point sources (i.e., industrial, municipal, and hospital wastewater treatment plants), nonpoint sources (such as atmospheric deposition and stormwater runoff), or unexpected scenarios [19,34,35]. For example, perfluorooctane sulfonyl compounds (PFOSs) are typical perfluorinated compounds mostly derived from textiles, leather antifouling coatings, semiconductor paper products, synthetic detergents, foam extinguishing agents, food additives, pesticides, and cosmetics. PFOSs have been widely detected in environmental media and the human body. Owing to the toxic effects of PFOSs, they were included in the Stockholm Convention in 2009 and have been gradually banned in numerous countries.

CECs are released into the surrounding environment during the production, use, and disposal of related products. For example, China's plastic ocean waste in 2011 was approximately 547–752,000 tons and it continued to expand at a rate of 4.55% per year until 2017, and then it had decreased to 257–353,000 tons by 2020 [36]. In the Hetao Plain of Inner Mongolia, China, the agricultural film remains in the soil, and the residue coefficient is as high as 40% [37,38]. CECs migrate, transform, agglomerate, degrade, and exhibit other environmental behaviors in environmental media, which affect their return to the environment. For example, water in the atmosphere inhibits the degradation of organophosphorus flame retardants caused by hydroxyl radicals [31]; perfluorooctanesulfonamides are converted to PFOS in earthworms; and perfluorohexane sulfonic acid and perfluorobutanesulfonate are converted in wheat [39]. Organophosphorus flame retardants exhibit biomagnification effects during food chain transfers [40]. When exposed to a mixture of o,p′-DDE and p,p′-DDE, the reduction in the ratio of estradiol to testosterone is greater than that observed with individual exposures. Northern bobwhite survival rates were lower when exposed to a mixture of TCDD and ethynyl estradiol or coumestrol compared to individual chemical exposures [41]. Smaller microplastics (i.e., 500 nm) can exacerbate the toxicity of POPs, where larger microplastics (30 μm) are less toxic, possibly due to size-dependent interactions between microplastics and POPs [42]. However, research on the effects of CECs mixtures is still in infancy, and current risk assessments are primarily based on concentration addition, which may lead to inaccurate risk estimates [43].

Currently, the primary focus of CECs removal technologies is on upgrading wastewater treatment plants. In line with this focus, membrane bioreactor (MBR) technology has shown improved removal performance of CECs before disinfection in conventional wastewater treatment plants [44,45]. In addition, activated carbon has demonstrated high adsorption capacity for CECs removal, although it has drawbacks such as slow adsorption kinetics and poor adsorption efficiency for hydrophilic pollutants [46,47]. Advanced oxidation processes based on ozone oxidation, ultraviolet radiation, gamma radiation, and electrooxidation have also been widely applied in CECs treatment [48–51].

CECs can persist in sediments for prolonged periods, resulting in the bioaccumulation of CECs in benthic organisms. Since the benthic organisms are consumed by fish, this food-web-driven bioaccumulation of CECs culminates in top predators having accumulated CECs to concentrations several orders of magnitude higher than that of the surrounding water [52,53]. The fate of CECs in soils may be influenced by interrelated processes [54]. Chemical substances with strong hydrophobicity or positive charges often exhibit high adsorption capacity, thereby reducing their potential for biodegradation, long-range transport, or plant accumulation [55–57]. Photodegradation of CECs represents a significant non-biological transformation process. For instance, under aerobic conditions, the half-life of bisphenol A in soil and sediment is estimated to be 3–37 days, whereas no degradation was observed in anaerobic or hypoxic estuarine sediments during a 120-day incubation experiment [58–60]. Microorganisms in soils also have the capability to degrade CECs, primarily occurring in the rhizosphere [61–63]. Consequently, it is necessary to conduct more detailed and systematic research on the pollution sources and environmental behaviors of CECs to provide a scientific basis for the accurate and scientific promotion of the prevention and control of pollutant health risks, and it is extremely important to treat and control their sources.

4. Exposure and Health Effects of CECs

Exposure to CECs occurs through diverse pathways, including industrial production, consumer goods, personal care products, plants and animals, food, and environmental media like soil. These CECs have a tendency to bioaccumulate in organisms and humans, resulting in threats to the environment and human health [64,65]. Scientific studies have highlighted their adverse impact on tissues, organs, and overall health, with a substantial portion of cancer risks linked to chemical use [66]. For example, persistent organic pollutants have been associated with intersexes, and chlorinated paraffins have shown toxic effects on zebrafish metabolism [67]. These contaminants enter the human body through multiple routes, including oral ingestion, respiratory inhalation, and skin absorption. Drinking water remains a primary source for most perfluorinated and polyfluoroalkyl substances (PFCs), while other sources contribute more to PFOA dust and air exposure [68]. Although studies indicate that micro/nanoplastics' main exposure pathway is ingestion, the contribution of other routes, like dust inhalation, should not be underestimated. However, accurately assessing human exposure to micro/nanoplastics is currently limited by detection methods [69,70].

The cumulative homeostasis of CECs in the human body reflects the synergistic effects of multiple exposure pathways. For example, with the gradual use of bisphenol S (BPS) as a substitute for bisphenol A, the highest levels of BPS are found in the urine of the Japanese population, followed by the United States, China, Kuwait, and Vietnam, and the levels of BPS in the urine of the rest of Asia are an order of magnitude lower than the levels of BPS elsewhere [71]. CECs are enriched in organisms and produce a variety of biological toxicities in organisms and human health, including endocrine disruption, growth and development toxicity, neurotoxicity, immunotoxicity, carcinogenicity, and teratogenicity [72]. Some studies have reported that photolysis will increase the toxic effect of polystyrene microplastics and polystyrene fragments after photodegradation will reduce the particle size and form persistent free radicals on the surface (such as CO· and COO·). Polystyrene fragments after photodegradation have the greatest impact on the growth inhibition and liver damage of grouper, followed by polystyrene fragments before photolysis and commercially available polystyrene particles [73]. It has also been reported that PFOA induces apoptosis in mouse hepatocytes, and the genotoxicity caused by the overproduction of reactive oxygen species (ROS) is due to the inhibition of complex I subunits in the electron transport chain and activation of the peroxide proliferation-activating receptor PPAR [74]. BPA also affects rat metabolism, with BPA exposure interfering with the biosynthesis of valine, leucine, and isoleucine and the metabolism of D-glutamine and D-glutamate, and low doses of BPA may have toxic effects on the nervous system [75]. Some studies have reported that PFOs exposure affects synaptic transmission, cell growth, and development, and mainly acts on calcium ion channels and produces neurotoxic effects [76].

Consumption of contaminated aquatic products is one of the most important pathways of human exposure to CECs; therefore, it is important and necessary to assess the human health risks caused by the intake of aquatic products. Some researchers found a significant positive association between urine levels of organophosphorus flame retardants (DPHP) in pregnant women and the risk of low birth weight in female infants. In the third trimester, the fetus is susceptible to the developmental toxicity of BDCIPP and BBOEP [77]. However, perfluorinated compounds (PFCs) have a serious impact on the occupationally exposed population, and the levels of PFCs in this population are much higher than those in the general population. Metabolomic studies have shown that 14 potential biomarkers are associated with oxidative stress, fatty acid beta-oxidation disorder, and kidney damage; however, the health risks of PFCs in occupational populations cannot be ignored, with as many as 4730 PFCs used in the global market [78,79]. According to incomplete statistics, 62% of CECs have environmental health hazards. For example, owing to the influence of new estrogen-like pollutants, the incidence of intersexes in wild barracudas in Bohai Bay has reached 50% [67]. In addition, the tetrabrombisphenol A derivative had a stronger neurotoxic effect than the parent compound, and exposure to this compound resulted in significant neuroethological changes in neonatal rats. The Journal of Science reports that antibiotic drugs in CECs increase the resistance of organisms and interfere with the homeostasis of organism populations and ecosystems [80]. Therefore, it is important to study the main driving factors and regulatory mechanism of EC transport for the treatment and control of CECs, and the joint ecological risk assessment of various pollutants needs to be further improved.

5. Risk Assessment and Management of CECs

Geographic visualization analysis of studies published in 2013, 2018, and 2023, and their correlations are shown in Figure 6. Globally, countries have the closest exchanges and cooperation in the field of POP research (Figure 6j–l), followed by antibiotics (Figure 6a–c), especially in Europe, the United States, and China, which have more extensive cooperative research, followed by Australia and South America.

Currently, the risk management and control of CECs is plagued by a few issues, primarily stemming from the lack of comprehensive laws and supervision. The absence of upper-level national laws on chemical management in countries like China hinders the regulation of CECs, leading to unclear numbers, unknown risks, and uncertainty regarding priority control of pollutants in specific industries and regions. Moreover, there is a weak foundation and insufficient reserves for effectively managing CECs, and innovation and governance capabilities need significant strengthening. Notably, existing environmental quality and industrial emission standards do not encompass CECs, despite their international significance. For instance, the "Surface Water Environmental Quality Standard" used in China lacks indicators for new POPs, such as perfluorinated and polyfluorinated compounds, which have been widely detected across various aquatic environments. Local management often focuses on conventional pollutants, with limited attention and supervision given to CECs and their potential health and environmental risks. Compared to conventional pollutants, CECs present more complex and hidden risks, posing greater challenges in terms of scientific understanding and effective governance. Consequently, addressing the risks associated with CECs necessitates urgent attention and comprehensive regulatory measures.

CECs in the environment pose a threat to human health. The derivation of environmental benchmarks and risk assessments of CECs is the only way to conduct risk management. Currently, aquatic water quality standards have been established for 35 pesticides, including organophosphorus pesticides, nicotine pesticides, acaricides, pyrethroids, carbamates, fungicides, and herbicides [81]. In addition, some studies have reported that a combination of the quantitative structure-activity relationship (QSAR), intermediate relationship model (ICE), and species sensitivity distribution model (SSD) can accurately predict the 5% species hazard concentration (HC_5), which is less than two times the actual experimental data concentration. The stratified ecological risk assessment method showed that 14.2% (HC_5) and 76.5% (HC1) of the surface water in China may present reproductive health risks, with the Yangtze River Basin being significantly higher than the other river basins [82]. It has been reported that 50 types of drugs and personal care products (PPCPs) are frequently detected in China's surface water, and after in-depth research, 12 PPCPs with greater risk to aquatic organisms have been selected using accurate probabilistic risk assessment methods [83]. Studies have also found that the ingestion of organophosphorus flame retardants through drinking water poses a potential carcinogenic risk, with a contribution rate of up to 72.4% [84]. In short, global research on the exposure and effects of CECs has made rapid progress, and it is urgent to determine the impact of CECs on the ecological environment and the risk assessment of human health, as well as to conduct comprehensive and in-depth mechanistic exploration and risk prevention and control of the environmental behavior, migration, transformation, and health effects of CECs.

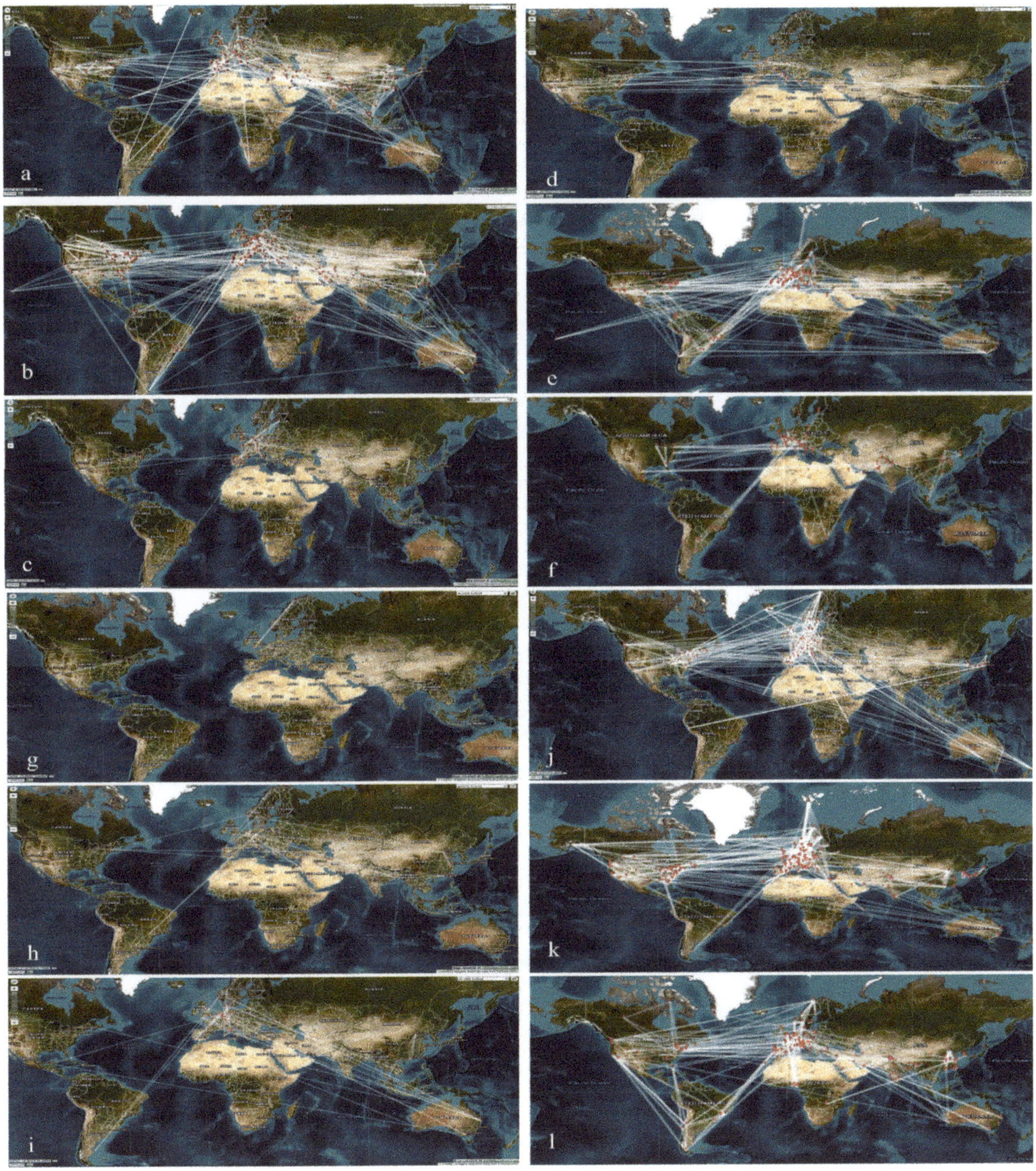

Figure 6. Geographical visualization of the literature published in 2013, 2018, and 2023 ((**a–c**): visualizations of antibiotics as the main topic; (**d–f**): visualizations with EDCs as the main inscription; (**g–i**): visualizations of microplastics for the title; (**j–l**): visualizations with POPs as the title. The white lines represent collaborative networks and the red dots represent research institutions).

Currently, in developed countries, there are environmental regulations addressing CECs. For example, the European Union has established an observation list [85] and implemented the Regulation on Registration, Evaluation, Authorization, and Restriction of Chemicals [86]. The United Kingdom has its own chemical investigation program and national implementation plans [87]. The United States has designated a candidate pollutant list that specifically focuses on unregulated contaminants in drinking water [88]. On 28 June 2022, China jointly issued the Ecological and Environmental Protection Plan for the Yellow River Basin and the Action Plan for Ecological Protection and Management of the Yellow River, focusing on the mainstream and major tributaries of the Yellow River to carry out environmental investigation, monitoring, and risk assessment of CECs. Various CECs, including endocrine disruptors, POPs, antibiotics, microplastics, and pesticides, have been detected in the agriculture surface water and sediments of the Yellow River Basin, among which the pollution levels of antibiotics and POPs are high according to statistical analysis of big data with the CiteSpace software (Figure 7). Microplastics have special carrier behavior and can absorb CECs such as antibiotics, PFAS, and endocrine disruptors from the environment [5]. With the ongoing battle against pollution, the focus of pollution control has shifted from the short-term and explicit risk management of conventional pollutants to the long-term and invisible risk management of CECs. However, current research has not organically combined the exposure and health impacts of CECs for the health risk assessment and management of CECs. Therefore, it is necessary to make scientific judgments based on scientific experiments and data; implement classification standards, quantitative risk assessment, differentiated assessment, and supervision; and provide scientific and technological support and systematic solutions to support the prevention and control of CECs.

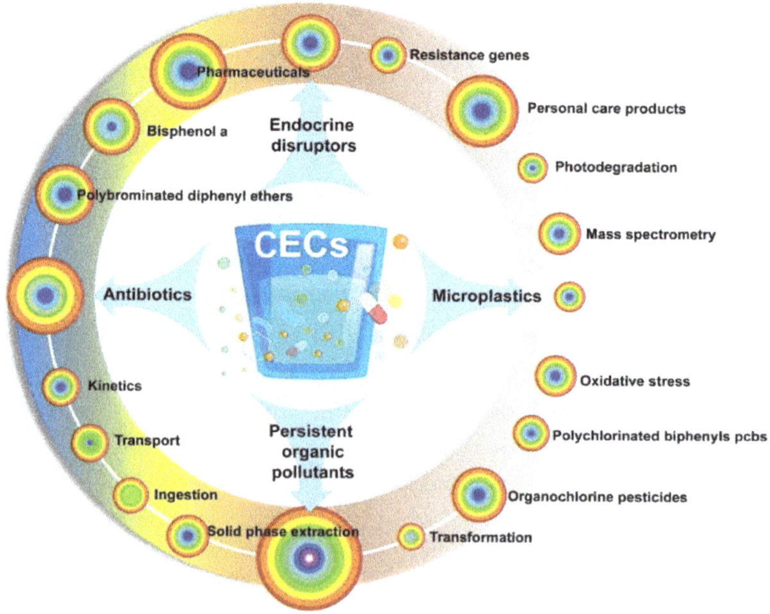

Figure 7. Various CECs in the agriculture-linked surface water and sediments of the Yellow River Basin.

Simultaneously, it is necessary to strengthen the top-level design, formulate CECs control strategies and action plans, determine the pollution bottom number, organize investigations and monitoring special actions, strengthen scientific research, and establish a technical system for the collaborative treatment of CECs and conventional pollutants. The control and treatment of CECs are related to social and economic development, ecological civilization construction, and people's lives and health, and are deep-seated environmental problems that must be solved as soon as possible. This is a long-term systematic project that should give full play to the leading and forcing role of ecological and environmental protection and coordinate the promotion and joint efforts of science and technology, economy, management, and policy to continuously improve environmental quality and promote the comprehensive green transformation of economic and social development.

6. Research Prospects

Currently, many CECs in the environment cannot be detected due to limitations in detection methods. Therefore, there is a need for the development of robust non-targeted monitoring approaches that can identify unknown compounds using minimal information. It is important to establish priority control lists for CECs to facilitate regulatory measures. Additionally, there should be a focus on the research and development of improved treatment methods for CECs in the environment, such as Advanced Oxidation Processes (AOPs) or other treatment technologies. To address the health risks posed by CECs, detailed research on their sources and behaviors is necessary. Overall, adopting a global approach based on screening, assessment, control, prohibition, reduction, and treatment will enable effective management of CECs and safeguard human health and the environment.

Author Contributions: W.F. wrote the manuscript and funding, Y.D. analyzed the data and drew pictures, F.Y. guided the manuscript, Q.M. modified the manuscript, and S.K.N. edited the manuscript. The manuscript was written through contributions of all authors. All authors have read and agreed to the published version of the manuscript.

Funding: This research was jointly supported by the Project of Science and Technology of the Inner Mongolia Province (2022YFHH0044) and the National Natural Science Foundation of China (grant numbers 42177400 and 41907338).

Data Availability Statement: Not applicable.

Conflicts of Interest: The authors declare no conflict of interest.

Abbreviations

AOPs	Advanced Oxidation Processes
BBOEP	butyl benzyl ortho-phthalate
BDCIPP	1,2-bis(2,4-dichlorophenyl)isopropanol
CECs	Contaminants of emerging concern
DDE	Dichlorodiphenylchloroethylene
DDT	dichlorodiphenyldichloroethane and its isomers
EDCs	Endocrine-Disrupting Chemicals
HCHs	hexachlorocyclohexane and isomers
HC_5	the 5% species hazard concentration
ICE	intermediate relationship model
MBR	Membrane bioreactor
MPs	microplastic
OCPs	organochlorine pesticides
PFCs	perfluorinated compounds
PFAS	per- and polyfluoroalkyl substances
PFOS	perfluorooctane sulfonate
POPs	persistent organic pollutants
PPAR	peroxide proliferation-activating receptor
PPCPs	personal care products
QSAR	quantitative structure-activity relationship
ROS	reactive oxygen species
SSD	species sensitivity distribution model
TCDD	2,3,7,8-tetrachlorodibenzo-p-dioxin

References

1. Song, F.; Li, T.; Hur, J.; Wu, F.; Meng, X. Eco-Colloidal Layer of Micro/Nanoplastics Increases Complexity and Uncertainty of Their Biotoxicity in Aquatic Environments. *Environ. Sci. Technol.* **2022**, *56*, 10547–10549. [CrossRef] [PubMed]
2. Li, L.; Li, F.; Deng, M.; Wu, C.; Zhao, X.; Song, K.; Wu, F. Microplastics Distribution Characteristics in Typical Inflow Rivers of Taihu Lake: Linking to Nitrous Oxide Emission and Microbial Analysis. *Water Res.* **2022**, *225*, 119117. [CrossRef] [PubMed]
3. Schnoor, J.L. Emerging Chemical Contaminants. *Environ. Sci. Technol.* **2003**, *37*, 375A. [CrossRef] [PubMed]
4. UNEP Inside the Push for a Global Chemicals Management Framework 2022. Available online: https://www.unep.org/news-and-stories/story/inside-push-global-chemicals-management-framework (accessed on 5 November 2023).
5. Wang, X.; Yu, X.; Wang, S.; Yin, X.; Qian, W.-X.; Lin, X.; Wu, Y.; Liu, C. Environmental Process, Effects and Risks of Emerging Contaminants in the Estuary-Coastal Environment. *Environ. Sci.* **2022**, *43*, 4810–4821. [CrossRef]
6. García-Fernández, A.J.; Espín, S.; Gómez-Ramírez, P.; Sánchez-Virosta, P.; Navas, I. Water Quality and Contaminants of Emerging Concern (CECs). In *Chemometrics and Cheminformatics in Aquatic Toxicology*; John Wiley & Sons, Ltd.: Hoboken, NJ, USA, 2021; pp. 1–21. ISBN 978-1-119-68139-7.
7. Meador, J.P.; Yeh, A.; Young, G.; Gallagher, E.P. Contaminants of Emerging Concern in a Large Temperate Estuary. *Environ. Pollut.* **2016**, *213*, 254–267. [CrossRef] [PubMed]
8. Kumar, A.; Sharma, G.; Kumari, A.; Guo, C.; Naushad, M.; Vo, D.-V.N.; Iqbal, J.; Stadler, F.J. Construction of Dual Z-Scheme g-C3N4/Bi4Ti3O12/Bi4O5I2 Heterojunction for Visible and Solar Powered Coupled Photocatalytic Antibiotic Degradation and Hydrogen Production: Boosting via I−/I3− and Bi3+/Bi5+ Redox Mediators. *Appl. Catal. B Environ.* **2021**, *284*, 119808. [CrossRef]
9. Liu, X.; Han, M.; Liu, Y.; Zhang, Z.; Lu, S.; Guo, X.; Xi, B.; Wu, F. Profiles and Potential Mobility of Antibiotic Resistance Genes in Different Bioelectrochemistry-Enhanced Constructed Wetlands. *Chem. Eng. J.* **2022**, *450*, 138005. [CrossRef]
10. Cao, X.; Huo, S.; Zhang, H.; Zhao, X.; Pang, C.; Ma, C.; Zheng, J.; Wu, F. Intensive Land-Based Activities Increase the Potential Risk of Benzo[α]Pyrene (BaP) to Aquatic Ecosystems and Human Health in Coastal Areas of China. *J. Clean. Prod.* **2022**, *371*, 133571. [CrossRef]
11. Ying, G.-G.; He, L.-Y.; Ying, A.J.; Zhang, Q.-Q.; Liu, Y.-S.; Zhao, J.-L. China Must Reduce Its Antibiotic Use. *Environ. Sci. Technol.* **2017**, *51*, 1072–1073. [CrossRef]
12. Fang, P.; Dai, H.; Zhuang, X.; Xie, H.; Luo, W.; Ren, M.; Zheng, J. Distribution of Antibiotic Resistance Genes in Typical Drinking Water Sources in the Lower Dongjiang River. *Ecol. Environ. Sci.* **2019**, *28*, 548–554.
13. Zhao, X.; Hu, Z.; Yang, X.; Cai, X.; Wang, Z.; Xie, X. Noncovalent Interactions between Fluoroquinolone Antibiotics with Dissolved Organic Matter: A 1H NMR Binding Site Study and Multi-Spectroscopic Methods. *Environ. Pollut.* **2019**, *248*, 815–822. [CrossRef] [PubMed]
14. Zhuo, L.; Wang, M.; Shi, Y.; Wang, Y.; Zhuang, X.; Xu, R.; Ren, M. Occurrence, Distribution, and Ecological Risk of Antibiotics in Surface Water of Typical Drinking Water Sources and Aquaculture in South China. *Asian J. Ecotoxicol.* **2019**, *14*, 164–175.
15. Hao, Y.; Xiong, S.; Wang, P.; Yang, R.; Pei, Z.; Li, Y.; Zhang, Q.; Jiang, G. Novel Brominated and Organophosphate Flame Retardants in the Atmosphere of Fildes Peninsula, West Antarctica: Continuous Observations from 2011 to 2020. *J. Hazard. Mater.* **2022**, *440*, 129776. [CrossRef] [PubMed]
16. da Costa Araújo, A.P.; da Luz, T.M.; Ahmed, M.A.I.; Ali, M.M.; Rahman, M.M.; Nataraj, B.; de Melo e Silva, D.; Barceló, D.; Malafaia, G. Toxicity Assessment of Polyethylene Microplastics in Combination with a Mix of Emerging Pollutants on Physalaemus Cuvieri Tadpoles. *J. Environ. Sci.* **2023**, *127*, 465–482. [CrossRef] [PubMed]
17. Ren, X.; Sun, Y.; Wang, Z.; Barceló, D.; Wang, Q.; Zhang, Z.; Zhang, Y. Abundance and Characteristics of Microplastic in Sewage Sludge: A Case Study of Yangling, Shaanxi Province, China. *Case Stud. Chem. Environ. Eng.* **2020**, *2*, 100050. [CrossRef]
18. Liu, X.; Wang, Z.; Wang, X.-L.; Li, Z.; Yang, C.; Li, E.-H.; Wei, H.-M. Status of Antibiotic Contamination and Ecological Risks Assessment of Several Typical Chinese Surface-Water Environments. *Environ. Sci.* **2019**, *40*, 2094–2100. [CrossRef]
19. Starling, M.C.V.M.; Amorim, C.C.; Leão, M.M.D. Occurrence, Control and Fate of Contaminants of Emerging Concern in Environmental Compartments in Brazil. *J. Hazard. Mater.* **2019**, *372*, 17–36. [CrossRef]
20. Visanji, Z.; Sadr, S.M.K.; Johns, M.B.; Savic, D.; Memon, F.A. Optimising Wastewater Treatment Solutions for the Removal of Contaminants of Emerging Concern (CECs): A Case Study for Application in India. *J. Hydroinformatics* **2019**, *22*, 93–110. [CrossRef]
21. Balakrishna, K.; Rath, A.; Praveenkumarreddy, Y.; Guruge, K.S.; Subedi, B. A Review of the Occurrence of Pharmaceuticals and Personal Care Products in Indian Water Bodies. *Ecotoxicol. Environ. Saf.* **2017**, *137*, 113–120. [CrossRef]
22. Blair, B.D.; Crago, J.P.; Hedman, C.J.; Klaper, R.D. Pharmaceuticals and Personal Care Products Found in the Great Lakes above Concentrations of Environmental Concern. *Chemosphere* **2013**, *93*, 2116–2123. [CrossRef]
23. Bonvin, F.; Rutler, R.; Chèvre, N.; Halder, J.; Kohn, T. Spatial and Temporal Presence of a Wastewater-Derived Micropollutant Plume in Lake Geneva. *Environ. Sci. Technol.* **2011**, *45*, 4702–4709. [CrossRef] [PubMed]
24. Weissinger, R.H.; Blackwell, B.R.; Keteles, K.; Battaglin, W.A.; Bradley, P.M. Bioactive Contaminants of Emerging Concern in National Park Waters of the Northern Colorado Plateau, USA. *Sci. Total Environ.* **2018**, *636*, 910–918. [CrossRef] [PubMed]
25. Masoner, J.R.; Kolpin, D.W.; Furlong, E.T.; Cozzarelli, I.M.; Gray, J.L.; Schwab, E.A. Contaminants of Emerging Concern in Fresh Leachate from Landfills in the Conterminous United States. *Environ. Sci. Process Impacts* **2014**, *16*, 2335–2354. [CrossRef] [PubMed]
26. Gao, Z.; Chen, Y.; Qin, Q.; Wang, R.; Dai, Z. Distribution Characteristics and Influencing Factors of Organochlorine Pesticides in Agricultural Soil from Xiamen City. *Int. J. Environ. Res. Public Health* **2023**, *20*, 1916. [CrossRef] [PubMed]

27. Ren, X.-M.; Yao, L.; Xue, Q.; Shi, J.; Zhang, Q.; Wang, P.; Fu, J.; Zhang, A.; Qu, G.; Jiang, G. Binding and Activity of Tetrabromobisphenol A Mono-Ether Structural Analogs to Thyroid Hormone Transport Proteins and Receptors. *Environ. Health Perspect.* **2020**, *128*, 107008. [CrossRef] [PubMed]
28. Meng, X.-Z.; Venkatesan, A.K.; Ni, Y.-L.; Steele, J.C.; Wu, L.-L.; Bignert, A.; Bergman, Å.; Halden, R.U. Organic Contaminants in Chinese Sewage Sludge: A Meta-Analysis of the Literature of the Past 30 Years. *Environ. Sci. Technol.* **2016**, *50*, 5454–5466. [CrossRef] [PubMed]
29. Feng, W.; Wang, T.; Zhu, Y.; Sun, F.; Giesy, J.P.; Wu, F. Chemical Composition, Sources, and Ecological Effect of Organic Phosphorus in Water Ecosystems: A Review. *Carbon Res.* **2023**, *2*, 12. [CrossRef]
30. Ji, G.; Gu, J.; Guo, M.; Zhou, L.; Wang, Z.; Shi, L.; Gu, A. A Systematic Comparison of the Developmental Vascular Toxicity of Bisphenol A and Its Alternatives in Vivo and in Vitro. *Chemosphere* **2022**, *291*, 132936. [CrossRef]
31. Liu, Q.; Zhou, Q.; Jiang, G. Nanomaterials for Analysis and Monitoring of Emerging Chemical Pollutants. *TrAC Trends Anal. Chem.* **2014**, *58*, 10–22. [CrossRef]
32. Zhong, H.; Liu, W.; Li, N.; Ma, D.; Zhao, C.; Li, J.; Wang, Y.; Jiang, G. Assessment of Perfluorohexane Sulfonic Acid (PFHxS)-Related Compounds Degradation Potential: Computational and Experimental Approaches. *J. Hazard. Mater.* **2022**, *436*, 129240. [CrossRef]
33. Gogoi, A.; Mazumder, P.; Tyagi, V.K.; Tushara Chaminda, G.G.; An, A.K.; Kumar, M. Occurrence and Fate of Emerging Contaminants in Water Environment: A Review. *Groundw. Sustain. Dev.* **2018**, *6*, 169–180. [CrossRef]
34. Abtahi, M.; Dobaradaran, S.; Torabbeigi, M.; Jorfi, S.; Gholamnia, R.; Koolivand, A.; Darabi, H.; Kavousi, A.; Saeedi, R. Health Risk of Phthalates in Water Environment: Occurrence in Water Resources, Bottled Water, and Tap Water, and Burden of Disease from Exposure through Drinking Water in Tehran, Iran. *Environ. Res.* **2019**, *173*, 469–479. [CrossRef] [PubMed]
35. Akhbarizadeh, R.; Dobaradaran, S.; Schmidt, T.C.; Nabipour, I.; Spitz, J. Worldwide Bottled Water Occurrence of Emerging Contaminants: A Review of the Recent Scientific Literature. *J. Hazard. Mater.* **2020**, *392*, 122271. [CrossRef] [PubMed]
36. Li, A.; Li, F.; Qiu, J.; Yan, C.; Liu, C.; Meng, F.; Li, Z.; Li, J.; Lang, Y.; Hu, H. Pollution Status, Biological Toxicity and Control Strategy of Microplastics in Water Environments: A Review. *J. Ocean. Univ. China* **2019**, *49*, 88–100.
37. Wang, F.; Wu, H.; Wu, W.; Wang, L.; Liu, J.; An, L.; Xu, Q. Microplastic Characteristics in Organisms of Different Trophic Levels from Liaohe Estuary, China. *Sci. Total Environ.* **2021**, *789*, 148027. [CrossRef] [PubMed]
38. Zhao, Z.-Y.; Wang, P.-Y.; Wang, Y.-B.; Zhou, R.; Koskei, K.; Munyasya, A.N.; Liu, S.-T.; Wang, W.; Su, Y.-Z.; Xiong, Y.-C. Fate of Plastic Film Residues in Agro-Ecosystem and Its Effects on Aggregate-Associated Soil Carbon and Nitrogen Stocks. *J. Hazard. Mater.* **2021**, *416*, 125954. [CrossRef] [PubMed]
39. Chen, M.; Qiang, L.; Pan, X.; Fang, S.; Han, Y.; Zhu, L. In Vivo and in Vitro Isomer-Specific Biotransformation of Perfluorooctane Sulfonamide in Common Carp (*Cyprinus carpio*). *Environ. Sci. Technol.* **2015**, *49*, 13817–13824. [CrossRef]
40. Wang, X.; Zhong, W.; Xiao, B.; Liu, Q.; Yang, L.; Covaci, A.; Zhu, L. Bioavailability and Biomagnification of Organophosphate Esters in the Food Web of Taihu Lake, China: Impacts of Chemical Properties and Metabolism. *Environ. Int.* **2019**, *125*, 25–32. [CrossRef]
41. McMurry, C.S.; Dickerson, R.L. Effects of Binary Mixtures of Six Xenobiotics on Hormone Concentrations and Morphometric Endpoints of Northern Bobwhite Quail (*Colinus virginianus*). *Chemosphere* **2001**, *43*, 829–837. [CrossRef]
42. Tang, Y.; Rong, J.; Guan, X.; Zha, S.; Shi, W.; Han, Y.; Du, X.; Wu, F.; Huang, W.; Liu, G. Immunotoxicity of Microplastics and Two Persistent Organic Pollutants Alone or in Combination to a Bivalve Species. *Environ. Pollut.* **2020**, *258*, 113845. [CrossRef]
43. Borgert, C.J.; Quill, T.F.; McCarty, L.S.; Mason, A.M. Can Mode of Action Predict Mixture Toxicity for Risk Assessment? *Toxicol. Appl. Pharmacol.* **2004**, *201*, 85–96. [CrossRef]
44. Caliman, F.A.; Gavrilescu, M. Pharmaceuticals, Personal Care Products and Endocrine Disrupting Agents in the Environment—A Review. *CLEAN—Soil Air Water* **2009**, *37*, 277–303. [CrossRef]
45. Spring, A.J.; Bagley, D.M.; Andrews, R.C.; Lemanik, S.; Yang, P. Removal of Endocrine Disrupting Compounds Using a Membrane Bioreactor and Disinfection. *J. Environ. Eng. Sci.* **2007**, *6*, 131–137. [CrossRef]
46. Órfão, J.J.M.; Silva, A.I.M.; Pereira, J.C.V.; Barata, S.A.; Fonseca, I.M.; Faria, P.C.C.; Pereira, M.F.R. Adsorption of a Reactive Dye on Chemically Modified Activated Carbons—Influence of pH. *J. Colloid Interface Sci.* **2006**, *296*, 480–489. [CrossRef] [PubMed]
47. Putra, E.K.; Pranowo, R.; Sunarso, J.; Indraswati, N.; Ismadji, S. Performance of Activated Carbon and Bentonite for Adsorption of Amoxicillin from Wastewater: Mechanisms, Isotherms and Kinetics. *Water Res.* **2009**, *43*, 2419–2430. [CrossRef] [PubMed]
48. Ikehata, K.; Jodeiri Naghashkar, N.; Gamal El-Din, M. Degradation of Aqueous Pharmaceuticals by Ozonation and Advanced Oxidation Processes: A Review. *Ozone Sci. Eng.* **2006**, *28*, 353–414. [CrossRef]
49. Rivera-Utrilla, J.; Sánchez-Polo, M.; Ferro-García, M.Á.; Prados-Joya, G.; Ocampo-Pérez, R. Pharmaceuticals as Emerging Contaminants and Their Removal from Water. A Review. *Chemosphere* **2013**, *93*, 1268–1287. [CrossRef] [PubMed]
50. Velo-Gala, I.; López-Peñalver, J.J.; Sánchez-Polo, M.; Rivera-Utrilla, J. Ionic X-Ray Contrast Media Degradation in Aqueous Solution Induced by Gamma Radiation. *Chem. Eng. J.* **2012**, *195–196*, 369–376. [CrossRef]
51. Vogna, D.; Marotta, R.; Napolitano, A.; Andreozzi, R.; d'Ischia, M. Advanced Oxidation of the Pharmaceutical Drug Diclofenac with UV/H2O2 and Ozone. *Water Res.* **2004**, *38*, 414–422. [CrossRef]
52. Pico, Y.; Belenguer, V.; Corcellas, C.; Diaz-Cruz, M.S.; Eljarrat, E.; Farré, M.; Gago-Ferrero, P.; Huerta, B.; Navarro-Ortega, A.; Petrovic, M.; et al. Contaminants of Emerging Concern in Freshwater Fish from Four Spanish Rivers. *Sci. Total Environ.* **2019**, *659*, 1186–1198. [CrossRef]

53. Previšić, A.; Rožman, M.; Mor, J.-R.; Acuña, V.; Serra-Compte, A.; Petrović, M.; Sabater, S. Aquatic Macroinvertebrates under Stress: Bioaccumulation of Emerging Contaminants and Metabolomics Implications. *Sci. Total Environ.* **2020**, *704*, 135333. [CrossRef]
54. Shi, Q.; Xiong, Y.; Kaur, P.; Sy, N.D.; Gan, J. Contaminants of Emerging Concerns in Recycled Water: Fate and Risks in Agroecosystems. *Sci. Total Environ.* **2022**, *814*, 152527. [CrossRef] [PubMed]
55. Fu, Q.; Sanganyado, E.; Ye, Q.; Gan, J. Meta-Analysis of Biosolid Effects on Persistence of Triclosan and Triclocarban in Soil. *Environ. Pollut.* **2016**, *210*, 137–144. [CrossRef] [PubMed]
56. Fu, Q.; Wu, X.; Ye, Q.; Ernst, F.; Gan, J. Biosolids Inhibit Bioavailability and Plant Uptake of Triclosan and Triclocarban. *Water Res.* **2016**, *102*, 117–124. [CrossRef] [PubMed]
57. Wu, X.; Ernst, F.; Conkle, J.L.; Gan, J. Comparative Uptake and Translocation of Pharmaceutical and Personal Care Products (PPCPs) by Common Vegetables. *Environ. Int.* **2013**, *60*, 15–22. [CrossRef] [PubMed]
58. Chang, B.-V.; Liu, J.-H.; Liao, C.-S. Aerobic Degradation of Bisphenol-A and Its Derivatives in River Sediment. *Environ. Technol.* **2014**, *35*, 416–424. [CrossRef] [PubMed]
59. Yang, Y.; Wang, Z.; Xie, S. Aerobic Biodegradation of Bisphenol A in River Sediment and Associated Bacterial Community Change. *Sci. Total Environ.* **2014**, *470–471*, 1184–1188. [CrossRef] [PubMed]
60. Yu, Y.; Liu, Y.; Wu, L. Sorption and Degradation of Pharmaceuticals and Personal Care Products (PPCPs) in Soils. *Environ. Sci. Pollut. Res.* **2013**, *20*, 4261–4267. [CrossRef] [PubMed]
61. Gao, Y.; Ren, L.; Ling, W.; Gong, S.; Sun, B.; Zhang, Y. Desorption of Phenanthrene and Pyrene in Soils by Root Exudates. *Bioresour. Technol.* **2010**, *101*, 1159–1165. [CrossRef]
62. LeFevre, G.H.; Hozalski, R.M.; Novak, P.J. Root Exudate Enhanced Contaminant Desorption: An Abiotic Contribution to the Rhizosphere Effect. *Environ. Sci. Technol.* **2013**, *47*, 11545–11553. [CrossRef]
63. Miya, R.K.; Firestone, M.K. Enhanced Phenanthrene Biodegradation in Soil by Slender Oat Root Exudates and Root Debris. *J. Environ. Qual.* **2001**, *30*, 1911–1918. [CrossRef] [PubMed]
64. Terrado, M.; Barceló, D.; Tauler, R. Multivariate Curve Resolution of Organic Pollution Patterns in the Ebro River Surface Water–Groundwater–Sediment–Soil System. *Anal. Chim. Acta* **2010**, *657*, 19–27. [CrossRef] [PubMed]
65. Esposito, G.; Prearo, M.; Renzi, M.; Anselmi, S.; Cesarani, A.; Barcelò, D.; Dondo, A.; Pastorino, P. Occurrence of Microplastics in the Gastrointestinal Tract of Benthic by–Catches from an Eastern Mediterranean Deep–Sea Environment. *Mar. Pollut. Bull.* **2022**, *174*, 113231. [CrossRef] [PubMed]
66. Wu, S.; Powers, S.; Zhu, W.; Hannun, Y.A. Substantial Contribution of Extrinsic Risk Factors to Cancer Development. *Nature* **2016**, *529*, 43–47. [CrossRef] [PubMed]
67. Gu, J.; Guo, M.; Yin, X.; Huang, C.; Qian, L.; Zhou, L.; Wang, Z.; Wang, L.; Shi, L.; Ji, G. A Systematic Comparison of Neurotoxicity of Bisphenol A and Its Derivatives in Zebrafish. *Sci. Total Environ.* **2022**, *805*, 150210. [CrossRef] [PubMed]
68. Zhu, Q.; Wang, M.; Jia, J.; Hu, Y.; Wang, X.; Liao, C.; Jiang, G. Occurrence, Distribution, and Human Exposure of Several Endocrine-Disrupting Chemicals in Indoor Dust: A Nationwide Study. *Environ. Sci. Technol.* **2020**, *54*, 11333–11343. [CrossRef] [PubMed]
69. Domenech, J.; Marcos, R. Pathways of Human Exposure to Microplastics, and Estimation of the Total Burden. *Curr. Opin. Food Sci.* **2021**, *39*, 144–151. [CrossRef]
70. Feng, W.; Deng, Y.; Cao, Y.; Liu, J.; Han, Y.; Liu, J.; Miao, Q.; Yang, F.; Zhu, Y.; Giesy, J.P. Biotechnology Remediation and Environmental Behavior of Microplastics in Soils: A Review. *Rev. Environ. Contam. Toxicol. (Former. Residue Rev.)* **2023**, *261*, 13. [CrossRef]
71. Liao, C.; Liu, F.; Alomirah, H.; Loi, V.D.; Mohd, M.A.; Moon, H.-B.; Nakata, H.; Kannan, K. Bisphenol S in Urine from the United States and Seven Asian Countries: Occurrence and Human Exposures. *Environ. Sci. Technol.* **2012**, *46*, 6860–6866. [CrossRef]
72. McRose, D.L.; Newman, D.K. Redox-Active Antibiotics Enhance Phosphorus Bioavailability. *Science* **2021**, *371*, 1033–1037. [CrossRef]
73. Wang, X.; Zheng, H.; Zhao, J.; Luo, X.; Wang, Z.; Xing, B. Photodegradation Elevated the Toxicity of Polystyrene Microplastics to Grouper (Epinephelus Moara) through Disrupting Hepatic Lipid Homeostasis. *Environ. Sci. Technol.* **2020**, *54*, 6202–6212. [CrossRef]
74. Li, K.; Sun, J.; Yang, J.; Roberts, S.M.; Zhang, X.; Cui, X.; Wei, S.; Ma, L.Q. Molecular Mechanisms of Perfluorooctanoate-Induced Hepatocyte Apoptosis in Mice Using Proteomic Techniques. *Environ. Sci. Technol.* **2017**, *51*, 11380–11389. [CrossRef] [PubMed]
75. Zeng, J.; Kuang, H.; Hu, C.; Shi, X.; Yan, M.; Xu, L.; Wang, L.; Xu, C.; Xu, G. Effect of Bisphenol A on Rat Metabolic Profiling Studied by Using Capillary Electrophoresis Time-of-Flight Mass Spectrometry. *Environ. Sci. Technol.* **2013**, *47*, 7457–7465. [CrossRef] [PubMed]
76. Liao, C.; Li, X.; Wu, B.; Duan, S.; Jiang, G. Acute Enhancement of Synaptic Transmission and Chronic Inhibition of Synaptogenesis Induced by Perfluorooctane Sulfonate through Mediation of Voltage-Dependent Calcium Channel. *Environ. Sci. Technol.* **2008**, *42*, 5335–5341. [CrossRef] [PubMed]
77. Luo, D.; Liu, W.; Wu, W.; Tao, Y.; Hu, L.; Wang, L.; Yu, M.; Zhou, A.; Covaci, A.; Xia, W.; et al. Trimester-Specific Effects of Maternal Exposure to Organophosphate Flame Retardants on Offspring Size at Birth: A Prospective Cohort Study in China. *J. Hazard. Mater.* **2021**, *406*, 124754. [CrossRef] [PubMed]

78. OECD. *Toward a New Comprehensive Global Database of Per- and Polyfluoroalkyl Substances (PFASs): Summary Reporton Updating the OECD 2007 List of Per- and Polyfluoroalkyl Substances (PFASs)*; OECD: Paris, France, 2018.
79. Lu, Y.; Gao, K.; Li, X.; Tang, Z.; Xiang, L.; Zhao, H.; Fu, J.; Wang, L.; Zhu, N.; Cai, Z.; et al. Mass Spectrometry-Based Metabolomics Reveals Occupational Exposure to Per- and Polyfluoroalkyl Substances Relates to Oxidative Stress, Fatty Acid β-Oxidation Disorder, and Kidney Injury in a Manufactory in China. *Environ. Sci. Technol.* **2019**, *53*, 9800–9809. [CrossRef] [PubMed]
80. Orive, G.; Lertxundi, U.; Brodin, T.; Manning, P. Greening the Pharmacy. *Science* **2022**, *377*, 259–260. [CrossRef] [PubMed]
81. Huang, P.; Liu, S.-S.; Wang, Z.-J.; Ding, T.-T.; Xu, Y.-Q. Deriving the Predicted No Effect Concentrations of 35 Pesticides by the QSAR-SSD Method. *Chemosphere* **2022**, *298*, 134303. [CrossRef]
82. Jin, X.; Wang, Y.; Jin, W.; Rao, K.; Giesy, J.P.; Hollert, H.; Richardson, K.L.; Wang, Z. Ecological Risk of Nonylphenol in China Surface Waters Based on Reproductive Fitness. *Environ. Sci. Technol.* **2014**, *48*, 1256–1262. [CrossRef]
83. Liu, N.; Jin, X.; Feng, C.; Wang, Z.; Wu, F.; Johnson, A.C.; Xiao, H.; Hollert, H.; Giesy, J.P. Ecological Risk Assessment of Fifty Pharmaceuticals and Personal Care Products (PPCPs) in Chinese Surface Waters: A Proposed Multiple-Level System. *Environ. Int.* **2020**, *136*, 105454. [CrossRef]
84. Li, J.; He, J.; Li, Y.; Liu, Y.; Li, W.; Wu, N.; Zhang, L.; Zhang, Y.; Niu, Z. Assessing the Threats of Organophosphate Esters (Flame Retardants and Plasticizers) to Drinking Water Safety Based on USEPA Oral Reference Dose (RfD) and Oral Cancer Slope Factor (SFO). *Water Res.* **2019**, *154*, 84–93. [CrossRef]
85. European Union. Commission Implementing Decision (EU) 2015/495 of 20 March 2015 Establishing a Watch List of Substances for Union-Wide Monitoring in the Field of Water Policy Pursuant to Directive 2008/105/EC of the European Parliament and of the Council (Notified under Document C(2015) 1756) Text with EEA Relevance. *Off. J. Eur. Union* **2015**, *78*, 40–42.
86. REACH. *European Chemicals Agency (ECHA)—How the European Union Regulates Chemicals*; REACH: Helsinki, Finland, 2016.
87. ALS. *Environmental Chemicals Investigation Programme 2—Overview*; ALS: Coventry, UK, 2015.
88. US EPA. *Contaminant Candidate List (CCL) and Regulatory Determination*; US EPA: Washington, DC, USA, 2016.

water

MDPI

Article

Features of the Formation of Strontium Pollution of Drinking Groundwater and Associated Health Risks in the North-West of Russia

Alexander I. Malov

N. Laverov Federal Center for Integrated Arctic Research of the Ural Branch of the Russian Academy of Sciences, 20 Nikolsky Ave., Arkhangelsk 163020, Russia; malovai@yandex.ru

Abstract: The purpose of this research was to determine the natural factors that contribute to maintaining the standard quality of fresh drinking groundwater in areas with high strontium content. Hazard index values for the consumption of water containing strontium were also calculated to assess the overall non-carcinogenic health risk from combined ingestion and dermal exposure. The results showed that the groundwater with strontium concentrations exceeding the maximum permissible concentrations had an increased correlation of strontium concentrations with total dissolved solids and celestite and gypsum saturation indices. A decrease in calcium content was recorded with a simultaneous increase in the concentration of magnesium and strontium. Reducing conditions in the aquifer were also favorable for the conservation of these waters. In waters of standard quality, all these factors did not appear, which indicates their formation in sediments with discretely located small inclusions of celestite and gypsum. These waters were characterized by a calcium bicarbonate composition, low total dissolved solids (TDS), and oxidizing conditions. Elevated radiocarbon contents indicate their relatively young age. In general, it was found that children in the study area are most vulnerable to risks. Fifty percent of wells supply drinking water that is unsafe for consumption. The water from about a third of the wells studied is dangerous for adults.

Keywords: drinking water; Sr; isotope–chemical composition; dating; health risks

Citation: Malov, A.I. Features of the Formation of Strontium Pollution of Drinking Groundwater and Associated Health Risks in the North-West of Russia. *Water* **2023**, *15*, 3846. https://doi.org/10.3390/w15213846

Academic Editors: Weiying Feng, Fang Yang and Jing Liu

Received: 4 October 2023
Revised: 1 November 2023
Accepted: 1 November 2023
Published: 3 November 2023

1. Introduction

Groundwater is one of the main components of the environment, without which the existence of living organisms, including humans, is impossible. Groundwater has a number of significant advantages over surface water as it is better protected from environmental pollution than surface water and should be used first [1]. However, due to their dynamism, they actively participate in the processes of interaction between water and rocks, transferring various chemical elements into solution, including those that negatively affect the quality of drinking water [2–5].

The greatest difficulties are presented by elements whose maximum permissible concentrations (MPC) are characteristic of natural waters (e.g., in $mg \cdot L^{-1}$: Fe 0.3, F 1.2–1.5, Be 0.0002, Se 0.01, As 0.05, Sr 7.0, Mn 0.1) [6]. The low maximum permissible concentrations of these elements are explained by their organoleptic (Fe, Mn) and sanitary–toxicological properties (F, Be, Se, As, Sr).

According to [7], animal studies have shown that the chemical similarity of strontium to calcium allows it to be exchanged for calcium in various biological processes; the most important of these is the replacement of calcium in the bones, which affects skeletal development, causing changes similar to those associated with rickets [8–11]. Strontium may reduce cartilaginous bone calcification in children and adolescents with greater effects than those in adults [12].

The study of strontium in relation to diseases with a predominant lesion of the osteoarticular system in the form of deforming chondroosteoarthritis began more than 170 years ago with Russian scientists' pioneering work on Urovskaya disease (the name was given according to the place of its first discovery: the Urov River near Lake Baikal in the south of Central Siberia) or Kashin–Beck disease (after the names of the doctors who first studied this disease) [13]. Later, it turned out that this disease is also common in Tadzhikistan [14], NW China [15], North Korea, and possibly in some areas of Africa [16,17]. However, despite a long period of study, the exact cause of this disease has not yet been established. Currently, more than 20 theories and hypotheses are being discussed in the scientific literature to explain the etiology and factors of the disease. The priority theory is biogeochemical, according to which the occurrence and course of the disease depend on environmental factors (i.e., a lack/excess of chemical elements or compounds). In particular, researchers have paid special attention to deficiencies of selenium and calcium and the low Ca:Sr ratio in groundwater and surface water, bottom sediments, soil, vegetation, animal bones, human teeth, and hair [18–21].

The relevance of studying fresh groundwater is due to the fact that: (1) drinking water is the most necessary and obligatory substance in the human diet, (2) groundwater with a high strontium content forms regional hydrogeochemical provinces in areas of widespread carbonate rocks in humid areas [11,22,23], and (3) groundwater with high strontium content is also widespread in arid zones due to the evaporative concentration of relatively shallow groundwater [2]. Significant research has been carried out in central Russia, Siberia, China, the US, and Australia [24–30]. Geological–hydrogeological, statistical, and thermodynamic analyses have shown that in a number of cases there is a relationship between the strontium content in water and the lithology of rocks, the age of groundwater, the total dissolved solids (TDS), and the degree of saturation of groundwater in relation to gypsum and celestite. However, the correlations were often somewhat weak [29].

In northwestern Russia, the highest Sr concentrations were established relatively recently in the carbonate aquifer in the mouth area of the Mezen River basin (see Figure 1b) [31]. Contamination of groundwater with strontium was also confirmed when drilling wells for water to the south in the cities of Karpogory, Bereznik, Shenkursk, and Velsk (see Figure 1a), and the implementation of a joint project by geochemists from Russia, Finland, and Norway, showed that over the entire area of distribution of Permian rocks P_1s, P_2u, P_2kz, and P_2t (see Figure 1a), even the surface water contains strontium at a level of 0.5 to 2 mg·L^{-1} [32].

However, it remains unclear why there is a significant variation in strontium concentrations in groundwater over short distances. At the same time, in our opinion, it is obvious that the conditions for the formation of groundwater with high and low strontium concentrations are different, and it is necessary to look for correlation dependencies separately for these two types of groundwater. Due to the fact that the MPC of strontium in Russia is 7 mg·L^{-1}, this value was chosen during the study to separate two groundwater clusters. The statistical study also showed that there is indeed a bimodal distribution of Sr concentrations in water, with the above two clusters being identified (see Figure S1, Table S1, and Section S1 of the Supplementary Materials). The Shapiro–Wilk test of normality indicated that there is a possibility of normal distributions in these two clusters (Table S2).

The purpose of this research was to determine natural factors that contribute to maintaining the standard quality of fresh drinking groundwater in areas where high strontium concentrations occur. To assess the overall non-carcinogenic health risk from combined ingestion and dermal exposure, hazard index (HI) values from consumption of water containing strontium were also calculated (see Section S4 of Supplementary Materials). Values below one indicate no significant non-cancer health risk. HI values greater than one reflect the potential for non-cancerous health outcomes, and this probability increases as HI values rise [33].

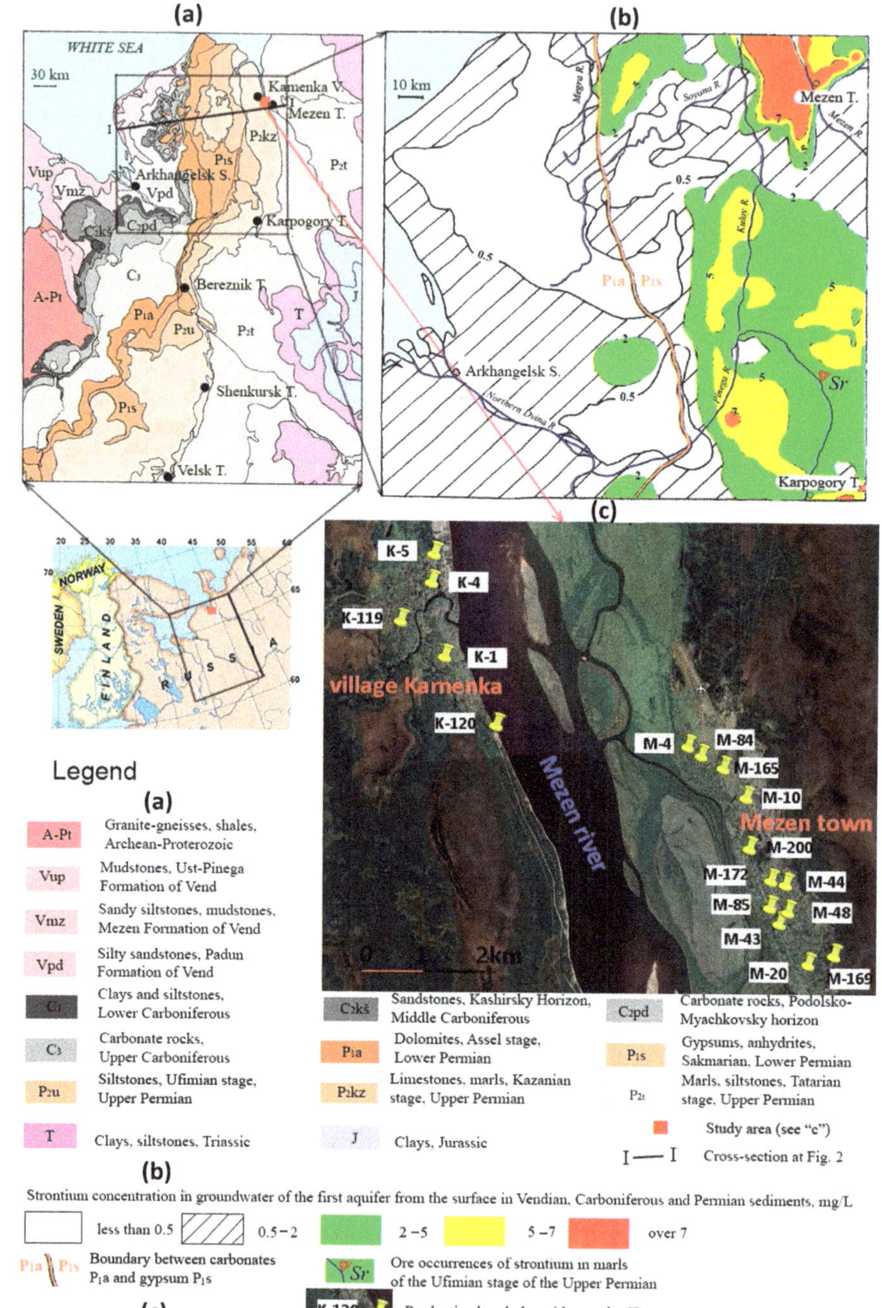

Figure 1. General location of the study site showing (**a**) geology, (**b**) strontium distribution in groundwater of the first aquifer from the surface (after [31]), and (**c**) sampling locations. I–I—cross-section at Figure 2.

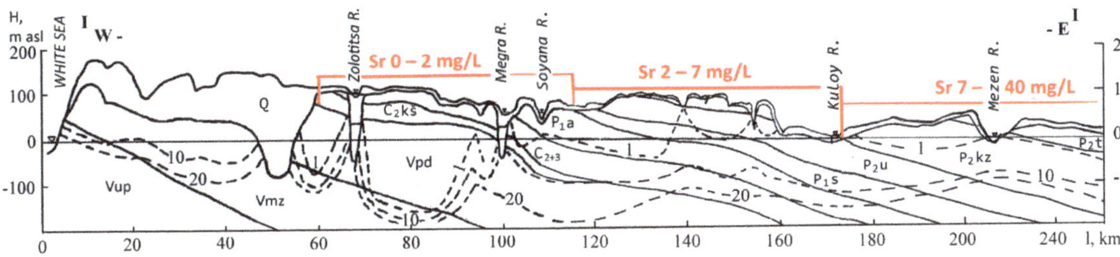

Figure 2. Schematic hydrogeological section along line I–I in Figure 1a. The dotted lines show the isolines of the total mineralization of groundwater (according to [31]); the red color shows the concentrations of strontium in fresh groundwater. Designations in Figure 1.

2. Geological Conditions

The research area is located in the junction zone of the Mezen syneclise and the eastern slope of the Baltic shield. Figure 1a shows the outcrop of this shield, composed of the most ancient Archean–Early Proterozoic (A-Pt) primary sedimentary rocks, transformed into intensely deformed and deeply metamorphosed gneisses, granite gneisses, amphibolites, and crystalline schists. At the beginning of the Riphean period (R) of the Late Proterozoic stage, a system of narrow extended grabens (aulacogens) was formed in this crystalline basement, which was filled with terrigenous, carbonate–terrigenous, and effusive-terrigenous strata up to several kilometers thick. Regional subsidence of the territory, which is associated with the formation of the Mezen syneclise, began in the Vendian (Ediacaran) period of the Late Proterozoic.

Sedimentation of the Ust Pinega (Vup) and Mezen (Vmz) formations (see Figure 2) proceeded in a relatively shallow epicontinental sea with weak hydrodynamics and a predominance of a reducing geochemical setting. Thin horizontal layering of the band type, accentuated by films of organic matter, probably arose during seasonal climate fluctuations. Against the background of calm and unidirectional subsidence, at times there were short-term rises of the bottom, accompanied by shallowing of the sea, as evidenced by the appearance of thin layers of siltstones and sandstones in the section. The formation of coarse-grained facies (probably deltaic or coastal-marine) in the Padun Formation (Vpd) is associated with the activation of multidirectional tectonic movements. From the end of the Vendian to the beginning of the formation of the Lower Carboniferous deposits, the region was mainly an area of removal and continental weathering. The significant proportion of kaolinite and cement in the terrigenous formations of the Vendian Padun Formation testifies to intense weathering processes [34,35].

In the Lower Carboniferous (C_1), the subsidence of the territory begins. The sediments are dominated by clays and siltstones, with interlayers of sandstones and conglomerates and thin interlayers of carbonate rocks: dolomites, limestones, and marls. The facies variability of the rocks of the Kashir horizon of the Middle Carboniferous (C_2kš), the absence of irregularities in the distribution of clastic material in terms of granulometric composition, poor roundness, and sorting of fragments, the presence of layers of gravelstones and conglomerates, and a sharp variability in thickness indicate the accumulation of these deposits in a turbulent coastal environment. The maximum distribution of the Middle Carboniferous Sea occurred during the Podolian time (C_2pd). Large areas accumulated dolomites and limestones. The fauna from these deposits testifies to a shallow, coastal zone of the marine basin with a salinity close to normal. In the Upper Carboniferous (C_3), the transgression reaches its maximum size. At this time, thick strata of dolomites had accumulated over most of the area.

The beginning of the Permian period was marked by a powerful Asselian transgression (P_1a), which manifested itself on a planetary scale. The marine basin was characterized by normal salinity and mostly shallow water, which is confirmed by the diversity of faunal remains. However, in some areas there was a regime in which the conditions of sediment deposition somewhat deviated from those of a normal sea basin, and here lagoons periodically formed, which are associated with the appearance of gypsum interlayers. The salinization and regression of the sea, which began in the Upper Asselian, continued most intensively in the Sakmarian time (P_1s), by the end of which the territory was a gradually salinizing lagoon. Here, at first, the accumulation of dolomites containing gypsum interlayers occurred, followed by the accumulation of thick strata (about 80 m) of gypsum and anhydrites.

The type of sediments of the Ufimian stage of the Upper Permian (P_2u), represented mainly by red-colored siltstones, indicates that their accumulation occurred in a lagoonal–continental setting under arid climate conditions. The gypsum content of the rocks is replaced by celestite ore occurrences in the direction from the base to the top of the formation (see Figure 1b). The accumulation of sediments of the Kazanian stage of the Upper Permian (P_2kz) is associated with successive transgressions of the sea and its regression. Therefore, marls and calcareous clays are replaced by clayey, sandy, and dolomitic limestones with inclusions of gypsum and anhydrite. The Tatarian stage (P_2t) is represented at the base of the section by marls and siltstones with horizontal or similar bedding, which indicates the existence of a basin whose conditions were close to marine ones. Subsequently, there was an increasing desalination of the saline lagoons with the formation of the sandy–silty part of the sediments.

This regime did not last long. Subsequently, the entire territory entered the path of the predominantly continental Mesozoic–Cenozoic stage of denudation leveling. During this period, a significant part of the Lower Permian gypsum anhydrite sequence, distributed at least 100–150 km west of the currently observed boundary of the Sakmarian stage of the Lower Permian, was eroded (see Figures 1 and 2) [31].

As shown in Figure 2, the maximum concentrations of strontium, at 7–40 $mg \cdot L^{-1}$, in fresh groundwater are confined to the distribution area of the Kazanian-stage carbonate deposits. To the west, in the area of accumulation of thick strata of gypsum and anhydrites, they decrease to 2–7 $mg \cdot L^{-1}$, and further to the west, terrigenous-carbonate rocks contain groundwater with minimal concentrations of strontium. In [31], it is shown that, in general, one can note the dependence of the strontium content in water on its concentrations in rocks: in the direction from east to west they decrease from 2400 to 10 $mg \cdot kg^{-1}$. The Sr content in sandstones is 200 $mg \cdot kg^{-1}$, and in carbonate rocks is 610 $mg \cdot kg^{-1}$ [36,37].

The waters with the highest concentrations of strontium are the most dangerous for consumption, since in addition to the strontium content above the MPC for fresh drinking water, Ca/Sr << 100 ratios are also observed, which can cause the occurrence of Urov endemia (Kashin–Beck disease) [2].

3. Materials and Methods

In June 2022, 17 samples of drinking water used for water supply to the population were taken from boreholes located in the estuary zone of the Mezen River basin (see Figure 1c) for the purpose of studying the isotopic chemical composition (see Tables 1–3). The field sample preparation was carried out as described in a previous paper [38]. The analytical procedures are also described in previous papers [38,39]. The calcium, strontium, magnesium, sodium, and potassium concentrations were determined with an uncertainty degree of 1–2% by using an atomic absorption spectrometer (AAS) (Perkin-Elmer 5100 PC, Turku, Finland). Alkalinity was measured by potentiometric titration with HCl using an automated titrator (Metrohm 716 DMS Titrino, Metrohm AG, Herisau, Switzerland) using the Gran method (detection limit 10^{-5} M, uncertainty at $\geq$0.5 $mmol \cdot L^{-1}$ 1–3% and at <0.5 $mmol \cdot L^{-1}$ 7%). The major anion concentrations (Cl^-, SO_4^{2-}) were measured by ion chromatography (HPLC, Dionex ICS 2000, ThermoFisher, Waltham, MA, USA) with an uncertainty of 2%.

Table 1. The physicochemical parameters of groundwater.

Sample ID	Location	H, m.a.s.l.	Sample Date// Depth (m)	T (°C)	pH (Unit)	Eh (mV)	O_2 ($mg \cdot L^{-1}$)	TDS ($mg \cdot L^{-1}$)	Sr ($mg \cdot L^{-1}$)
K-120	N 65.87238 E 44.13921	6	18 June//40–60	5.3	8.03	151	3.0	410	12.09
K-1	N 65.88087 E 44.12091	22	18 June//40–60	7.1	7.99	155	8.0	446	8.23
M-200	N 65.85261 E 44.23201	11	21 June//40–60	4.8	7.90	−60	2.8	469	1.91
M-84	N 65.86652 E 44.21601	13	20 June//40–60	5.3	7.77	73	3.2	473	0.76
M-10	N 65.86009 E 44.23067	10	19 June//40–60	5.5	7.87	−121	2.1	538	17.10
M-20	N 65.83675 E 44.26264	28	21 June//40–60	4.0	7.89	−103	2.2	562	1.50
M-85	N 65.84367 E 44.23960	10	19 June//40–60	5.6	7.82	8	6.0	646	26.31
M-4	N 65.86776 E 44.20915	11	20 June//40–60	4.2	7.45	66	0.6	656	1.02
M-169	N 65.83541 E 44.25338	19	21 June//40–60	5.0	7.78	−58	3.4	663	6.60
M-48	N 65.84318 E 44.24603	12	22 June//40–60	5.3	7.86	−78	2.8	669	4.62
M-44	N 65.84731 E 44.24515	17	22 June//40–60	5.3	7.67	−6	3.0	705	2.40
M-43	N 65.84137 E 44.24300	9	19 June//40–60	6.5	7.61	−109	0.6	731	32.00
M-165	N 65.86820 E 44.22311	17	20 June//40–60	4.6	7.56	58	2.0	752	1.50
M-172	N 65.84790 E 44.23991	13	22 June//40–60	8.2	7.64	38	5.6	780	2.70
K-4	N 65.89190 E 44.11613	10	17 June//40–60	6.0	7.28	−41	4.5	803	40.11
K-5	N 65.89627 E 44.11808	11	17 June//40–60	5.7	7.06	−116	0.0	857	27.14
K-119	N 65.88612 E 44.10291	12	17 June//40–60	5.0	7.67	−16	3.1	979	39.06

Sample ID	Na^+ ($mg \cdot L^{-1}$)	Ca^{2+}	Mg^{2+}	K^+	Cl^-	SO_4^{2-}	HCO_3^-	Ca^{2+}/Sr	Water Type [a] (-)
K-120	24.0	48.3	16.8	3.61	24.8	16.0	265	4.0	Ca-Mg-HCO_3
K-1	18.9	58.1	20.5	3.73	21.2	26.7	289	7.1	Ca-Mg-HCO_3
M-200	23.1	61.3	21.1	3.49	10.6	7.1	341	32.0	Ca-Mg-HCO_3
M-84	19.8	74.1	15.7	3.60	14.2	15.6	330	98.0	Ca-HCO_3
M-10	27.4	64.5	25.2	3.68	38.9	8.3	353	3.8	Ca-Mg-HCO_3
M-20	49.0	65.3	18.7	3.71	10.6	9.2	404	43.0	Ca-Na-HCO_3
M-85	35.4	61.3	29.6	3.91	17.7	85.1	387	2.4	Ca-Mg-HCO_3
M-4	38.9	107.0	9.1	4.25	15.9	56.5	424	107.0	Ca-HCO_3
M-169	55.9	74.9	25.1	4.16	24.8	6.5	465	11.0	Ca-Na-Mg-HCO_3
M-48	39.8	83.0	29.8	3.78	24.8	15.7	467	18.0	Ca-Mg-HCO_3
M-44	54.5	105.0	11.8	4.22	10.6	4.4	512	44.0	Ca-Na-HCO_3
M-43	80.0	49.3	25.1	4.31	33.6	72.9	433	2.2	Na-Ca-HCO_3
M-165	63.9	111.0	21.3	4.62	37.2	36.0	476	74.0	Ca-Na-HCO_3
M-172	72.4	111.0	7.9	3.90	33.6	44.0	494	41	Ca-Na-HCO_3
K-4	43.7	85.0	33.4	4.13	54.9	52.3	489	2.1	Ca-Mg-HCO_3
K-5	84.0	85.4	21.9	6.41	65.8	108.0	458	3.2	Ca-Na-HCO_3
K-119	170.0	58.9	37.2	11.10	158.0	123.0	382	1.5	Na-Mg-Ca-HCO_3-Cl

[a] Cations and anions with a content higher than 25 mg-eq·% are listed in descending order.

Table 2. The mineral saturation indices (SI) of groundwater.

Sample ID	Dolomite	Calcite	Strontianite	Anhydrite	Gypsum	Celestite
K-120	0.88	0.42	1.87	−2.98	−2.56	−2.57
K-1	0.90	0.43	1.64	−2.70	−2.28	−2.54
M-200	0.91	0.44	0.98	−3.25	−2.84	−3.76
M-84	0.58	0.38	0.45	−2.83	−2.42	−3.81
M-10	0.98	0.44	1.92	−3.19	−2.78	−2.76
M-20	1.03	0.54	0.95	−3.13	−2.72	−3.77
M-85	0.84	0.33	2.02	−2.23	−1.82	−1.59
M-4	−0.03	0.27	0.30	−2.16	−1.75	−3.18
M-169	1.08	0.53	1.52	−3.26	−2.85	−3.31
M-48	1.33	0.64	1.44	−2.85	−2.44	−3.10
M-444	0.75	0.60	1.02	−3.29	−2.88	−3.92
M-43	0.38	0.09	1.96	−2.40	−1.98	−1.57
M-165	0.79	0.50	0.69	−2.39	−1.98	−3.25
M-172	0.44	0.55	0.99	−2.28	−1.87	−2.88
K-4	0.16	0.04	1.77	−2.34	−1.93	−1.66
K-5	−0.59	−0.24	1.32	−2.02	−1.61	−1.51
K-119	0.79	0.25	2.14	−2.19	−1.77	−1.35

Table 3. Characteristics of the isotopic composition of groundwater and parameters used to estimate the residence time of groundwater in the aquifer.

Sample ID	^{14}C (pmc)	$\delta^{13}C$ (‰)	^{14}C [a] (pMC)	$^{14}C_0$ [b] (pmc)	$^{14}C_0$ [c]	$^{14}C_0$ [d]	$^{14}C_0$ [e]
K-120	44.04 ± 0.47	−10.6	42.79 ± 0.46	45.00	*NC*	60.42	85
K-1	45.4 ± 0.41	−10.7	44.1 ± 0.4	45.36	*NC*	60.99	85
M-200	50.15 ± 0.53	−9.1	48.56 ± 0.51	39.64	*NC*	51.87	85
M-84	65.02 ± 0.58	−13.2	63.49 ± 0.57	54.29	50.47	76.95	85
M-10	57.45 ± 0.66	−8.7	55.58 ± 0.64	38.21	*NC*	49.59	85
M-20	54.57 ± 0.54	−13.3	53.29 ± 0.53	54.64	53.98	75.81	85
M-85	49.32 ± 0.61	−11.8	49.02 ± 0.59	49.29	*NC*	67.26	85
M-4	74.47 ± 0.66	−14.6	72.92 ± 0.65	59.29	63.17	83.22	85
M-169	60.53 ± 0.65	−15.7	59.4 ± 0.64	63.21	90.87	89.49	85
M-48	58.3 ± 0.54	−15.8	57.23 ± 0.5	63.57	93.62	90.06	85
M-44	61.39 ± 0.60	−13.5	59.98 ± 0.59	55.36	52.79	76.95	85
M-43	44.64 ± 0.55	−9.9	43.29 ± 0.53	42.50	*NC*	56.43	85
M-165	72.2 ± 0.67	−12.0	70.32 ± 0.65	50.00	24.61	68.4	85
M-172	62.87 ± 0.56	−15.2	61.64 ± 0.55	61.43	77.44	86.64	85
K-4	50.24 ± 0.44	−9.8	48.71 ± 0.43	42.14	*NC*	55.86	85
K-5	59.23 ± 0.54	−13.3	57.84 ± 0.53	54.64	21.96	75.81	85
K-119	29.53 ± 0.49	−8.6	28.56 ± 0.47	37.86	*NC*	49.02	85
Sample ID	**^{14}C Age [b] (year BP)**	**^{14}C Age [c]**	**^{14}C Age [d]**	**^{14}C Age [e]**		**U (ppb)**	**$^{234}U/^{238}U$ (unit)**
K-120	424 ± 104	*NC*	2675 ± 163	6171 ± 144		0.130	1.12
K-1	214 ± 213	*NC*	2529 ± 169	5968 ± 204		0.075	1.13
M-200	modern	*NC*	236 ± 235	5027 ± 187		0.243	2.78
M-84	modern	modern	1337 ± 187	2237 ± 82		0.649	1.67
M-10	modern	*NC*	modern	3472 ± 102		0.026	1.63
M-20	208 ± 207	modern	2836 ± 84	4014 ± 128		0.094	1.65
M-85	modern	*NC*	2628 ± 138	5143 ± 163		0.225	2.08
M-4	modern	modern	842 ± 78	1049 ± 121		0.607	1.81
M-169	570 ± 76	3588 ± 109	3468 ± 91	2898 ± 109		0.027	3.48
M-48	804 ± 99	4369 ± 131	4540 ± 240	3336 ± 106		0.108	3.43
M-44	modern	modern	1801 ± 92	2810 ± 65		1.633	1.54
M-43	modern	*NC*	1820 ± 84	6130 ± 136		0.042	2.06
M-165	modern	modern	modern	1259 ± 76		0.994	1.23
M-172	modern	1632 ± 83	2791 ± 61	2605 ± 118		1.885	1.57
K-4	modern	*NC*	817 ± 88	4935 ± 92		0.078	2.26
K-5	modern	modern	2007 ± 104	3186 ± 140		0.085	1.59
K-119	2424 ± 272	*NC*	4662 ± 206	9828 ± 285		0.014	3.94

[a] normalized radiocarbon concentration; [b] initial carbon isotopic composition and calibrated age according to the Pearson model; [c] same for Mook model; [d] same for Ferronsky model; [e] same for Vogel model; *NC* not calculated.

Estimates of the residence time of groundwater in the aquifer were made using carbon isotopes ^{14}C and ^{13}C [40,41]. Two models were used to determine the initial radiocarbon content in the groundwater recharge area ($^{14}C_0$). For isotopic exchange between solid carbonate and total dissolved inorganic carbon (TDIC), the Ingerson and Pearson model was used [42]. For isotope exchange between soil CO_2 and TDIC, the Mook model [43,44] was used (see [38,45]).

Due to the fact that calculations using these models showed a very young age of groundwater, Ferronsky and Polyakov's [46] model to account only for the dissolution of carbonates was also used:

$$^{14}C_0 = -5.7\delta^{13}C \quad (1)$$

In addition, according to Vogel's [47] generalization that the radiocarbon activity of freshly formed groundwater in many parts of the world averages 85 ± 5% of the radiocarbon activity of a modern wood standard, the simplest model was also used:

$$^{14}C_0 = 85pmc \quad (2)$$

Calib Rev 8.1.0 was used to calibrate the radiocarbon ages [48,49].

4. Results

4.1. The Physicochemical Parameters of Groundwater

Physicochemical parameters characterizing the water composition of the Upper Permian Kazan carbonate aquifer (P_2kz) are shown in Table 1 and Section S2 of the Supplementary Materials and illustrated by Piper diagrams in Figure 3.

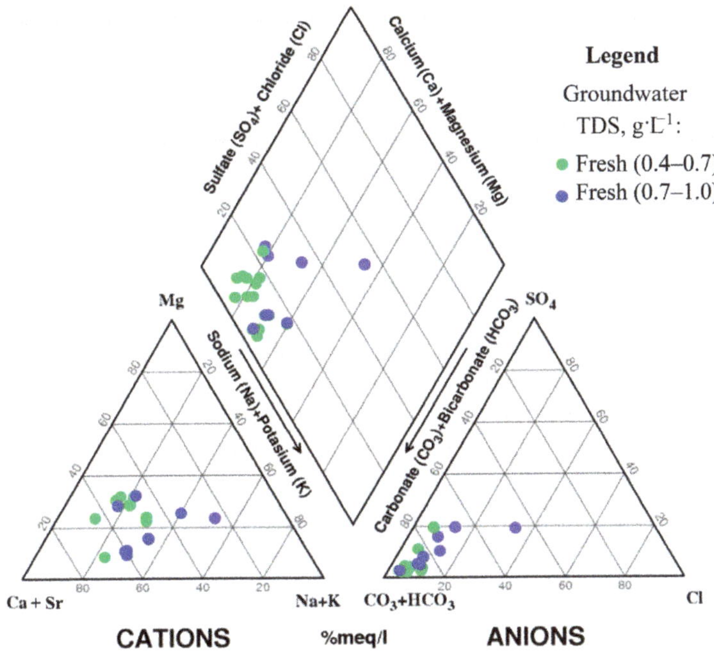

Figure 3. Piper diagrams.

TDS in fresh water ranged from 410 to 979 mg·L^{-1}. The Ca-Mg-HCO_3 composition was typical for low-mineralized waters. With an increase in their mineralization, the chemical composition changed to Ca-Na-HCO_3 and Na-Mg-Ca-HCO_3-Cl. As the composition of the groundwater changed, its pH also changed from the range of 7.45–8.03 (average value = 7.84) to the range of 7.06–7.67 (average value = 7.47). Sr concentrations ranged from 0.76 to 40.11 mg·L^{-1}, regardless of the overall water composition (Table 1). The water saturation indices (SI) for gypsum and anhydrite were largely negative (Table 2), and there was a deficiency of SO_4^{2-} relative to Ca^{2+} (Figure 4a), which indicated relatively small amounts of Ca^{2+} transferred into water due to the dissolution of gypsum. The exception was the most mineralized sample K-119, in which the Ca^{2+}:SO_4^{2-} ratio was close to unity (1.15).

In low-mineralized waters, the ratio (Ca^{2+} + Mg^{2+}):HCO_3^- is close to unity. As the degree of water mineralization increases, a deficiency (Ca^{2+} + Mg^{2+}) is observed (Figure 4b). At the same time, the ratios Na^+:HCO_3^- (Figure 4c) and Na^+:Ca^{2+} (Figure 4d) increase. Sea coasts are characterized by an observed increase in the concentration of chlorine in groundwater up to 158 mg·L^{-1} (see Figures 1a and 2). However, an excess of sodium compared to chlorine (Figure 4f) and negative values of chlor-alkali indices (Figure 4e) indicate the possibility of Na^+ passing into solution due to the cation exchange of alkaline earth elements for alkaline ones [50,51]. Additional sources of sodium can be aluminosilicates such as albite and oligoclase.

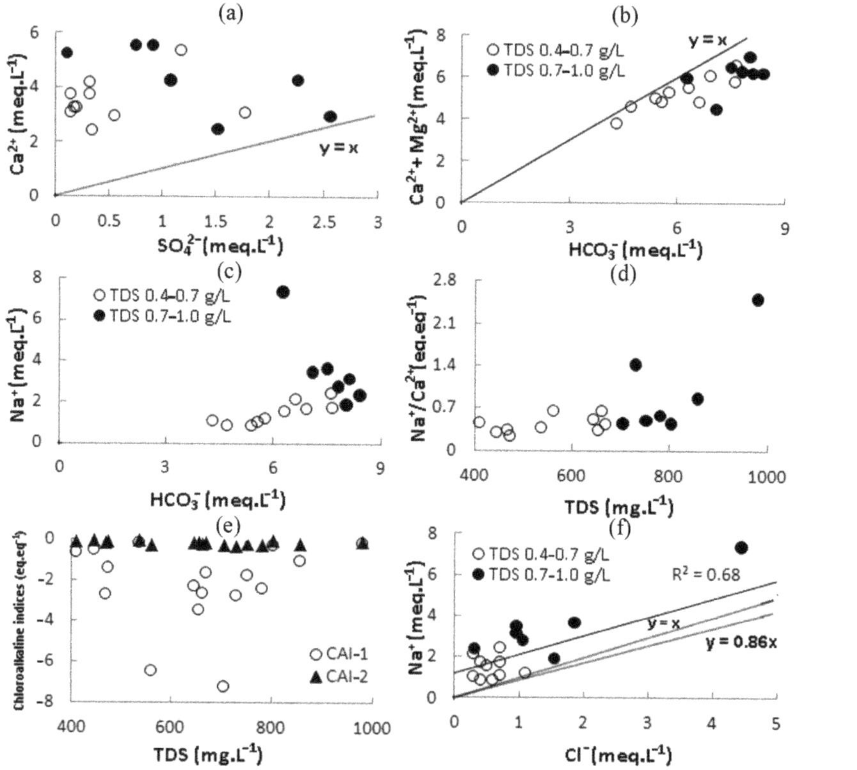

Figure 4. Diagrams showing element concentrations (meq·L^{-1}) and their ratios (eq·eq^{-1}) in the fresh groundwater of the study area with TDS groups; Ca^{2+} vs. $SO_4{}^{2-}$ (**a**), (Ca^{2+} + Mg^{2+}) vs. $HCO_3{}^-$ (**b**), Na^+ vs. $HCO_3{}^-$ (**c**), Na^+/Ca^{2+} vs. TDS (**d**), Chloroalkaline indices vs. TDS (**e**) and Na^+ vs. Cl^- (**f**).

4.2. Isotopic Parameters of Groundwater

The results of determinations of carbon and uranium isotopes in groundwater of the Upper Permian Kazan carbonate aquifer (P_2kz) are presented in Table 3.

As can be seen from the table, the Pearson and Mook models generally provide values for the residence time of groundwater in the aquifer that can be characterized as "modern." Four samples showed values ranging from 1632 ± 83 to 4540 ± 240 years. Calculations using the Ferronsky model showed groundwater age values comparable to those determined by the Mook model approximately up to 4662 ± 206 years, and according to the Vogel model they were twice as high.

In general, for all models, one can see a tendency to correlate the age of groundwater with radiocarbon concentrations in groundwater (Figure S2), and we considered it more logical to use the ^{14}C (pmc) values when analyzing changes in the strontium concentration depending on water age (see Section 4.5).

The ^{14}C values varied from 74.47 ± 0.66 to 29.53 ± 0.49 pmc. They did not correlate with TDS. A more detailed description is given in Section 4.5.

Uranium concentrations were maximum (0.2–1.9 μg·L^{-1}) under oxidizing conditions, where uranium was in the 6^+ form; under reducing conditions, uranium passed into the 4^+ state, and its content, as a rule, does not exceed 0.1 μg·L^{-1} (Figure S3a). The maximum concentrations of uranium tended to be in more alkaline conditions than the minimum concentrations (pH 7.5–8; Figure S3b), where they were in the composition of uranyl-carbonate complexes. The maximum concentrations of uranium also predominated in younger waters (^{14}C = 60–80 pmc) when compared to the minimal concentrations

(^{14}C = 20–60 pmc; Figure S3c). In waters with maximum uranium concentrations, where rock dissolution processes predominate, the ratio of uranium isotopes ^{234}U:^{238}U was minimal (1–2); in reducing conditions where recoil loss factors predominate, it rose to 4 (Figure S3d). Under oxidizing conditions, elevated uranium concentrations correlated with TDS (R^2 = 0.59; Figure S3e) and the main TDS-determining ions—Na^+ (R^2 = 0.73; Figure S3f), HCO_3^- (R^2 = 0.75; Figure S4a), and Ca^{2+} (R^2 = 0.67; Figure S4b)—but did not correlate with SO_4^{2-}, Cl^-, or Mg^{2+} (Figure S4c–e). Under reducing conditions, there was no correlation between the minimum concentrations of uranium, TDS, and basic ions.

4.3. Dependence of Strontium on the Acid–Base and Redox Properties of Fresh Water in the Upper Permian Kazan Carbonate Aquifer (P_2kz)

Elevated strontium concentrations showed a tendency to increase simultaneously with a decrease in pH to neutral values (Figure 5a), although the significance level for this dependency was above 5% (coefficient of determination R^2 = 0.44, coefficient of correlation r = –0.67, significance value p = 0.07; see Section S3 of the Supplementary Materials). For low strontium concentrations there was no correlation (R^2 = 0.09, r = 0.31, p = 0.42).

The dependence of the SI values for calcite and dolomite on pH was well pronounced for groundwater with high strontium concentrations (R^2 = 0.93–0.88, r = 0.97–0.94, p = 0). At the same time, the supersaturation of water in relation to these minerals was replaced by undersaturation, which was characterized by a change in SI from 0.4 and 1 to –0.2 and –0.6, respectively, as it approached neutral pH values (Figure 5b). There was practically no tendency to increase Sr in this direction (Figure 5g) (R^2 = 0.35 and 0.18, r = –0.59 and –0.43, p = 0.12 and 0.29, respectively).

A correlation between saturation indices for dolomite and pH was observed in groundwater with low strontium concentrations (R^2 = 0.64, r = 0.8, p = 0.01), with SI changing from 1.3 to –0.03. There was no correlation between SI values for calcite and pH (R^2 = 0.19, r = 0.45, p = 0.22), with SI changing from 0.6 to 0.3. The concentration of Sr increased in the direction of increasing the supersaturation of water with respect to calcite and dolomite (R^2 = 0.30–0.34, r = 0.56–0.58, p = 0.12–0.1), which may indicate other sources of Sr in the water.

The SI values for strontianite did not correlate with pH (R^2 = 0.28 and 0.34, r = 0.53 and 0.58, p = 0.18 and 0.1 for water samples with high and low strontium concentrations, respectively; Figure 5c). The same can be said about the concentration of Sr in water with a high content of it. Low strontium concentrations increased towards increasing the water supersaturation with strontianite from SI 0.3 to 1.5 (Figure 5h) (R^2 = 0.79, r = 0.89, p = 0.01).

Negative saturation indices for gypsum and celestite were recorded throughout the entire range of their values. For water with high strontium contents they increased as they approached neutral pH values from –2.8 to –1.6 and –1.4, respectively (Figure 5b,c) (R^2 = 0.49–0.45, r = –0.7 to –0.67, p = 0.05 –0.07). The concentrations of strontium increased in a similar manner (Figure 5g,h) (R^2 = 0.48–0.74, r = 0.7–0.86, p = 0.06–0.006). For samples with low strontium contents, SI increased from −2.9 and −3.9 to −1.8 and −2.9, respectively (Figure 5b,c) (R^2 = 0.59–0.18, r = –0.76 to –0.42, p = 0.02–0.26); however, they were not correlated with Sr (Figure 5g,h) (R^2 = 0.14–0.12, r = –0.37 to 0.35, p = 0.32–0.36).

There was a dependence of TDS on pH (R^2 = 0.56, r = –0.75, p = 0.03 in water samples with high strontium contents and R^2 = 0.36, r = –0.6, p = 0.09 in water samples with low strontium contents), which was expressed with an increase in TDS as the pH approached neutral values in the P_2kz carbonate aquifer (Table 1). Elevated strontium concentrations showed a high positive correlation with TDS (Figure 5d) (R^2 = 0.82; r = 0.9, p = 0.002). Low strontium concentrations did not correlate with TDS (R^2 = 0.09, r = 0.3, p = 0.43).

As in the case of pH, for groundwater with a high content of strontium, a slight decrease in the supersaturation of water in calcite and dolomite was found as the degree of water mineralization increased (R^2 = 0.49–0.31; Figure 5e). Saturation indices for strontianite did not correlate with TDS (R^2 = 0.01; Figure 5f). The most pronounced undersaturation of water was with respect to gypsum and celestite (Figure 5e,f), and this undersaturation decreased as TDS increased from SI = −2.8 to −1.6 and from −2.8 to −1.4, respectively (R^2 = 0.64–0.73, r = 0.8 –0.89, p = 0.02–0.003). For groundwater with a low content of strontium, there was no correlation of TDS with saturation indices for all minerals except celestite (Figure 5e,f).

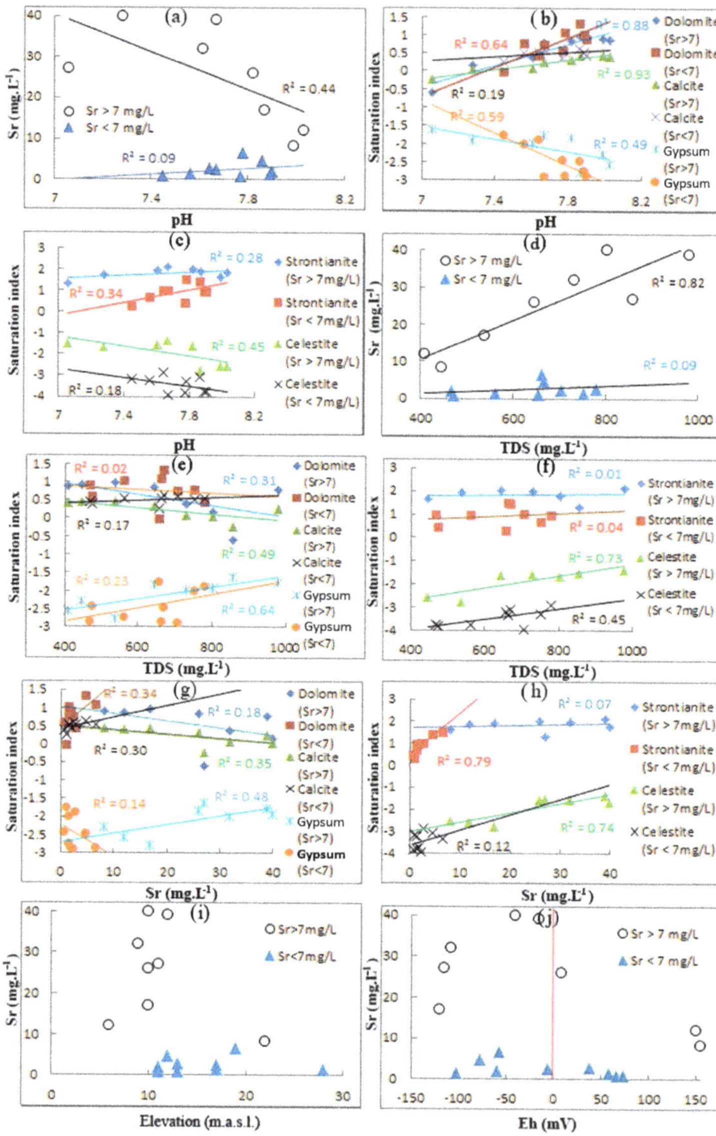

Figure 5. Graphs of strontium concentrations in water as a function of pH (**a**), TDS (**d**), SI (**g**,**h**), elevation (**i**), and Eh (**j**); graphs of groundwater saturation indices depending on pH (**b**,**c**) and TDS (**e**,**f**). Red line is Eh = 0.

Groundwater with high Sr concentrations tended to predominate in the lowlands near the river (Figure 5i) and in the reducing conditions of the aquifer (Figure 5j). Figure 6a shows the correlation between the maximum Sr concentrations and the minimum U concentrations characteristic of a reducing environment in aquifers [52] (see Table 2). Figure 6b also demonstrates the presence of two different correlation trends for the maximum and minimum Sr concentrations with ^{234}U:^{238}U activity ratios, with the same correlation coefficients (R^2 = 0.64). A similar trend (without separation of water types according to Sr concentrations) was noted by Plechacek et al. [30].

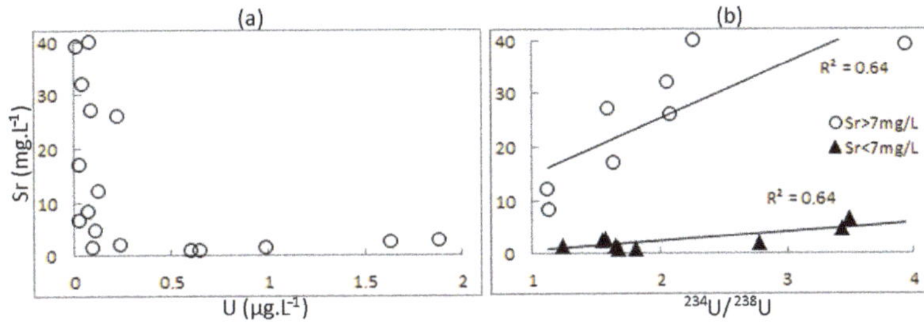

Figure 6. Graphs of strontium concentrations in water as a function of U (**a**) and $^{234}U/^{238}U$ (**b**) in the fresh groundwater.

4.4. Strontium vs. Chemical Composition of the Fresh Groundwater in the Upper Permian Kazan Carbonate Aquifer

Figures 5d and 7a,b show that increased strontium concentrations were correlated with TDS, Mg^{2+}, and HCO_3^- (R^2 = 0.82–0.69, r = 0.91–0.83, p = 0.002–0.01). There was a trend of increasing Sr concentrations with increasing SO_4^{2-}, Na^+, and Cl^- (Figure 7c–e) (R^2 = 0.5–0.38, r = 0.71–0.62, p = 0.05–0,1). There was virtually no correlation with Ca^{2+} (R^2 = 0.16; Figure 7f).

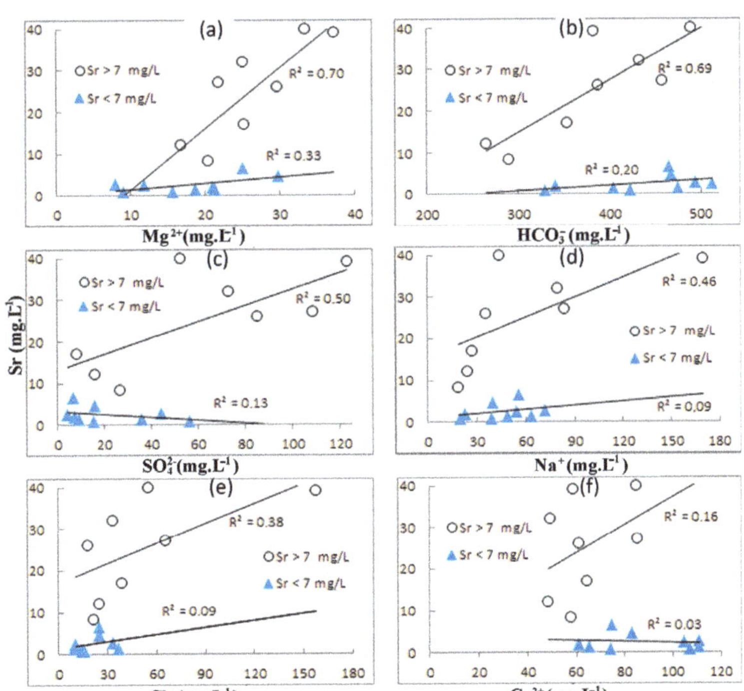

Figure 7. Graphs of strontium concentrations in water as a function of Mg^{2+} (**a**), HCO_3^- (**b**), SO_4^{2-} (**c**), Na^+ (**d**), Cl^- (**e**), and Ca^{2+} (**f**).

For these waters, a close dependence of the contents of sulfates and sodium on the total amount of dissolved substances was also recorded (R^2 = 0.77–0.72) (Figure 8a). The dependence of TDS on the contents of chlorides, bicarbonates, and magnesium was somewhat lower (R^2 = 0.62–0.52), and the correlation with calcium was practically absent (R^2 = 0.23).

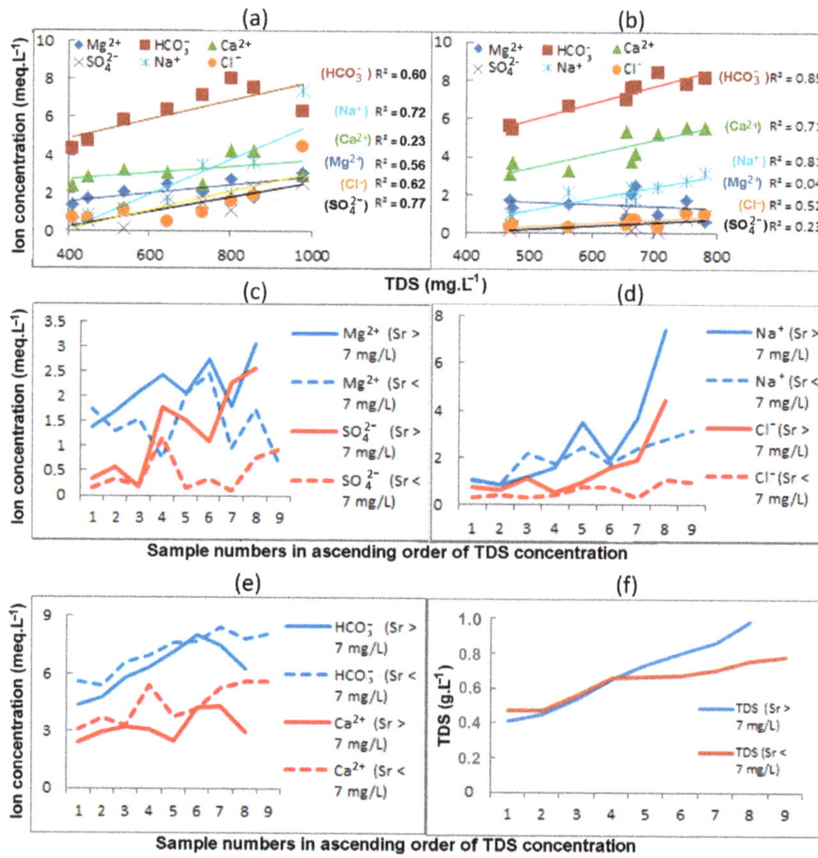

Figure 8. Diagrams of the dependence of the contents of the main cations and anions on the total amount of dissolved substances in water with excesses of the PMC for strontium (**a**) and in water of standard quality (**b**); the ratio of the concentrations of main cations and anions in waters of these two types (**c**–**f**).

Figure 8c,d show that the contents of Mg^{2+}, $SO_4{}^{2-}$, Na^+, and Cl^- in water with strontium concentrations above the MPS are noticeably higher than in water of standard quality; the concentrations of $HCO_3{}^-$ and Ca^{2+} show the opposite pattern (Figure 8e).

In addition, in water of standard quality there was no correlation between strontium contents and the ionic composition and total amount of dissolved substances (Figures 5d and 7). At the same time, there was a close dependence of TDS on the contents of bicarbonates, sodium, and calcium (R^2 = 0.89–0.71; Figure 8b). It was somewhat lower in relation to chlorides (R^2 = 0.52) and was practically absent in relation to sulfates and magnesium (R^2 = 0.23 and 0.04).

4.5. Dependence of Strontium Content on Water Age

Figure 9a shows a decrease in Ca^{2+} concentrations with decreasing ^{14}C (pmc) values, that is, increasing age of water. At the same time, the concentrations of Sr and Mg^{2+} increased, and in Figure 9c,d a clear increase in Mg^{2+}:Ca^{2+} and Sr:Ca^{2+} can be seen. It should be noted that these graphs were built for the entire data set (17 water samples). For water with high (8 samples) and low (9 samples) Sr content, a weak correlation was found between ^{14}C and Ca^{2+} (Figure 9b) (R^2 = 0.30–0.58, r = 0.55–0.76, p = 0.16–0.02); there is no correlation between ^{14}C and Sr (R^2 = 0.07–0.08, r = –0.27 to –0.28, p = 0.52 to 0.46).

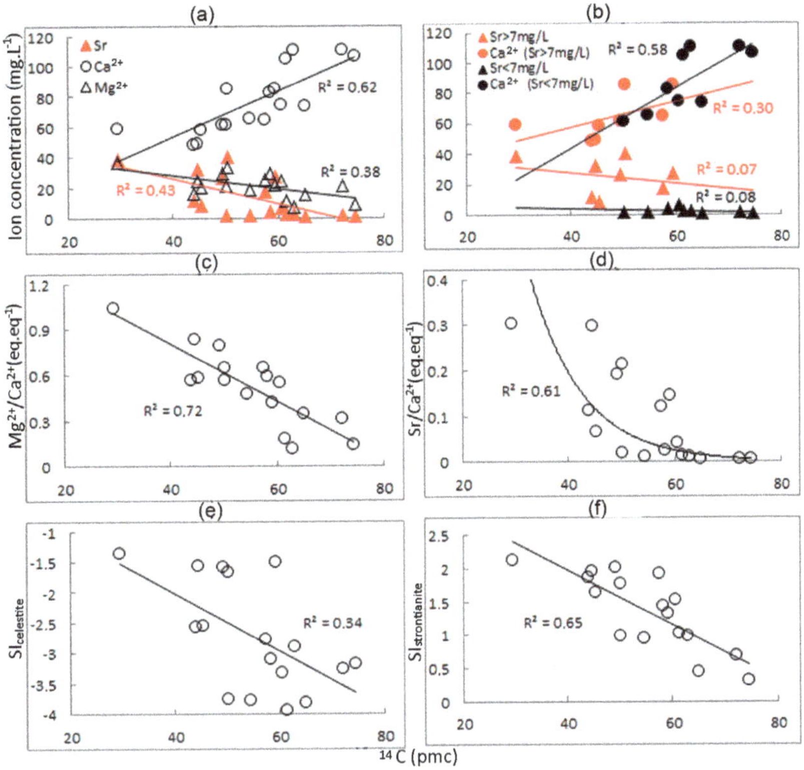

Figure 9. ^{14}C (pmc) vs. (**a**) strontium, calcium, and magnesium; (**b**) strontium and strontium in high and low-Sr water; (**c**) Mg^{2+}:Ca^{2+}; (**d**) Sr:Ca^{2+}; (**e**) $SI_{celestite}$; (**f**) $SI_{strontionite}$.

Graphs constructed from the entire data set (17 water samples) (Figure 9e,f) also show a correlation of SI for strontianite with the age of water (R^2 = 0.65) and a weak correlation of SI for celestite with the age of water (R^2 = 0.34). In water of standard quality (9 samples), there was also a correlation of SI for strontianite with the age of water (r = –0.61, p = 0.047) and no correlation of SI for celestite with the age of water (r = 0.41, p = 0.27). In water exceeding the MPC for strontium, the correlation of the saturation indices of strontianite and celestite with the age of the water was not established (r = –0.64 and –0.27, p = 0.08 and 0.52).

5. Discussion

5.1. Groundwater with Strontium Concentrations above the MPC

We have shown above that an increase in strontium concentrations in fresh groundwater was observed simultaneously with an increase in the amount of dissolved substances and a decrease in pH to approximately neutral values. Similar trends were also noted by

Krainov [2] and Macgrove [29]. However, the correlation of strontium and pH in our case was insignificant (r = −0.67, p = 0.07) compared to the correlation of strontium and the amount of dissolved substances (r = 0.9, p = 0.002). As a result, there was a clear increase in the values of $SI_{celestite}$ and SI_{gypsum} simultaneously with an increase in the content of dissolved substances in water (r = 0.89–0.8, p = 0.003–0.02), while the correlation of $SI_{celestite}$ and SI_{gypsum} with pH was noticeably less pronounced (r = −0.67 and −0.7, p = 0.07 and 0.05). There was also a clear increase in strontium concentration along with an increase in $SI_{celestite}$ and SI_{gypsum} (r = 0.86–0.7, p = 0.006–0.06), while water saturation with calcite and dolomite had little effect on the increase in strontium content. A relationship between $SI_{celestite}$ and $SI_{strontianite}$ and the age of groundwater was noted.

A correlation of magnesium with strontium and total mineralization of groundwater (r = 0.84 and 0.75, p = 0.01 and 0.03) and no correlation of calcium with strontium and total mineralization of groundwater (r = 0.4 and 0. 48, p = 0.32 and 0.23) were found for strontium-rich groundwater. In addition, it was found that with increasing water age, there was a decrease in calcium concentration with a parallel increase in strontium and magnesium concentrations and an increase in magnesium–calcium and strontium–calcium ratios.

Thus, it can be assumed that it is not an increase in magnesium concentrations as a result of the dedolomitization of rocks [23,29,53–55], but a decrease in the calcium concentrations (Figures 9a and 10) that causes an increase in the content of strontium, which replaces it.

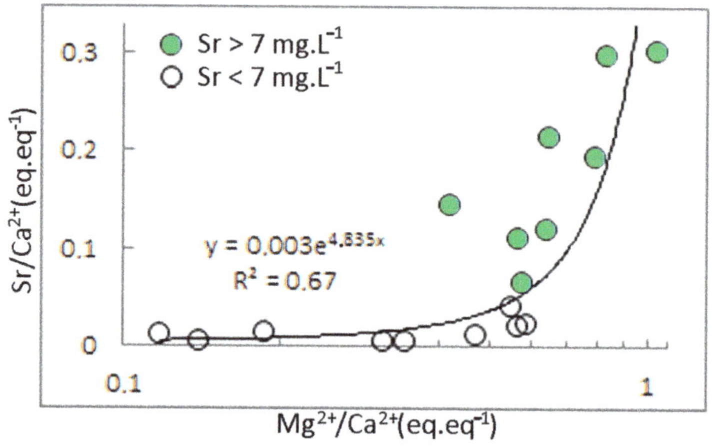

Figure 10. $Sr:Ca^{2+}$ vs. $Mg^{2+}:Ca^{2+}$ plot.

However, a separate analysis of age-related changes in the concentrations of Ca and Sr in waters with strontium contents above and below the MPC (Figure 9b) shows less certain dependencies, and therefore it is advisable to check them in the future using more extensive material through periodical samplings.

The correlation of strontium with bicarbonate ions is also significant (r = 0.83, p = 0.01), since bicarbonate ions are one of the main components of low-mineralized water. However, in more mineralized waters they are partially "replaced" by sulfate ions and chloride [24,25] (see Figure 8a,c–e). At the same time, as shown above, calcium concentrations increased in proportion to the concentrations of bicarbonate ions, also in low-mineralized waters, and then were partially "replaced" by sodium (see Figure 8e,d).

Based on the established correlation of the concentrations of strontium (Figure 5d), chlorine ions, and sodium (Figure 8d) with the total mineralization of groundwater and the tendency of increasing strontium concentrations with increasing sodium and chlorine content, it can be assumed that there is an influence of the upwelling of salt waters from the lower substage of the Kazan aquifer (see Figure 2) to increase the strontium content in drinking water. According to [26], the contents of TDS and Sr in saline groundwater are 3100 and 9 ppm, respectively. That is, water mineralization up to 700–1000 ppm can be obtained by mixing one part of salt water with 4–8 parts of fresh water (400 ppm). However, to achieve a Sr concentration in this mixture of 7 ppm, its concentration in fresh water should already be about 6.6 ppm: $(9 \times 1 + 6.6 \times 6)/7 = 7$. That is, the effect of salt water is insignificant in relation to the increase in strontium concentration [56].

The correlation of strontium with sulfate ions and a significant increase in the concentration of sulfate ions with an increase in TDS, as well as the correlation of strontium with $SI_{celestite}$ and SI_{gypsum} indicate the formation of high concentrations of strontium due to the dissolution of these minerals.

The occurrence of the collected water samples with a high content of strontium in the zone of reducing conditions is possibly due to the difficult water exchange in this area and its relatively weak flushing, as a result of which celestite inclusions are preserved there. Other researchers have noted a similar trend [29,30].

The regularity of this factor is also justified by the increase in strontium concentrations in boreholes gravitating to the lowlands near the river (see Figure 1c, Kamenka village and Figure 11).

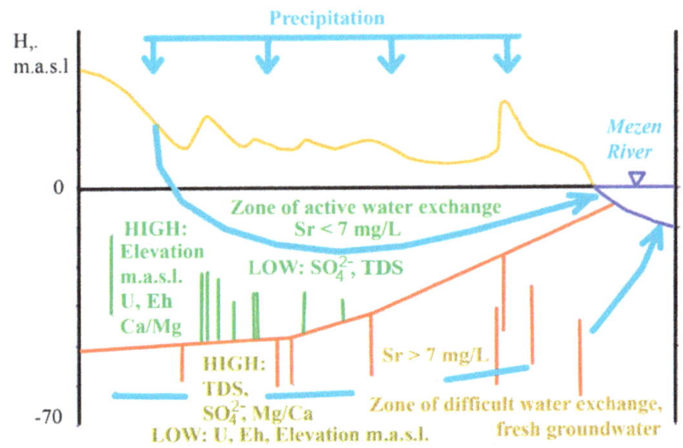

Figure 11. Scheme of the formation of strontium contamination. Green and red vertical lines are groundwater sampling intervals.

5.2. Groundwater with Strontium Content below the MPC

Strontium concentrations in strontium-poor groundwater did not correlate with pH or TDS. Their average values fluctuated around 2.6 $mg \cdot L^{-1}$. Water with strontium content below the MPC was characterized by reduced saturation in relation to $SrSO_4$ and $CaSO_4$, and strontium did not correlate with $SI_{celestite}$ and SI_{gypsum}. Water was supersaturated in $CaCO_3$, $CaMg(CO_3)_2$, and $SrCO_3$; as its supersaturation decreased, the strontium content in water also decreased.

Strontium concentrations in strontium-poor groundwater also did not correlate with major groundwater ions. The concentrations of most ions were noticeably lower than in groundwater with a high content of strontium. The exceptions are bicarbonate ions and calcium. The SO_4^{2-} content was most significantly reduced (Figure 8c), and there was no dependence of strontium concentrations on sulfate ($r = -0.36$, $p = 0.34$), which indicates the formation of strontium due to leaching of discretely distributed small inclusions of $SrSO_4$ and $CaSO_4$. Chloride and sodium concentrations were also reduced, indicating less upwelling of the salty groundwater. But assessments of its influence showed that the upwelling effect of brackish waters on increasing strontium concentrations is significant

(44%) [56]. The dedolomitization processes had virtually no effect on Sr concentrations (Figure 10).

Thus, the main factors characteristic of groundwater with strontium content above the MPC (correlation of Sr with TDS and $SI_{celestite}$ and SI_{gypsum} and the effect of dedolomitization) were not manifested in groundwater with strontium content below the MPC.

5.3. Estimation of Exposure and Human Health Risk

Data on the carcinogenicity of strontium are very limited, and there is insufficient information to assess its carcinogenic potential due to the lack of adequate studies on long-term chronic exposure [57].

The assessment of non-carcinogenic risk to human health from contact with groundwater was carried out in accordance with the procedure described by the US Environmental Protection Agency [33,57,58]. A deterministic approach was used for two routes of exposure (ingestion and water through the skin) to two subpopulations (adults and children). It is important to emphasize that strontium is not volatile, thus inhalation exposure is not likely, and therefore inhalation was not included in the risk assessment models in this study.

The methodology for assessing exposure and risk to human health, as well as the main parameters and values used for deterministic exposure calculations, are provided in Section S4 Supplementary Materials.

The calculation results are shown in Figure 12:

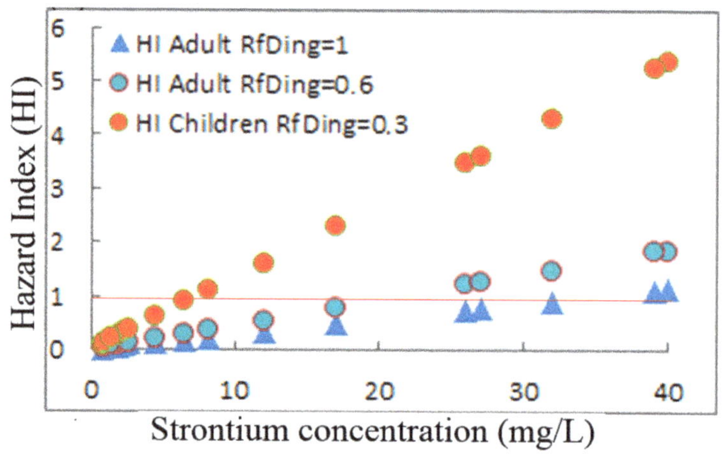

Figure 12. Hazard index (HI) to assess the overall non-carcinogenic health risk from combined ingestion and skin exposure of investigated fresh drinking waters.

Values of HI lower than one indicate no significant non-cancer health risk. Values of HI greater than one depict an existing likelihood of non-cancer health effects occurring and the probability increases as the values rise (see Section S4 Supplementary Materials).

Children in the area are most vulnerable to risks. Fifty percent of the wells contain water that is dangerous for consumption by children of about nine years of age weighing 30 kg. For younger children, the situation is even worse.

For adults weighing up to 70 kg, water from about a third of the studied wells is dangerous.

6. Conclusions

The purpose of this research was to determine natural factors that contribute to maintaining the standard quality of fresh drinking groundwater in areas with high strontium content. The results showed that:

(1) For waters with strontium concentrations above the MPC, the following features are characteristic:
- Correlations of strontium with TDS and saturation indices for celestite and gypsum, indicating the formation of these waters in sediments with high contents of celestite and gypsum.
- An increase in strontium–calcium and magnesium–calcium ratios during the period when groundwater is in the aquifer, associated with the process of dedolomitization.
- Correlation of strontium concentrations with the concentrations of the main ions, except for calcium.
- Reducing conditions in the aquifer, indicating difficult water exchange in the aquifer, promoting the preservation of strontium-containing minerals.

(2) For waters with strontium concentrations below the MPC, the following features are characteristic:
- Strontium concentrations do not correlate with TDS, $SI_{celestite}$, and SI_{gypsum}, indicating the formation of water in sediments with discretely located small inclusions of celestite and gypsum.
- The processes of dedolomitization practically do not affect the growth of strontium concentrations.
- Oxidizing conditions and active water exchange in the aquifer are favorable for the formation of water with standard quality.

In general, areas of development of groundwater of a Ca-HCO_3 composition with reduced values of the total content of dissolved substances and increased values of the oxidation-reduction potential will be favorable for obtaining drinking water of standard quality.

The assessment of non-carcinogenic risk to human health from contact with groundwater has shown that children in this area are currently the most vulnerable to risks. Fifty percent of wells contain water that is dangerous to drink.

Supplementary Materials: The following supporting information can be downloaded at: https://www.mdpi.com/article/10.3390/w15213846/s1, S1. Statistical distribution of strontium concentration values in drinking groundwater in the North-West of Russia; S2. Summary statistics on compositions of drinking groundwater in the North-West of Russia; S3. Correlation matrices on compositions of drinking groundwater in the North-West of Russia; S4. Estimation of exposure and human health risk; Figure S1. Histograms (a) all 17 samples, (b) 9 samples with strontium contents less than 7 mg/L, (c) 8 samples with strontium contents more than 7 mg/L; Figure S2. Plots of the radiocarbon/residence time in the fresh groundwater Pearson model (a), Ferronsky model (b), and Vogel model (c); Figure S3. Plots of the U concentration in the fresh groundwater vs. Eh (a), pH (b), ^{14}C (c), $^{234}U/^{238}U$ (d), TDS (e), Na^{2+} (f). (Ox)—oxidizing conditions in the aquifer, (Red)—reducing conditions in the aquifer; Figure S4. Plots of the U concentration in the fresh groundwater vs. HCO_3^- (a), Ca^{2+} (b), SO_4^{2-} (c), Cl^- (d), Mg^{2+} (e). (Ox)—oxidizing conditions in the aquifer, (Red)—reducing conditions in the aquifer. Table S1. Summary statistics on strontium concentration values. Table S2. Test of Normality Shapiro-Wilk. Table S3. Parameters and values used for deterministic exposure calculations. Refs [59–64] cited in Supplementary Materials.

Funding: This research was funded by the Russian Science Foundation, grant number 23-27-10004, https://rscf.ru/project/23-27-10004/. (accessed on 10 September 2023).

Data Availability Statement: Data are available upon reasonable request from the corresponding author.

Conflicts of Interest: The author declares no conflict of interest.

References

1. Zektser, I.S. Groundwater as a Component of the Environment. In *Geology and Ecosystems*; Zektser, I.S., Marker, B., Ridgway, J., Rogachevskaya, L., Vartanyan, G., Eds.; Springer: Boston, MA, USA, 2006; pp. 91–105. [CrossRef]
2. Krainov, S.R.; Ryzhenko, B.N.; Shvets, V.M. *Geochemistry of Groundwater. Fundamental, Applied and Environmental Aspects*; CenterLitNefteGaz: Moskow, Russia, 2012; p. 672. (In Russian)

3. Shen, J.; Schäfer, A. Removal of fluoride and uranium by nanofiltration and reverse osmosis: A review. *Chemosphere* **2014**, *117*, 679–691. [CrossRef]
4. Vaiopoulou, E.; Gikasb, P. Regulations for chromium emissions to the aquatic environment in Europe and elsewhere. *Chemosphere* **2020**, *254*, 126876. [CrossRef]
5. He, X.; Li, P.; Shi, H.; Xiao, Y.; Guo, Y.; Zhao, H. Identifying strontium sources of flowback fluid and groundwater pollution using 87Sr/86Sr and geochemical model in Sulige gasfield, China. *Chemosphere* **2022**, *306*, 135594. [CrossRef] [PubMed]
6. *SanRaR 2.1.4.1074-01*; Drinking Water. Hygienic Requirements for Water Quality of Centralized Drinking Water Supply Systems. Quality Control. Sanitary and Epidemiological Rules and Regulations Approved by the Chief State Sanitary Doctor of the Russian Federation on 26 September 2001. Russian Federation: Moscow, Russia, 2001. (In Russian)
7. Amata, R.; Diamond, G.L.; Dorsey, A.; Fransen, M.E. *Toxicological Profile for Strontium*; U.S. Department of Health and Human Services, Public Health Service, Agency for Toxic Substances and Disease Registry: Atlanta, GA, USA, 2004.
8. Bartley, J.C.; Reber, E.F. Toxic effects of stable strontium in young pigs. *J. Nutr.* **1961**, *75*, 21–28. [CrossRef] [PubMed]
9. Colvin, L.B.; Creger, C.R. Stable strontium and experimental bone anomalies. *Fed. Proc. Fed. Am. Soc. Exp. Biol.* **1967**, *26*, 416.
10. Colvin, L.B.; Creger, C.R.; Ferguson, T.M.; Crookshank, H.R. Experimental epiphyseal cartilage anomalies by dietary strontium. *Poult. Sci.* **1972**, *51*, 576–581. [CrossRef] [PubMed]
11. Keesari, T.; Sabarathinam, C.; Sinha, U.K.; Pethaperumal; Thilagavathi, R.; Kamaraj, P. Fate and transport of strontium in groundwater from a layered sedimentary aquifer system. *Chemosphere* **2022**, *307*, 136015. [CrossRef] [PubMed]
12. Khandare, A.L.; Validandi, V.; Rajendran, A.; Singh, T.G.; Thingnganing, L.; Kurella, S.; Nagaraju, R.; Dheeravath, S.; Vaddi, N.; Kommu, S.; et al. Health risk assessment of heavy metals and strontium in groundwater used for drinking and cooking in 58 villages of Prakasam district, Andhra Pradesh, India. *Env. Geochem. Health* **2020**, *42*, 3675–3701. [CrossRef]
13. Zamana, L.V.; Rikhvanov, L.P.; Soktoev, B.R.; Baranovskaya, N.V.; Epova, E.S.; Solodukhina, M.A.; Mikhailova, L.A.; Kopylova, Y.G.; Khvashchevskaya, A.A. New data on chemical composition of natural waters in the area of distribution of Urov (Kaschin–Beck) disease (Transbaikal region). *Bull. Tomsk Polytech. Univ. Geo Assets Eng.* **2019**, *330*, 121–133. (In Russian)
14. Höllriegl, V.; München, H.Z. Strontium in the environment and possible human health effects. *Encycl. Environl. Health* **2011**, *10*, 268–275. [CrossRef]
15. Wang, Z. A historic overview of research and control on Kashin-Beck Disease in China. *Chin. J. Endem.* **1999**, *18*, 161–163.
16. Levander, O.A. Selenium. In *Trace Elements in Human and Animal Nutrition*; Mertz, W., Ed.; Academic Press Inc.: Orlando, FL, USA, 1986; Volume 2, pp. 209–279.
17. Malaisse, F.; Mathieu, F. (Eds.) *Big Bone Disease, a Multidisciplinary Approach of Kashin–Beck Disease in Tibet Autonomous Region (PR China)*; Les Presses Universitaires de Gembloux: Gembloux, Belgium, 2008; p. 148.
18. Vinogradov, A.P. Geochemical investigations in the area of Urov endemia. *Dokl. Akad. Nauk* **1939**, *23*, 64–67. (In Russian)
19. Ermakov, V.V.; Gulyaeva, U.A.; Tyutikov, S.F.; Kuzmina, T.G.; Safonov, V.A. Biogeochemistry of calcium and strontium in the landscapes of Eastern Transbaikalia. *Geochem. Int.* **2017**, *55*, 1105–1117. [CrossRef]
20. Ermakov, V.; Bech, J.; Gulyaeva, U.; Tyutikov, S.; Safonov, V.; Danilova, V.; Roca, N. Relationship of the mobile forms of calcium and strontium in soils with their accumulation in meadow plants in the area of Kashin–Beck endemia. *Env. Geochem. Health* **2020**, *42*, 159–171. [CrossRef] [PubMed]
21. Rikhvanov, L.P.; Soktoev, B.R.; Baranovskaya, N.V.; Ageeva, E.V.; Belyanovskaya, A.I.; Deriglazova, M.A.; Yusupov, D.V.; Epova, E.S.; Solodukhina, M.A.; Zamana, L.V.; et al. Integrated geochemical studies of the components of the natural environment in the endemic regions of Transbaikalia. *Bull. Tomsk. Polytech. Univ. Geo Assets Eng.* **2021**, *332*, 7–25. (In Russian)
22. Vsevolozhsky, V.A. *Fundamentals of Hydrogeology*; Publishing House of Moscow State University: Moscow, Russia, 2007; p. 448. (In Russian)
23. Bui, D.T.; Khosravi, K.; Karimi, M.; Busico, G.; Khozani, Z.S.; Nguyen, H.; Mastrocicco, M.; Tedesco, D.; Cuoco, E.; Kazakis, N. Enhancing nitrate and strontium concentration prediction in groundwater by using new data mining algorithm. *Sci. Total Environ.* **2020**, *715*, 136836. [CrossRef]
24. Limantseva, O.A.; Ryzhenko, B.N. Model for Sr accumulation in the Carboniferous deposits of the Moscow artesian basin. *Geochem. Int.* **2008**, *46*, 935–944. [CrossRef]
25. Limantseva, O.A.; Ryzhenko, B.N.; Cherkasova, E.V. Model for the formation of fluorine-bearing rocks in the Carboniferous deposits of the Moscow artesian basin. *Geochem. Int.* **2007**, *45*, 900–917. [CrossRef]
26. Ivanova, N.I. Strontium distridution patterns in groundwater and aquifer host rocks in the southeastern part of the Severnaya Dvina artesian basin. *Mosc. Univ. Geol. Bull.* **2014**, *69*, 258–266. [CrossRef]
27. Ivanova, I.S.; Shvartsev, S.L.; Pokrovsky, O.S. Distribution of strontium in groundwater of the upper hydrodynamic zone of the Sredneobsky artesian basin (Tomsk region). In *Geological Evolution of the Interaction of Water with Rocks*; Zamana, L.V., Shvartsev, S.L., Eds.; Publishing House of BSC SB RAS: Ulan-Ude, Russia, 2018; pp. 110–114.
28. Kaleem, M.; Naseem, S.; Bashir, E.; Shahab, B.; Rafiq, T. Discrete geochemical behavior of Sr and Ba in the groundwater of Southern Mor Range, Balochistan, a tracer for igneous and sedimentary rocks weathering and related environmental issues. *Appl. Geochem.* **2021**, *130*, 104996. [CrossRef]
29. Musgrove, M. The occurrence and distribution of strontium in U.S. groundwater. *Appl. Geochem.* **2021**, *126*, 104867. [CrossRef]
30. Plechacek, A.; Scott, S.R.; Gotkowitz, M.B.; Ginder-Vogel, M. Strontium and radium occurrence at the boundary of a confined aquifer system. *Appl. Geochem.* **2022**, *142*, 105332. [CrossRef]

31. Malov, A.I. *Groundwater of the South-East White Sea: Formation, Role in Geological Processes*; UB RAS: Yekaterinburg, Russia, 2003; p. 234. Available online: https://www.elibrary.ru/item.asp?id=1947446429 (accessed on 10 September 2023). (In Russian)
32. Salminen, R.; Chekushin, V.; Bogatyrev, I.; Fedotova, E.; Tomilina, O.; Zhdanova, L.; Tenhola, M.; Glavatskikh, S.; Gregorauskiene, V.; Kashulina, G.; et al. *Geochemical Atlas of the Eastern Barents Region*; Elsevier: Amsterdam, The Netherlands, 2004; p. 548.
33. USEPA. *Risk Assessment Guidance for Superfund. Volume I: Human Health Evaluation Manual (Part E, Supplemental Guidance for Dermal Risk Assessment)*; Final. EPA/540/R/99/005; U.S. Environmental Protection Agency: Washington, DC, USA, 2004.
34. Stankovsky, A.F.; Verichev, E.M.; Grib, V.P.; Dobeiko, I.P. Vendian of the southeastern White Sea. *Izv. Acad. Sci. USSR. Ser. Geol.* **1981**, *2*, 78–87. (In Russian)
35. Stankovsky, A.F.; Verichev, E.M.; Dobeiko, I.P. Vendian of the southeastern White Sea. In *Vendian System. Historical-Geological and Paleontological Substantiation. Vol. 2. Stratigraphy and Geological Processes*; Sokolov, B.S., Fedonkin, M.A., Eds.; Nauka: Moscow, Russia, 1985; pp. 67–76. (In Russian)
36. Turekian, K.K. *Chemistry of the Earth*; Holt, Rinehart and Winston Inc.: New York, NY, USA, 1972; p. 131.
37. Turekian, K.K.; Wedepohl, K.H. Distribution of the elements in some major units of the earth's crust. *Geol. Soc. Am. Bull.* **1961**, *72*, 175–192. [CrossRef]
38. Malov, A.I.; Sidkina, E.S.; Ershova, D.D.; Cherkasova, E.V.; Druzhinin, S.V. Time regularities of strontium concentration in drinking groundwater distant from the sea coast. *Env. Geochem. Health* **2023**, *45*, 8097–8118. [CrossRef] [PubMed]
39. Malov, A.I.; Sidkina, E.S.; Ryzhenko, B.N. Model of the Lomonosov diamond deposit as a water–rock system: Migration Species, Groundwater Saturation with Rock-Forming and Ore Minerals, and Ecological Assessment of Water Quality. *Geochem. Int.* **2017**, *55*, 1118–1130. [CrossRef]
40. Münnich, K.O. Messungen des ^{14}C-Gehaltes von hartem Grundwasser. *Naturwissenschaften* **1957**, *44*, 32–34. [CrossRef]
41. Münnich, K.O. Isotopendatierung von Grundwasser. *Naturwissenschaften* **1968**, *55*, 3–11. [CrossRef]
42. Ingerson, E.; Pearson, F.J. Estimation of age and rate of motion of groundwater by the ^{14}C method. In *Recent Researches in the Fields of Atmosphere, Hydrosphere and Nuclear Geochemistry*; Marusen: Tokyo, Japan, 1964; pp. 263–283.
43. Mook, W.G. On the reconstruction of the initial ^{14}C content of groundwater from the chemical and isotopic composition. In Proceedings of the Eighth International Conference on Radiocarbon Dating 1, Lower Hutt, New Zealand, 18–25 October 1972; Royal Society of New Zealand: Wellington, New Zealand, 1972; pp. 342–352.
44. Mook, W.G. The dissolution-exchange model for dating groundwater with ^{14}C. In *Interpretation of Environmental Isotope and Hydrochemical Data in Groundwater Hydrology*; IAEA: Vienna, Austria, 1976; pp. 213–225.
45. Han, L.-F.; Plummer, N. A review of single-sample-based models and other approaches for radiocarbon dating of dissolved inorganic carbon in groundwater. *Earth Sci. Rev.* **2016**, *152*, 119–142. [CrossRef]
46. Ferronsky, V.I.; Polyakov, V.A. Isotopes of the Earth's Hydrosphere. Springer: Amsterdam, The Netherlands, 2012. [CrossRef]
47. Vogel, J.C. Carbon-14 dating of groundwater. In Proceedings of the Symposium on Use of Isotopes in Hydrology, Vienna, Austria, 9–13 March 1970; IAEA: Vienna, Austria, 1970; pp. 225–237.
48. Reimer, P.J.; Austin, W.E.N.; Bard, E.; Bayliss, A.; Blackwell, P.G.; Ramsey, C.B.; Butzin, M.; Cheng, H.; Edwards, R.L.; Friedrich, M.; et al. The IntCal20 Northern Hemisphere radiocarbon age calibration curve (0–55 ka cal BP). *Radiocarbon* **2020**, *62*, 725–757. [CrossRef]
49. Stuiver, M.; Reimer, P.J.; Reimer, R.W. CALIB 8.2 [WWW Program]. Available online: http://calib.org (accessed on 16 February 2021).
50. Schoeller, H. Qualitative evaluation of groundwater resources. In *Methods and Techniques of Groundwater Investigation and Development*; UNESCO: Paris, France, 1967; Volume 33, pp. 44–52.
51. Schoeller, H. Geochemistry of groundwater. In *Groundwater Studies—An International Guide for Research and Practice*; UNESCO: Paris, France, 1977; Volume 15, pp. 1–18.
52. Steffanowski, J.; Banning, A. Uraniferous dolomite: A natural source of high groundwater uranium concentrations in northern Bavaria, Germany? *Env. Earth Sci.* **2017**, *76*, 508. [CrossRef]
53. Plummer, L.N. Defining reactions and mass transfer in part of the Floridan Aquifer. *Water Resour. Res.* **1977**, *13*, 801–812. [CrossRef]
54. Back, W.; Hanshaw, B.B.; Plummer, L.N.; Rahn, P.H.; Rightmere, C.T.; Rubin, M. Process and rate of dedolomitization: Mass transfer and ^{14}C dating in a regional carbonate aquifer. *Geol. Soc. Am. Bull.* **1983**, *94*, 1415–1429. [CrossRef]
55. Musgrove, M.; Banner, J.L. Controls on the spatial and temporal variability of vadose dripwater geochemistry: Edwards aquifer, central Texas. *Geochem. Cosmochim. Acta* **2004**, *68*, 1007–1020. [CrossRef]
56. Malov, A.I. Features of the formation of high strontium concentrations in drinking groundwater near sea coast. *Dokl. Earth Sci.* **2023**, *512*, 898–901. [CrossRef]
57. USEPA. *Health Effects Support Document for Strontium*; EPA 820-P-14-0012014; U.S. Environmental Protection Agency: Washington, DC, USA, 2014. Available online: www.epa.gov/safewater/ccl/pdf/Strontium.pdf (accessed on 10 September 2023).
58. USEPA. *Supplemental Guidance for Developing Soil Screening Levels for Superfund Sites*; OSWER 9355.4-24; U.S. Environmental Protection Agency: Washington, DC, USA, 2002.
59. Gerba, C.P. *Risk Assessment. Environmental and Pollution Science*, 3rd ed.; Elsevier: Amsterdam, The Netherlands, 2019; pp. 541–563. [CrossRef]

60. Means, B. Risk-Assessment Guidance for Superfund, Volume 1. In *Human Health Evaluation Manual, Part A*; Interim Report Final No. PB-90–155581/XAB; EPA-540/1–89/002; Environmental Protection Agency; Office of Solid Waste and Emergency Response: Washington, DC, USA, 1989. Available online: https://www.osti.gov/biblio/7037757 (accessed on 18 January 2022).
61. Ondayo, M.A.; Watts, M.J.; Hamilton, E.M.; Mitchell, C.; Mankelow, J.; Osano, O. Artisanal gold mining in Kakamega and Vihiga counties, Kenya: Potential human exposure and health risk. *Env. Geochem. Health* **2023**, *45*, 6543–6565. [CrossRef] [PubMed]
62. USEPA. Regional Screening Levels (RSLs). Calculator. Available online: https://epa-prgs.ornl.gov/cgi-bin/chemicals/csl_search (accessed on 3 August 2023).
63. Zhang, H.; Zhou, X.; Wang, L.; Wang, W.; Xu, J. Concentrations and potential health risks of strontium in drinking water from Xi'an, Northwest China. *Ecotoxicol. Environ. Saf.* **2018**, *164*, 181–188. [CrossRef] [PubMed]
64. Zimoch, I.; Łobos, E. Evaluation of health risk caused by chloroform in drinking water. *Desalin. Water Treat.* **2015**, *57*, 1027–1033. [CrossRef]

water MDPI

Article

Spatial Variations and Distribution Patterns of Soil Salinity at the Canal Scale in the Hetao Irrigation District

Zhiyuan Hu [1,2,†], Qingfeng Miao [1,2,†], Haibin Shi [1,2,*], Weiying Feng [3,*], Cong Hou [1,2], Cuicui Yu [1,2] and Yunfang Mu [1,2]

1 College of Water Conservancy and Civil Engineering, Inner Mongolia Agricultural University, Hohhot 010018, China; huzhiyuan@emails.imau.edu.cn (Z.H.); imaumqf@imau.edu.cn (Q.M.); nndhoucong@emails.imau.edu.cn (C.H.); yucuicui1112@emails.imau.edu.cn (C.Y.); 2021202060038@emails.imau.edu.cn (Y.M.)
2 High Efficiency Water-Saving Technology and Equipment and Soil and Water Environment Effect in Engineering Research Center of Inner Mongolia Autonomous Region, Hohhot 010018, China
3 School of Materials Science and Engineering, Beihang University, Beijing 100191, China
* Correspondence: shb@imau.edu.cn (H.S.); fengweiying@buaa.edu.cn (W.F.)
† These authors contributed equally to this work.

Abstract: Soil salinization is a major factor impacting global crop yields. To explore the spatial distribution and influencing factors of soil water and salt in typical canals of the Hetao irrigation district, regional soil information was monitored at fixed locations. In this study, classical statistics, geostatistics, and spatial autocorrelation methods were used to conduct quantitative analyses of soil salt content, water content, soil particle size distribution, and groundwater depth. The variation coefficient of the soil salt content in the 20–40 and 40–60 cm soil layers was between 10% and 100%, which corresponds to a medium degree of variation; the other soil layers had strong degrees of variation. The soil moisture content in each layer varied moderately. The gold coefficients of soil salt content and water content were less than 0.25, and the Z value was greater than 0, showing a strong spatial correlation and certain spatial agglomeration characteristics, which were mainly affected by structural factors in the study area. The distribution patterns of soil water and salt were affected by soil particle size. Sand content decreased with increasing depth, soil salt was negatively correlated with sand content, and soil water was positively correlated with sand content. Soil salinity was significantly affected by groundwater depth and increased with decreasing groundwater depth, following an exponential relationship. When the groundwater depth exceeded 1.7 m, the soil salt content exhibited small changes with groundwater depth. The results of this study could play a guiding role in terms of understanding the degree of soil salinization surrounding canals in the Hetao irrigation area and adjusting land management strategies over time.

Keywords: salinization; soil particle diameter; spatial variations; groundwater depth; ordinary Kriging

Citation: Hu, Z.; Miao, Q.; Shi, H.; Feng, W.; Hou, C.; Yu, C.; Mu, Y. Spatial Variations and Distribution Patterns of Soil Salinity at the Canal Scale in the Hetao Irrigation District. *Water* **2023**, *15*, 3342. https://doi.org/10.3390/w15193342

Academic Editor: Micòl Mastrocicco

Received: 17 August 2023
Revised: 20 September 2023
Accepted: 21 September 2023
Published: 23 September 2023

1. Introduction

As the global food demand and the world population rapidly increase, cultivated soil conditions are gradually deteriorating, leading to a decline in its potential [1–3]. Soil salinization is a major factor affecting global crop yield reduction; more than 8×10^8 hm^2 of land worldwide is affected by salinization, accounting for more than 6% of the total land area [4,5]. Currently, about 20% of arable land is salinized due to unscientific agricultural practices, and it is projected that by 2050, half of the world's arable land will be affected by salinization. Thus, there is a pressing need to better understand soil salinization processes and develop agricultural practices that will enable the production of the necessary amount of food to feed humanity while minimizing soil salinization and other degradation processes [3,6].

The main causes of soil salinization in coastal areas were found to be the intensive exploitation of aquifers and the continuous rise of sea level leading to seawater intrusion [7,8].

In contrast, the Hetao irrigation area in Inner Mongolia was identified as a typical salinization mega-irrigation area in the middle and upper reaches of the Yellow River. This area is situated in an inland arid region with low rainfall and high evaporation (the evaporation ratio in this region exceeds 10), resulting in significant salt accumulation [9,10]. Since the 1960s, extensive scientific studies and preventive measures have been undertaken by domestic and foreign scholars regarding water and salt migration, salinization prevention, and environmental effects [11,12]. One approach utilized to address the issue of saline water intrusion in coastal areas involves the abstraction–recharge process [13]. Additionally, in the Hetao irrigation area of Inner Mongolia, measures such as improving canal head engineering and implementing mandatory water-saving measures have been adopted [14]. While these actions enhance water supply security, they can disrupt the regional water balance system and impact the hydrological cycles of farmland within the irrigated area, thus profoundly affecting regional agriculture. Hence, conducting a quantitative study on the spatial distribution of soil salt is crucial in order to discern the patterns of soil salinization change. Such research carries significant implications for the scientific management and rational utilization of saline land [15–18].

Based on classical statistical theory and geostatistical theory, there have been numerous research studies and applications focused on the theory of regional spatial variation [19]. These studies have explored the heterogeneity and spatiotemporal variability of the riparian undercurrent zone. However, it is important to note that the current methods employed only capture the variability of sediment hydraulic conductivity in the riparian undercurrent zone through calculated values using functions, while they are unable to provide a detailed representation of the spatial variability. Multi-fractal analysis has been used to reveal the changing trends of soil moisture content and conductivity and multi-scale soil properties in small areas [20,21], with several studies indicating that soil does not possess ideal fractal characteristics. Instead, it only exhibits fractal characteristics within specific spatial scales. However, studying soil salinization at large scales requires significant manpower investment and time costs. With the development of geographic information technology, remote sensing technology, and machine learning methods, landscape changes caused by salinization in a region can be more efficiently and directly reflected [22]. This advancement is convenient for studying the macro-level causes of this phenomenon. In recent years, studies have shown that the spatial interpolation method provides good accuracy in predicting soil salinity [23]. But it has been found to have shortcomings in quantifying the degree of spatial correlation. To overcome this limitation, researchers have turned to the use of Moran's I index in spatial clustering recognition methods. This index has proven to be effective in reflecting the overall degree of spatial correlation and evaluating the significance of independent locations [24]. However, in the context of the modernization of irrigation areas, existing studies have mostly focused on the spatiotemporal variations in soil salinity at the irrigation and farmland scales; there is a lack of medium-scale studies [25]. Moreover, because soil properties are dominated by continuous changes in space, certain spatial structural features can only be displayed at specific sampling scales. Small-scale research should not be blindly extrapolated to a larger scale [26]. Zhang Wencong [14] quantitatively studied the dynamic changes in farmland soil water and groundwater in different irrigation periods at the canal level in the Hetao irrigation district using the water balance method but did not reveal the spatial characteristics or rules of salt variation in different soil layers.

Identifying the spatial variation characteristics and distribution patterns of water and salt in the Hetao irrigation area at the canal scale could allow for the creation of a basic reference system to reveal the spatiotemporal dynamic characteristics and scale transformation features of soil attributes. In this study, typical canals in the Hetao irrigation district were selected as the study area, and the spatial distribution patterns of soil water and salt were evaluated after spring irrigation and at an early stage of crop growth. The degree of soil salinization within a typical canal was determined to provide scientific guidance for subsequent accurate improvement and sustainable development of saline-alkali land.

2. Materials and Methods

2.1. Overview of the Study Area

2.1.1. Physical Geography and Climatic Conditions

The research area is located in Wuyuan County, Bayannaoer City, Inner Mongolia Autonomous Region, China (41°6′44″ N–41°7′41″ N, 108°19′48″ E–108°20′49″ E). Its geographic location is shown in Figure 1. From the Yihe trunk canal in the south, to the highway in the north, to a new canal in the east, and to the left second branch canal in the west, the terrain is relatively flat. This area is 1022–1025 m above sea level and covers an area of 2.344 km^2. Corn, sunflowers, beets, and tomatoes are grown in this area. Its annual average solar radiation is 153.44 cal/cm^2, its annual sunshine duration is 2702 h, its average temperature is 17.76 °C, its accumulated temperature is 3569 °C, its frost-free period is 117–136 days its average annual amount of rainfall is 153.1 mm, and its average annual groundwater depth is 1.86 m; it has a middle-temperate continental climate. The area's average soil organic matter content is 9.5 g/kg, and its total nitrogen content is 1.2 g/kg. The meteorological data used in this study were obtained from the National Meteorological Information Center and are shown in Figure 2.

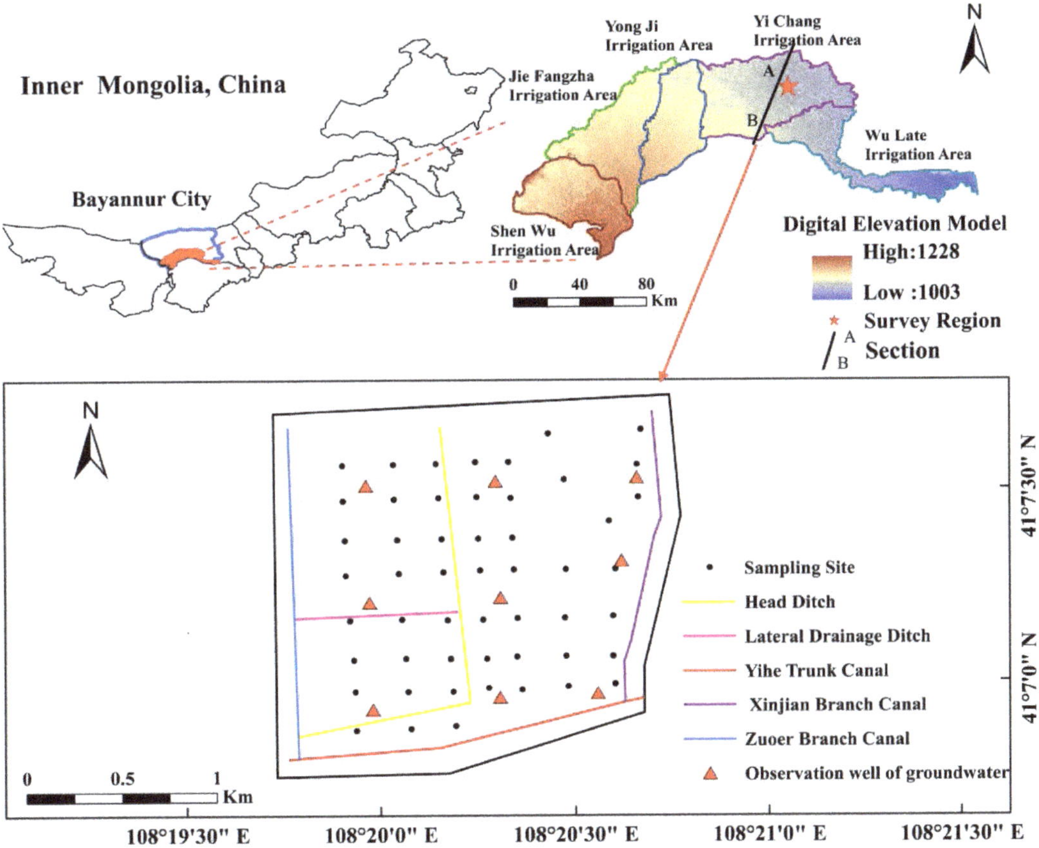

Figure 1. The distribution of the study area and sampling points.

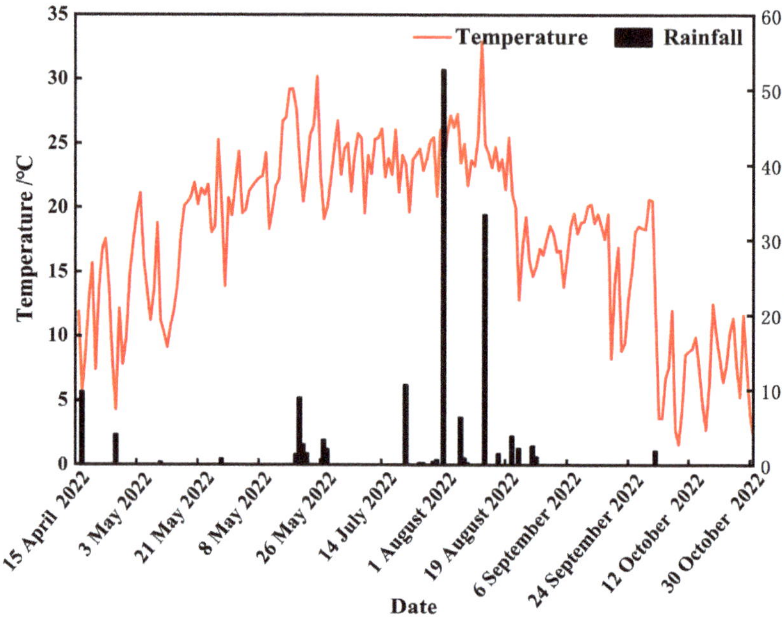

Figure 2. Rainfall and temperature during the growth period of 2022.

2.1.2. Hydrogeological Framework

According to the geological survey, the Hetao Irrigation District is a Mesozoic–Cenozoic fault basin that formed in the late Jurassic. It is part of the Ordos platform syncline of the North China Platform in terms of its geological structure (Figure 3a). The sediments in the irrigation area are mainly derived from the river sediments of Langshan and the ancient Yellow River. According to its sedimentary structural conditions, an alluvial lake aquifer system with a fine-grained phase as its main component was formed in the Hetao Basin. The sediment particles are mainly medium-grain sand and coarse sand. The groundwater flows from west to east, with a large permeability coefficient and strong runoff conditions. As shown in Figure 3b, the aquifer in the irrigation area has complex geometry and multi-layer overlap. On the basis of Quaternary stratification, the hydrogeological characteristics of the aquifer, and its development and utilization value according to its burial conditions, it is divided into Upper Pleistocene (Qp_3)–Holocene (Q_h) and Middle Pleistocene aquifer systems (Q_{p2}). The Qp3-Qh aquifer has the greatest thickness and the widest distribution. The Q_{p2} aquifer is widely distributed in the piedmont and uplift areas, and its burial depth is relatively shallow.

2.2. Sampling Point Layout and Data Acquisition

2.2.1. Sampling Point Layout

Monitoring points were arranged in accordance with a 200 × 200 m grid in the test area, and 52 soil-water- and -salt-sampling points were slightly adjusted for the special terrain and features (Figure 1). Sampling began in April 2022, when the maize and sunflowers were at the seedling stage. The field coordinates were determined using a handheld global positioning system, and the sampling depth ranged from the surface to a depth of 100 cm. The total sampling depth was divided into six layers (0–10, 10–20, 20–40, 40–60, 60–80, and 80–100 cm), and the excavated samples were collected by layer with a soil drill (stainless steel 1 m) and placed in sealed bags.

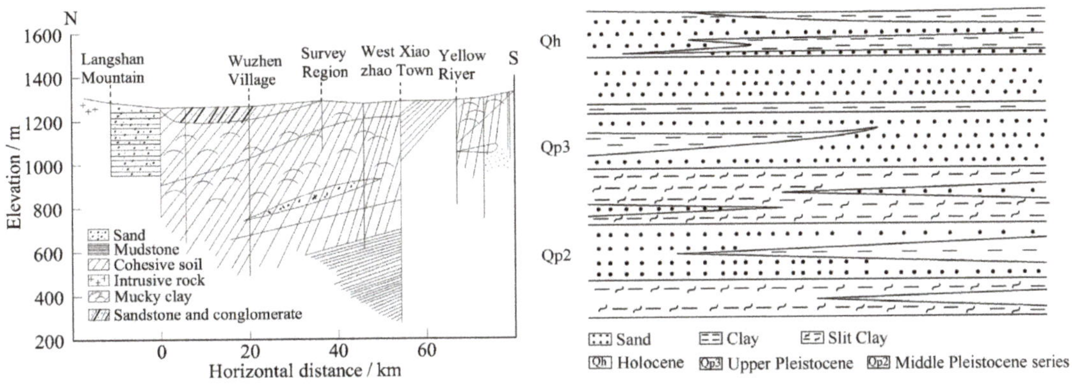

(**a**) Hydrogeological profile (**b**) Aquifer schematic diagram

Figure 3. Schematic diagrams of hydrogeological conditions in Hetao Irrigation District.

2.2.2. Data Measurements

A drying method [27] was used to determine the soil moisture content, and soil conductivity was measured simultaneously. After air drying and grinding, the conductivity of the soil extract in a 1:5 ratio of soil to distilled water was tested using an electrical conductivity meter (INESA DDS-307A, Shanghai, China) after being passed through a 2 mm screen [28]. The depth of the groundwater was measured every seven days using the plumb–plumb method [14].

2.2.3. Soil Texture

A laser particle size analyzer was used to test the particle size distributions of the ground and sieved soil samples (HELOS & RODOS, Sympatec, Clausthal, Germany). The U.S. Department of Agriculture soil particle grading standard [29,30] was used for the analysis. The soil particle sizes were divided into clay (0–2 μm), silt (2–50 μm), and sand (50–2000 μm). The soil textures in the study area were divided into five categories: loam, sandy loam, silty loam, and silty loam (Figure 4).

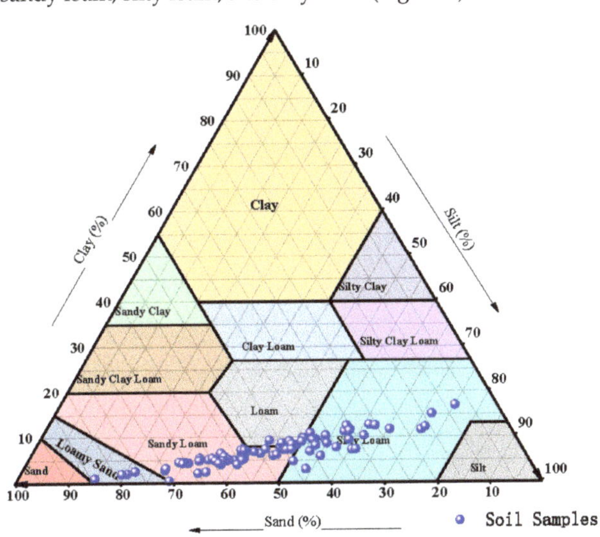

Figure 4. Soil textures in the study area.

2.3. Research Methods

2.3.1. Semi-Variance Function

The semi-variance function is a basic tool used in geostatistics [31]. The spatial variation structure of the regionalized variable is represented by half of the variance between sampling points and is calculated as follows (1):

$$\gamma(h) = \frac{1}{2N(h)} \sum_{i=1}^{N(h)} [Z(x_i) - Z(x_i + h)]^2 \quad (1)$$

In Formula (1), $N(h)$ is the lag distance equal to the logarithm of point h; $Z(x_i)$ is the measured value of the variable at point x_i; and $Z(x_i + h)$ is the measured value of the variable deviating from h at point x_i.

The fitting accuracy of the semi-variance function was characterized using the determination coefficient (R^2) and residual sum of squares (RSS), in which R^2 was between 0 and 1; the larger the value, the better the fitting effect. The smaller the RSS value, the better the simulation effect [32].

2.3.2. Ordinary Kriging

Kriging, or spatial local interpolation, is an unbiased optimal estimation method for regionalized variables in a limited region. It utilizes semi-variance theory and structural analysis, making it a key component of geostatistics [1]. Suppose there are n measurement points in the neighborhood of the estimation point x_0, that is, $x_1, x_2, \ldots, x_n$; accordingly, the ordinary Kriging formula used was as follows:

$$Z(x_0) = \sum_{i=1}^{n} \lambda_i Z(x_i) \quad (2)$$

In Formula 2, $Z(x_i)$ is the observation value selected for Kriging near the estimate point x_0; $Z(x_0)$ is the Kriging estimate at x_0; and λ_i is the weight of the ith observation with respect to the estimated point x_0, which is obtained from the Kriging equations established via the semi-variance function.

Testing the normal distribution of data has the same premise as using the Kriging method for the spatial analysis of soil characteristics. The Kriging method is effective only when the analyzed data follow a normal distribution; otherwise, there may be a proportional effect. After conducting the Kolmogorov–Smirnov test, it was found that the soil salt content and water content datasets did not conform to a normal distribution; after natural logarithm conversion, they showed an approximate normal distribution, which met the requirements of geostatistical analysis. Therefore, the data adopted for variance function calculations were obtained after logarithmic conversion [33].

2.3.3. Spatial Autocorrelation Method

Spatial autocorrelation analysis, derived from biometrics, is a spatial statistical method that tests for significant spatial dependence between sample elements and describes the spatial distribution of geographical phenomena in a study area. This is the premise behind and basis of linear regression analysis [34]. This method was carried out with the spatial autocorrelation index Moran's I, which was used to analyze the spatial autocorrelation of soil salinity. The formulae used are as follows:

$$I = \frac{\sum_{i=1}^{n} \sum_{j=1}^{n} w_{ij}(x_i - x)(x_j - x)}{S^2 \sum_{i=1}^{n} \sum_{j=1}^{n} w_{ij}} \quad (3)$$

$$s^2 = \frac{1}{n} \sum_{i=1}^{n} (x_i - x)^2 \quad (4)$$

In Equations (3) and (4), I is the spatial autocorrelation index, and w_{ij} is the spatial weight matrix between regions i and j. This study was based on the first-order ROOK weight matrix of the common boundary, where the adjacent space is 1 and the non-adjacent space is 0. x_i is the observed value of a variable [1] at sample point I, $\bar{x}$ is the mean value of the variable, x_j is the observed value at sample point j, s^2 is the variance of the variable, and n is the total number of variables. Moran's I was converted to a standard normal distribution using the following formula:

$$Z(I) = \frac{1 - E(I)}{\sqrt{var(i)}} \quad (5)$$

In Formula (5), $E(I)$ is the theoretical expected value, and $var(i)$ is the variance. When $Z(I) = 0$, that is, when I is equal to its mathematical expectation, the observed values show an independent random distribution, corresponding to Moran's I scatter plot. The observed values were evenly distributed in four quadrants but were not significant in the Local Indicators of Spatial Association (LISA) cluster plot. When $Z(I) > 0$, there is a positive spatial correlation in the region, corresponding to the distribution of observed values in quadrants 1 and 3 on the Moran's I scatter plot and spatial similarity in the LISA cluster plot. When $Z(I) < 0$, there is a negative spatial correlation within the region, which corresponds to the distribution of observed values in quadrants 2 and 4 on Moran's I scatter plot and presents spatial heterogeneity in the LISA cluster plot [35].

2.4. Data Processing and Analysis

Origin (2018, OriginLab, Northampton, MA, USA) software was used to analyze the data and draw maps, and GS + 9.0 (Youwan Technology, Beijing, China) software was used to perform logarithmic transformation on data sources that did not conform to a normal distribution. A semi-variance function model was obtained after calculation, simulation, and testing. In ArcGIS (10.8, Esri, CA, USA) software, ordinary Kriging interpolation was used to draw spatial distribution characteristic maps of the soil salinity and groundwater depth. The spatial analysis module of Geoda (1.20.0.36, Luc Anselin, Chicago, IL, USA) software was used to calculate Moran's I index and Z value and to draw a Moran's I index scatter plot.

3. Results and Analysis

3.1. Analysis of Soil Salinization Characteristics

Based on the data from the samples collected on 8 June 2022, we conducted a classical statistical analysis of the salt and water content of the soil layers at 0–10, 10–20, 20–40, 40–60, 60–80, and 80–100 cm. The corresponding statistical characteristics are listed in Table 1. Overall, the average soil moisture content of each layer increased with increasing depth. The average soil salt content decreased with increasing depth in the surface layers (0–40 cm) and increased with increasing depth in the deep layers (40–100 cm). The ratios of the maximum to minimum salt content in each soil layer were 99.06, 34.00, 21.98, 20.86, 38.46, and 39.92 in the 0–10, 10–20, 20–40, 40–60, 60–80, and 80–100 cm layers, respectively, while the variation in salt ranges were 50.01, 17.16, 10.70, 10.92, 22.10, and 25.69 g/kg, respectively. There was an evident accumulation of surface soil salt, and the soil salt variability was high in the vertical direction.

It is generally believed that when the coefficient of variation is less than 10%, there is weak variability, while values of 10–100% indicate moderate variability and those greater than 100% indicate strong variability. The coefficient of variation of the soil salt content in each layer was strong, except in the 20–40 and 40–60 cm soil layers. The soil moisture content in each layer had a moderate degree of variation. The variation coefficients of soil salt content and soil water content in the 0–60 cm soil layer decreased layer by layer, indicating that the spatial variability of soil water and salt decreased with increasing soil depth. This may have been due to the influence of factors affecting the spatial variations of the soil water and soil salt content (such as climate, soil texture, and human activities) weakening with increasing soil depth. In addition, the coefficient of variation of the soil salt content was higher than that of the soil water content at the same depth, indicating that the spatial variability of the soil salt content was stronger than that of the soil water content.

Table 1. Statistical analysis of soil salinity.

	Depth/cm	Minimum	Maximum	Average Value	Standard Deviation	Variance	Skewness	Kurtosis	Coefficient of Variation/%
Soil salt content (g/kg)	0–10	0.51	50.52	4.64	8.11	65.77	4.41	21.96	1.75
	10–20	0.52	17.68	2.90	2.94	8.67	2.89	11.57	1.01
	20–40	0.51	11.21	2.58	2.21	4.89	2.19	5.22	0.86
	40–60	0.55	11.47	2.95	2.26	5.09	1.68	3.19	0.77
	60–80	0.59	22.69	3.63	4.12	16.95	3.05	11.13	1.13
	80–100	0.66	26.35	3.59	4.03	16.20	3.89	19.93	1.12
Soil moisture content (%)	0–10	13.71	44.65	21.12	4.84	23.43	2.68	10.72	0.23
	10–20	15.60	38.54	22.94	3.68	13.56	1.56	5.79	0.16
	20–40	8.06	31.98	24.19	3.64	13.28	−1.61	6.76	0.15
	40–60	10.76	54.74	26.48	5.79	33.47	1.88	12.58	0.21
	60–80	13.42	62.92	28.48	6.77	45.89	3.25	14.75	0.23
	80–100	22.17	69.21	20.01	7.10	50.37	4.24	21.34	0.35

Analysis of soil salt and moisture content using traditional statistical methods can only be used to summarize soil conditions. It cannot reflect the characteristics impacting local change; that is, it can produce an estimate of the overall situation, but cannot quantitatively describe the randomness, structure, independence, or correlation of the distribution of soil salt and water. To solve these problems, it is necessary to use further geostatistical methods to analyze and discuss the structure of spatial variation [33].

3.2. Analysis of Spatial Variation Characteristics of Soil Salinization

The geostatistical analysis module of GS + 9.0 software was used to carry out a semi-variance function analysis of salt and water content. The specific fitting results and parameters are listed in Table 2. The salt and water proportions of each layer of soil were well fitted by the index model, with an RSS range of 3×10^{-6}–0.127 and an R^2 range of 0.419–0.927, both of which reached significant levels. The ratio of the bullion value (C_0) to the abasement value (C_0 + C) is the bullion coefficient; the size of the bullion value indicates the degree of spatial correlation between soil properties. A bullion coefficient of less than 0.25 indicates a strong spatial correlation that is mainly affected by structural factors (climate, parent material, terrain, and soil type). A bullion coefficient between 0.25 and 0.75 indicates a moderate special correlation that is affected by structural and random factors. A bullion coefficient greater than 0.75 indicates a weak spatial correlation that is mainly affected by random factors (fertilization, tillage measures, planting systems, and other human activities). A bullion coefficient close to 1 indicates that there is constant variation across the entire scale [23]. The coefficients for salt content in the 0–60 cm soil layer ranged from 0.001 to 0.223, showing a strong spatial correlation that was mainly affected by structural factors. The bullion coefficient for the soil water content ranged from 0.001 to 0.031, showing a strong spatial correlation. In this study, the spatial variation of soil salt and water content was mainly affected by structural factors.

3.3. Soil Salt Content and Water Content Are Spatially Autocorrelated

To further study the degree of spatial dependence and the local distribution pattern of soil salinization in the test area, the test area was latticed at a scale of 200 × 200 m to improve research accuracy. As shown in Figure 5, Moran's *I* was used as an index to measure the spatial autocorrelation of soil salt content (SSC) and soil moisture (SM). The Geoda platform was used to analyze the global and local spatial autocorrelation of soil salt content and conductivity, and a Moran's *I* scatter plot was obtained.

Table 2. Semi-variance function types and model parameters of salinity factors in saline soil.

Index	Depth/cm	Model	Nugget (C_0)	Sill ($C_0 + C$)	Partial Sill $C_0/(C_0 + C)$	RSS	R^2	Range/km
Soil salt content	0–10	Exponential	0.019	1.763	0.011	0.102	0.927	0.031
	10–20	Exponential	0.032	0.826	0.039	0.069	0.826	0.018
	20–40	Exponential	0.086	0.639	0.135	0.072	0.721	0.015
	40–60	Exponential	0.133	0.596	0.223	0.089	0.546	0.011
	60–80	Exponential	0.001	0.752	0.001	0.127	0.419	0.004
	80–100	Exponential	0.113	0.675	0.167	0.012	0.663	0.005
Soil moisture content	0–10	Exponential	0.0002	0.047	0.004	0.0003	0.799	0.012
	10–20	Exponential	0.001	0.029	0.031	0.0001	0.764	0.015
	20–40	Exponential	0.010	14.940	0.001	0.001	0.619	0.009
	40–60	Exponential	0.001	0.636	0.002	0.291	0.856	0.034
	60–80	Exponential	0.0001	0.043	0.002	0.001	0.679	0.008
	80–100	Exponential	0.0001	0.036	0.003	0.0003	0.731	0.008

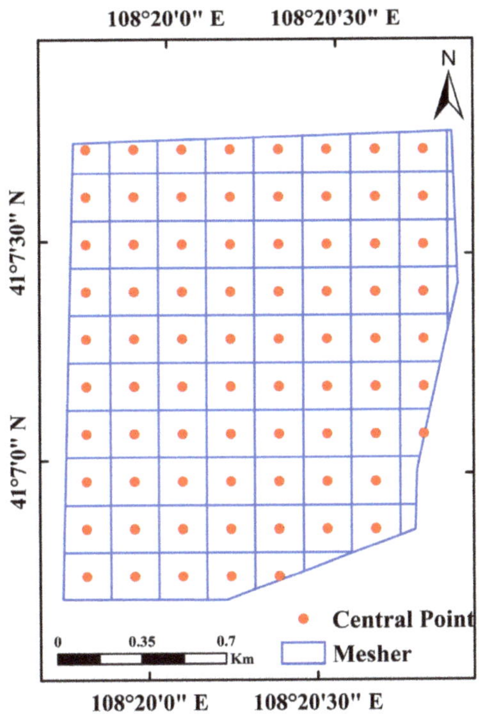

Figure 5. A grid map of the study area.

To further study the degree of spatial dependence and the local distribution patterns of soil salinization in the test area, the test area was latticed at a scale of 200 × 200 m to improve research accuracy. Table 3 presents the statistical results of the indicators from each layer of saline soil compared with Moran's *I* at a water content depth of the same soil layer. The salt content in the 0–100 cm soil layer ranged from 0.09 to 0.19, and the *Z* value ranged from 1.59 to 2.62. Moran's *I* and *Z* values of the soil water content ranged from 0.43–0.48 and 5.51–5.85, respectively. The *Z* values of the overall soil salt and water content in the test area were both greater than 0 ($p < 0.05$), and the spatial positive correlation was highly significant among all soil layers, indicating that the soil salt and water content had a strong spatial dependence. As presented in Figures 6 and 7, there was a positive correlation between the salt content in the 0–100 cm soil layers, but the accumulation phenomenon was not significant relative to the moisture content. Except for one sample point, the scattered points of most of the sample points fell in the first and third quadrants, indicating a spatial agglomeration trend. These results show that the spatial distribution of the soil salt and water content in the test area was not random but regular, showing specific spatial agglomeration characteristics; these spatial agglomeration characteristics were determined by the spatial heterogeneity of the structural factors in the areas in which the sample points were located. The Geoda platform was used to analyze the global and local spatial autocorrelations of soil salt content and moisture, and a Moran's *I* scatter plot was obtained.

Table 3. Statistics of indicators of each layer of saline soil.

Depth/cm	Moran'*I*		*Z*		*P*	
	Soil Salt Content (g/kg)	Soil Moisture Content (%)	Soil Salt Content (g/kg)	Soil Moisture Content (%)	Soil Salt Content (g/kg)	Soil Moisture Content (%)
0~10	0.16	0.46	1.61	5.61	<0.05	<0.05
10~20	0.11	0.46	1.58	5.64	<0.05	<0.05
20~40	0.09	0.46	1.55	5.58	<0.05	<0.05
40~60	0.16	0.48	2.14	5.85	<0.05	<0.05
60~80	0.19	0.43	2.62	5.21	<0.05	<0.05
80~100	0.10	0.45	1.84	5.51	<0.05	<0.05

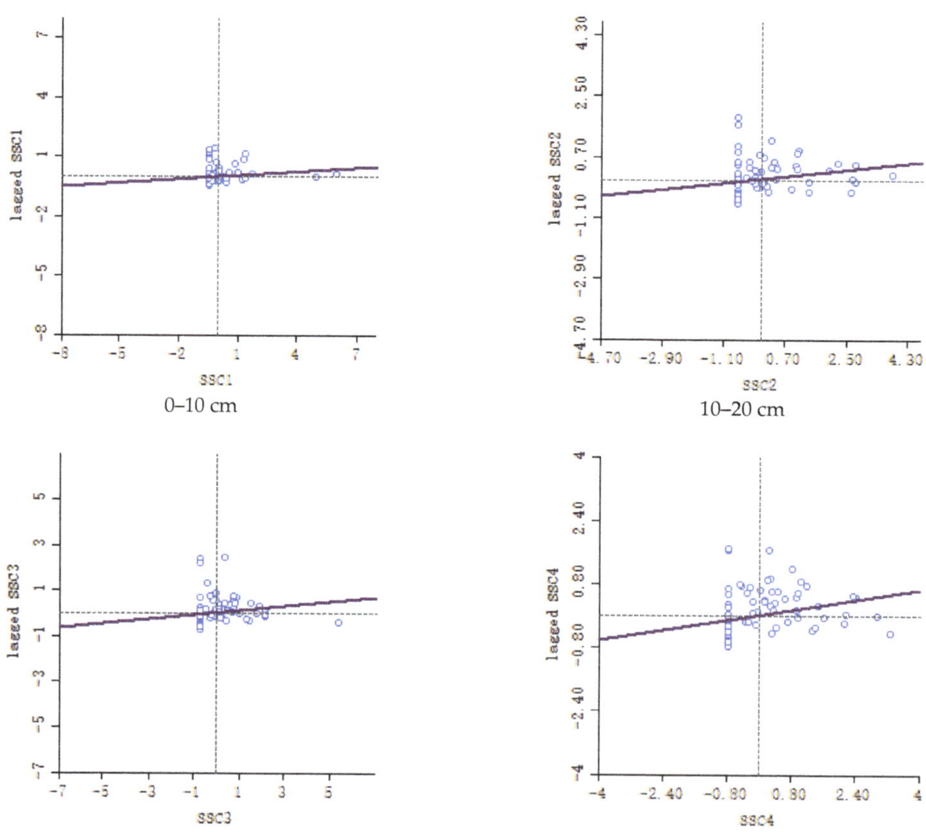

Figure 6. *Cont.*

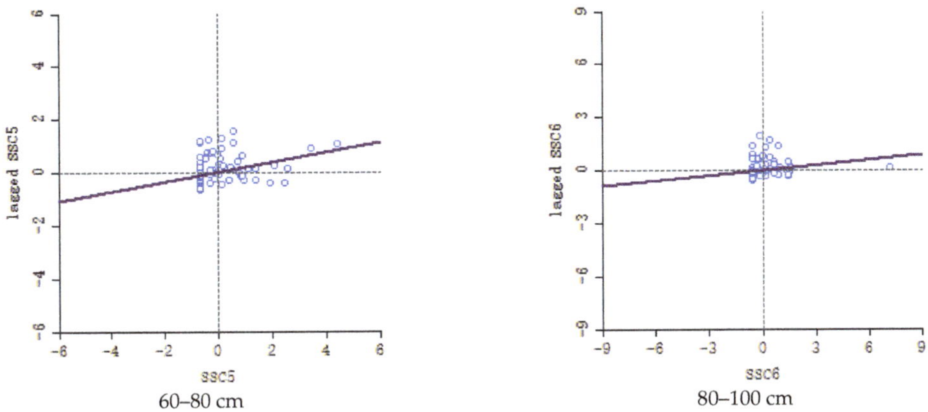

Figure 6. Moran's *I* scatter plots of the soil salt content.

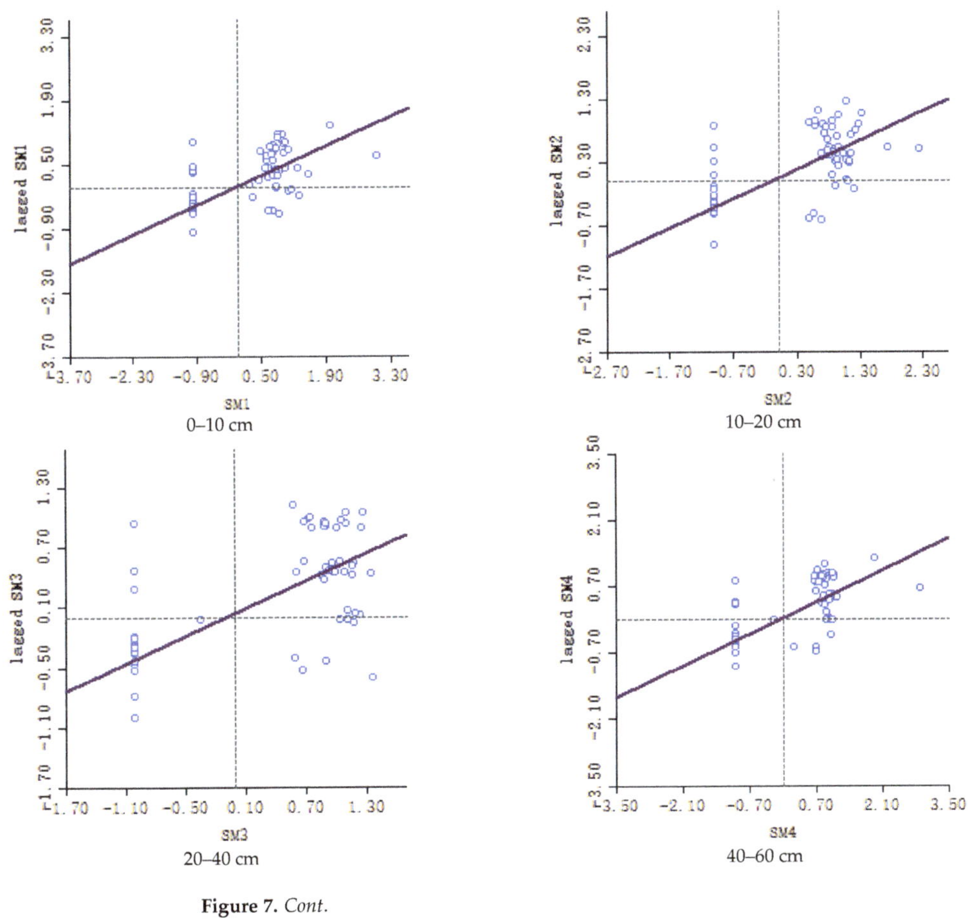

Figure 7. *Cont.*

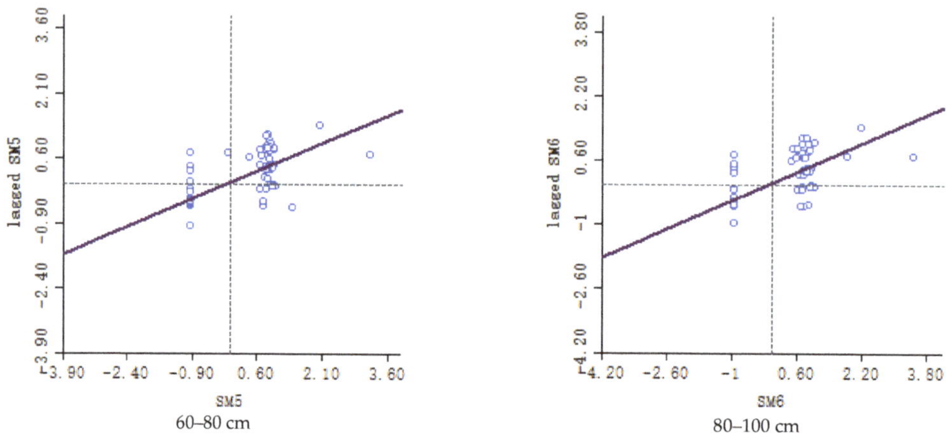

Figure 7. Moran's *I* scatter plots of the soil moisture content.

3.4. Spatial Distribution Characteristics of Soil Salt and Water Content

The optimal fitting model and parameters are listed in Table 2. The semi-variance function types and model parameters of the salinity factors in saline soil and the results obtained via ordinary Kriging interpolation are shown in Figures 8 and 9, respectively. From the perspective of spatial distribution, areas with high soil salt content appeared in the wasteland in the transition from cultivated land to villages in the northeast of the study area and the unseeded cultivated land in the southwest of the study area. Areas with a low soil salt content were distributed in the northwest and southeast of the study area where the salt content in the surface and deep soil samples was significantly different. This difference was likely due to the lack of vegetation cover and shallow groundwater depth in these areas. Capillary water in the vadose zone rises under the prolonged action of strong evaporation, and soluble salts in the deep soil and groundwater are brought to the surface, resulting in increased salinity. Fields in the northeast and southwest were irrigated and planted in mid-May, resulting in relatively low salinity.

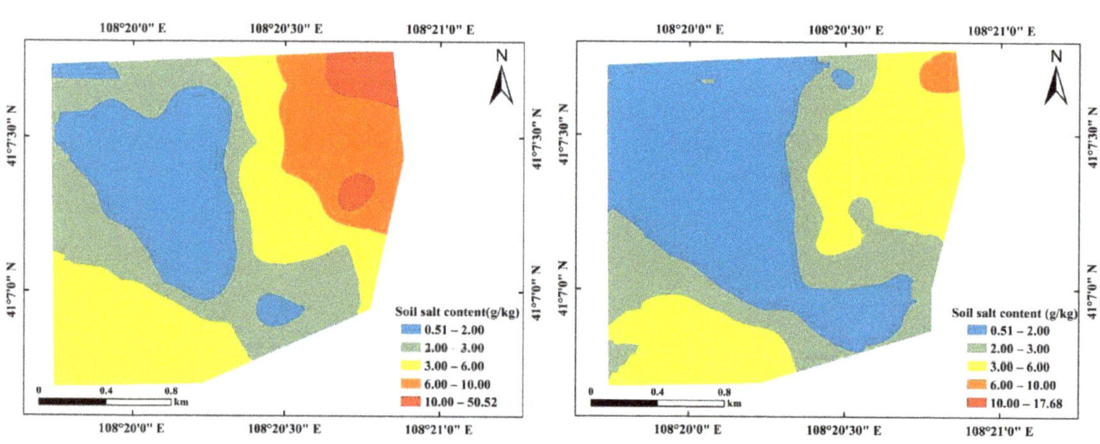

Figure 8. *Cont.*

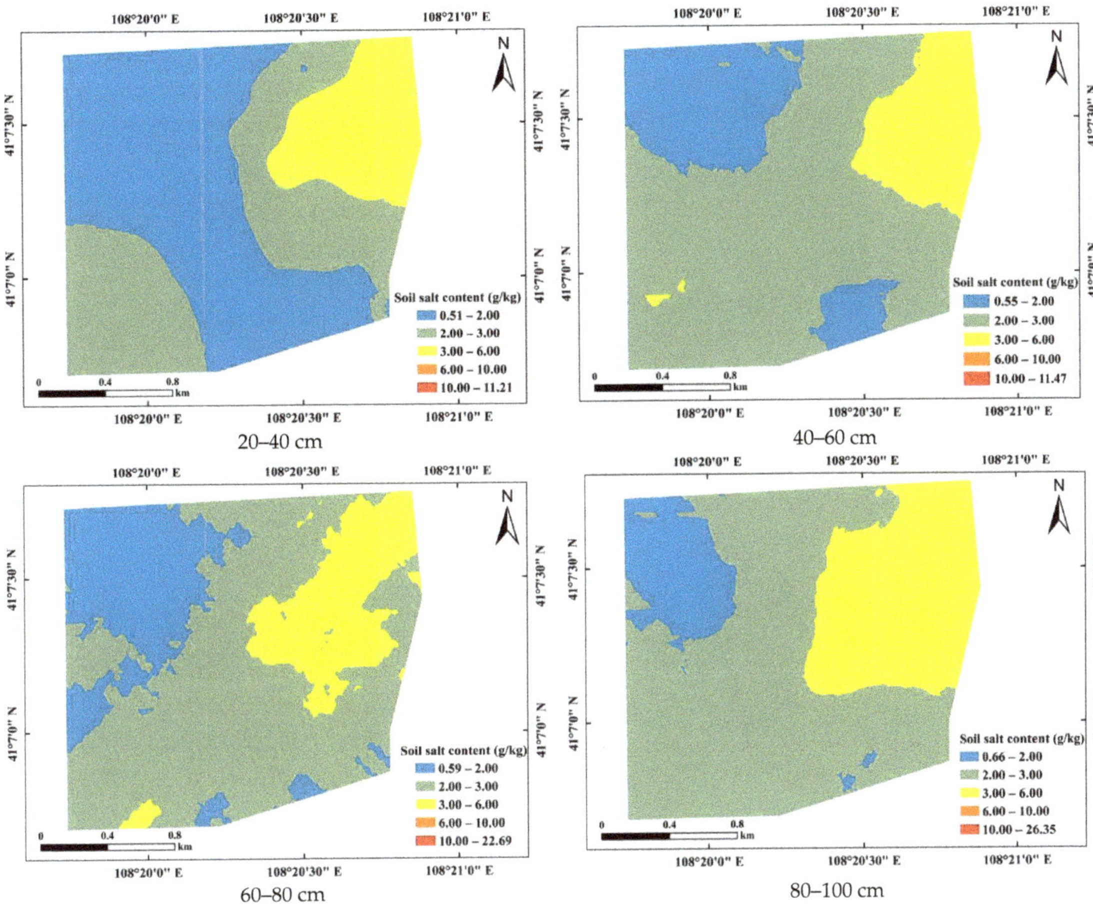

Figure 8. The spatial distribution of the soil salt content.

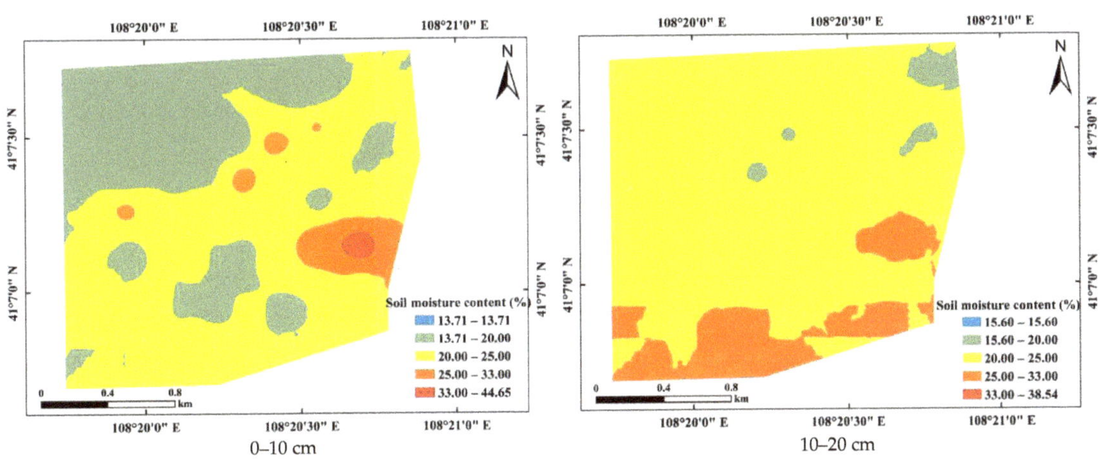

Figure 9. *Cont.*

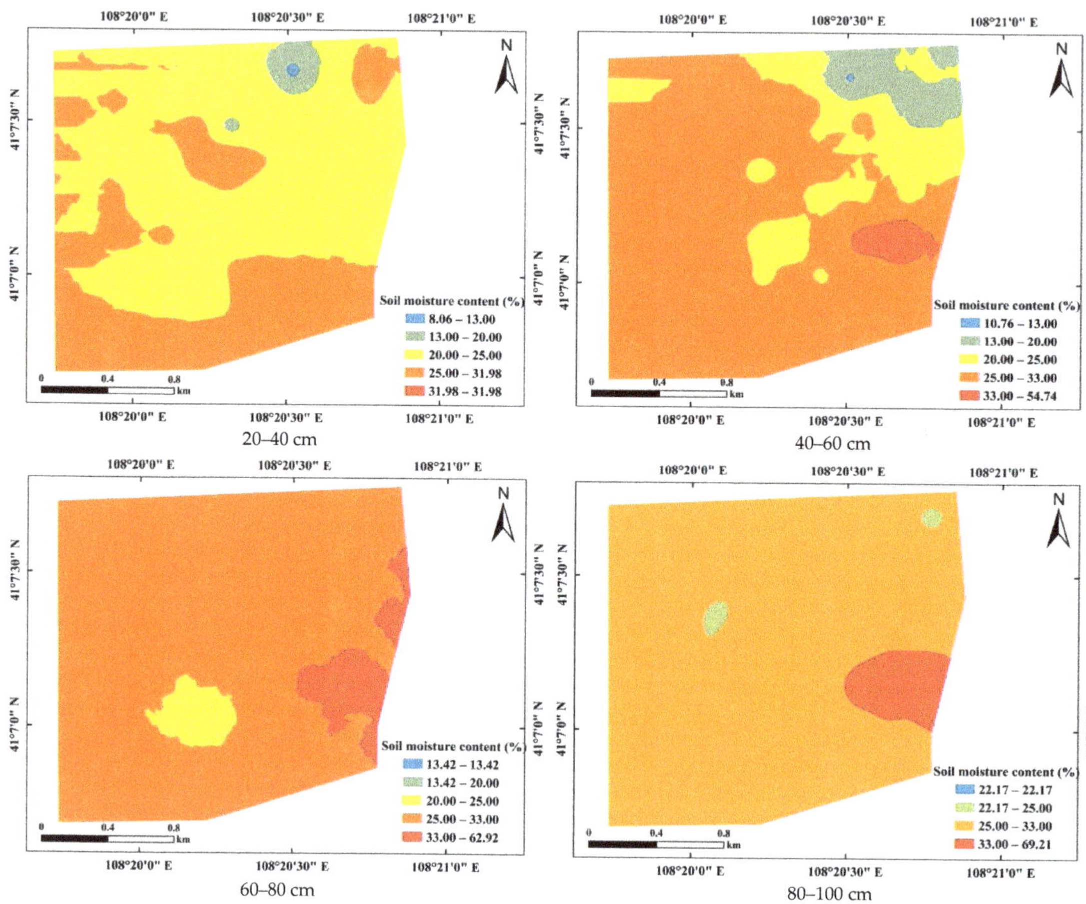

Figure 9. The spatial distribution of the soil moisture content.

According to the classification standard for salinized soil [23], soil is divided into non-saline (salt content less than 2 g/kg; secure crop growth), mildly saline (salt content of 2–3 g/kg; mild crop inhibition), moderately saline (salt content of 3–6 g/kg; moderate crop inhibition), severely saline (6–10 g/kg; severe crop inhibition), and saline soil (salt content greater than 10 g/kg; a crop-damaging amount of salt). As shown in Figure 10a, the proportion of areas with different salt and water levels ranges in the different soil layers in the test area increased layer by layer by 18%. The soil salt content increased by 9% layer by layer from 0 to 40 cm in the soil with a salt content of $\leq$2 g/kg and by 2% and 6% for the soil with a salt content of between 2 and 3 g/kg, respectively. The areas with a 3–6 g/kg soil salt content were 33%, 29%, and 15% decreased layer by layer, respectively. Severely saline soils with 6–10 g/kg soil salt content appeared at the soil layers of 0–10 cm and 10–20 cm, accounting for 12% and 1% of samples, respectively. A soil salt content of >10 g/kg was only observed in samples at a depth of 0–10 cm, accounting for 5% of the total area and located mostly in the local wasteland in the transition area from cultivated land to villages.

As shown in Figure 10b, with an increase in soil depth, soil water content gradually increased. Low-water-content areas of 13% to 20% occurred in the 0–10, 10–20, 20–40, and 40–60 cm soil layers, accounting for 33%, 3%, 2%, and 6% of the total land, respectively. The proportions of the area with a water content of 20–25% were 60%, 82%, 60%, 18%, 3%, and 1%, respectively. The proportions of the area with a water content of 25–33% were 6%,

15%, 38%, 74%, 89%, and 92%. The high-value area with a water content greater than 33% was located in deep soil, and a change in the area's proportions was not evident [24,36]. Sampling was carried out at the beginning of the growth period in this experiment, and groundwater and its salt moved upward continuously due to strong evaporation, which was the main source of the recharge of the soil water and salt content in the study area.

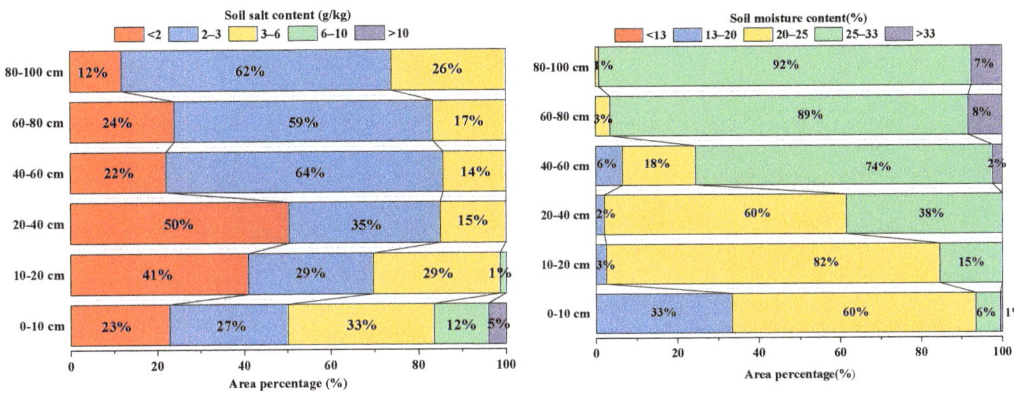

(**a**) Soil salt content (**b**) Soil moisture content

Figure 10. The proportionate areas of different soil salt and water content intervals in different layers.

3.5. Effect of Groundwater Depth on Soil Salt Content in Different Layers

This study shows that soil salt content and soil salinization status are greatly affected by the depth of groundwater. The shallower the groundwater level, the greater the evaporation and the more serious the soil salt accumulation. The soil salt content was lower in areas with deeper groundwater levels. This shows that the groundwater depth in the study area restricted the soil salt content ("salt comes with water, salt goes with water"), and soil salt is transported to the surface soil layer by groundwater through the process of evaporation. In addition, in some parts of the study area, spring irrigation significantly impacted groundwater depth, resulting in shallow groundwater levels. Even if the groundwater salt content is low, more water will be carried via evaporation into the soil, resulting in soil salt accumulation. Therefore, the soil will not be salinized if the groundwater is controlled at a depth that does not allow salt to enter the shallower soil through evaporation.

The spatial distribution of shallow groundwater depth in the study area is shown in Figure 11 through interpolation. It can be observed that the groundwater depth changes significantly before and after spring irrigation. The groundwater burial depth on the side of the new channel is obviously less than that on the side of the second channel on the left, likely due to the fact that only some farmers in the area use the new channel for spring irrigation. Furthermore, the drainage ditches in the study area have an impact on the change in groundwater burial depth. Figure 11 demonstrates that the groundwater burial depth in the right region of the study area decreases rapidly, in part, after spring irrigation. This test also reveals that the groundwater burial depth of the drainage ditches exhibits a similar trend of change, indicating its effective role in drainage. In terms of water quality, the EC value of the water in the ditch before spring irrigation was 2.7 dS/m, which increased to 3.5 dS/m after spring irrigation, resulting in a 29.6% increase in salt value.

Through an analysis of the relationship between soil salinity and groundwater depth, we observed that with shallower groundwater depths, the relationship between the two became exponential (Figure 12); the correlation was good, the R^2 value was no less than 0.9421, and the fitting degree of these two values was high. Guan Xiaoyan [36,37] found that the local groundwater depth in the area of the present study was greater than 1.5 m. When the electrical conductivity value was less than 0.8 dS/m, the yield of corn, wheat, and sunflower was not affected [22]. When the underground water depth was between

1 and 1.7 m, the soil salt content changed greatly with the underground water depth. When the underground water depth was greater than 1.7 m, the overall soil salt content was low, and the salt content of each layer changed slightly. Therefore, to effectively control soil salinization, the underground water depth should be maintained below 1.7 m.

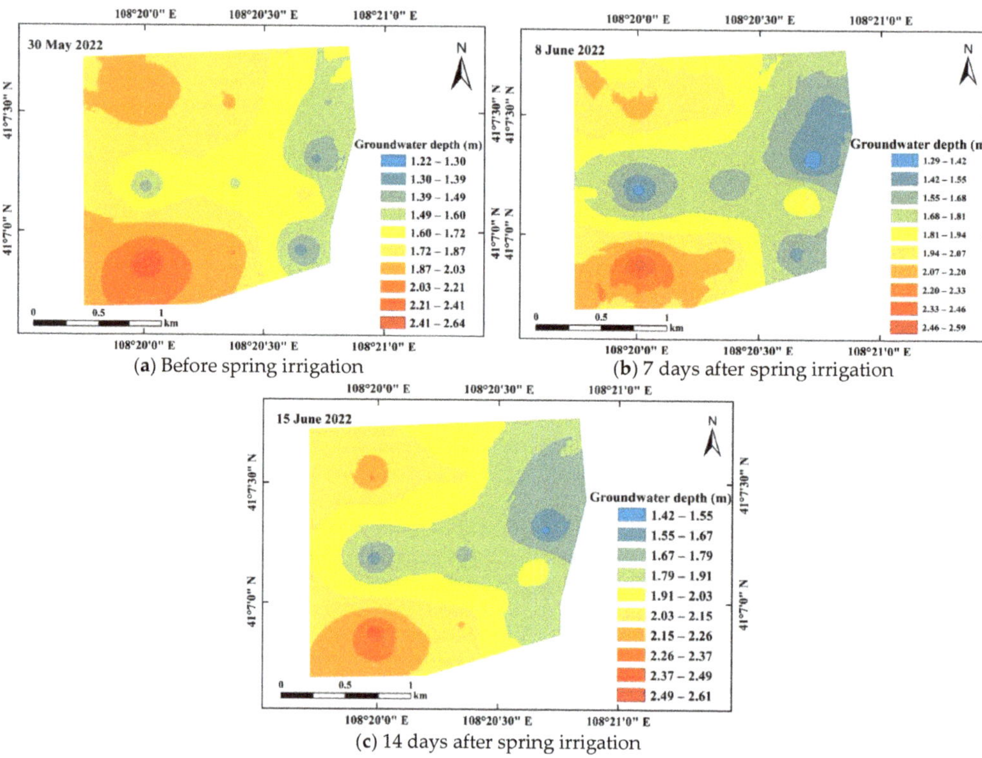

Figure 11. Groundwater depth in different periods in the study area.

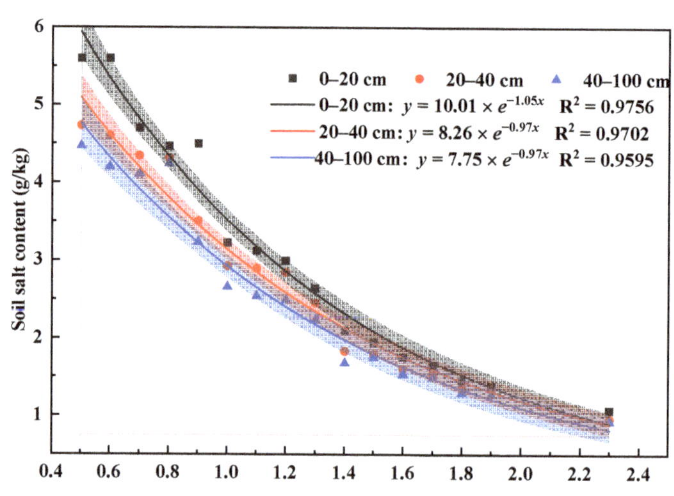

Figure 12. The relationship between soil salt content and groundwater depth.

4. Discussion

4.1. Variability Mechanism of Soil Water and Salt

According to the classical statistical analysis of the soil salt factors in the study area, there is an obvious phenomenon of surface accumulation in the surface soil [12]. The soil water content increases layer by layer in the 0–40 cm deep soil layer, while the soil salt content decreases layer by layer. From the perspective of salt content variation, the maximum salt content of the surface soil is 99 times greater than the minimum value, with a variation range of 50.01 g/kg. This is due to the study area being located in arid and semi-arid regions in the Hetao irrigation area, with high groundwater levels and little rainfall. In this area, spring irrigation is usually used to "wash salt" before crop planting. However, the leaching of salt is limited when irrigation water infiltrates the soil. As a result, when the downward leaching reaches a certain depth, pore water flow reverses upward to supplement the evaporation requirement after irrigation stops. This causes salt to gather on the soil surface again, forming a high-salt zone with strong variability, which aligns with the research conclusions of some scholars in the Hetao irrigation district [37,38].

According to the results of the geostatistical analysis of the study area, soil water and salt in the 0–100 cm soil layer showed a strong spatial correlation, primarily influenced by structural factors. The irrigation areas in Hetao primarily rely on yellow water diversion, which has a high salinity, thereby resulting in the mixing of soil ions. Consequently, the chemical composition of groundwater is altered, and the precipitation behavior of certain minerals, particularly carbonate minerals, is induced. Such behavior impacts the pore microstructure of the soil and leads to changes at the macroscopic scale such as alterations in soil particle size distribution [8]. This study's findings demonstrate a decrease in soil salinity as the sand content increases, whereas an increase in soil salinity is observed with higher clay and silt content [3]. The relationship between soil water and soil clay is negative, whereas it is positive for clay and silt content. This is mainly due to the large surface area of soil clay and silt, which constitute the active segment of the soil, facilitating effective water penetration and solute leaching [39,40]. Conversely, sand is relatively inert, thus impeding deep seepage and solute transport. This circumstance promotes an increase in the reaction surface area and the production of more mineral precipitation [41]. Maize and sunflowers were the primary crops planted in the study area. Although crop cultivation and management in the area were performed relatively scientifically (mulching, formula fertilization, etc.), the soil organic matter content in the study area was only 9.5 g/kg, meaning the soil had insufficient fertility. Considering the local situation, this cultivated land should probably be subjected to a straw-crushing treatment before future crop planting. This treatment can effectively improve the decomposition rate of organic residues and the synthesis rate of organic matter, promote the soil-ripening process, increase the silt and clay content in the soil, and play a role in soil moisture conservation [42].

The results of Zhang Wencong's quantification of the conversion between soil water and groundwater at the scale of canals in the Hetao Irrigation District showed that the levels of recharge of soil water into groundwater over a 2-year growth period were 207.73 mm and 236.94 mm [14]. Thus, it can be observed that the bidirectional exchange between soil water and shallow groundwater was frequent during the growth period. This frequent water exchange indicates that there is significant hydrological exchange between surface water and shallow groundwater in the Hetao irrigation area. Furthermore, it is important to note that soil salinization generally occurs in areas where the groundwater depth is low. This occurs when salt-rich pore water from groundwater continuously evaporates at or near the surface, leading to salt precipitation [37]. The key to preventing soil salinization is to control groundwater levels. Therefore, the area's irrigation and drainage system should be improved, and technical agricultural measures such as reasonable irrigation should be taken to prevent the increase in the groundwater level and soil salt return. Generally, when the groundwater depth is more than 2–3 m, the accumulation of salt generated via soil evaporation is small [37,38]. However, the results of this study show that when the groundwater depth is more than 1.7 m, the soil salt content is low, and the soil salt content changes little in each layer, which does not affect the yield of corn and sunflower in the region. It is important to note that this area's groundwater depth varies seasonally, so controlling the groundwater depth in the study area is not the only standard with which to control soil salinization. Soil salinization should be controlled in various aspects by combining groundwater composition and other soil improvement methods [43].

4.2. Spatial Correlation of Soil Water and Salt

Kriging interpolation is the most commonly used interpolation method in geostatistics. It can effectively estimate the attribute value of an unsampled area. Some scholars at home and abroad have used inverse-distance-weighted interpolation to study soil spatial variation [23,24,32,44]. However, when the variation coefficient of the predicted object is large, Kriging interpolation has higher accuracy. According to Michael J. Friedel, hydrogeological units can generate variation functions for Kriging interpolation with sufficient support, which can improve the spatial density of spectral remote sensing images [45]. Kriging is the most commonly used interpolation method in geological statistics. Some scholars have used the inverse-distance-weighted interpolation method to study soil spatial variation. However, some studies have indicated that the accuracy of Kriging is higher when the coefficient of variation of the predicted object is large [23,24,32,44,46–48]. The combination of ArcGIS and geostatistics can be used to effectively monitor spatial variations in regional soil salt factors [2,20,46–50]. Based on the regionalized variable theory, GS+9 software was used to carry out semi-variance function fitting. We found that the spatial variation in the soil salt and water content at the canal scale was best represented by an exponential model. The ecological processes associated with spatial change on a small scale may be consistent with the conclusions of some factors related to land use, such as crop type, soil texture, and tillage methods, as reported by Guo Xudong et al. [51]. Studies have shown that if there is a large variation in soil properties within a small spatial range, the nugget value will be larger than the absent value, and Moran's I index is a more robust method to use [35]. Our study area lies between the scale of an irrigation area and the scale of farmland with flat terrain. There was no large wasteland as the object of salt discharge for the cultivated land; therefore, it did not exhibit dry salt discharge [52], and drainage and salt discharge depended on the drainage channels in the area. Future research should investigate how different spatial arrangements of cultivated land and wasteland affect the spatial variability of soil water and salt under varying irrigation and drainage conditions [53].

In this study, it was found that the spatial autocorrelation of soil water and salt at the canal scale is good [23,24], indicating a strong dependence relationship (Figures 5 and 6). Consequently, ordinary Kriging interpolation provides reliable results for soil environmental parameters, rendering them valuable for land salinization management. It is worth noting that the spatial variability of soil is influenced by the variation of the sampling scale. As the sampling scale increases, the complexity of soil water and salt variation increases, potentially because of homogeneity at a small scale. In the future, the research scope of spatial correlation should gradually be expanded to include varying scales, such as canals, branch canals, and irrigation areas [2,3].

In the study area, soil water and salt showed a spatial concentration trend with a strong dependence relationship [29,46,54,55]. This showed that the evaluation of soil salinization is a complex problem, and factors such as crops, soil texture, canal system layout, and groundwater should be considered [56–58]. The effects of different planting times and canal system layouts on the spatial variation of soil salinity were investigated, and the effects of different particle size distributions and groundwater depths on soil salinity were studied. The timely determination of the influence of the above factors on the spatial variation and distribution pattern of soil salinity has important reference value for soil salinization prevention and can be used as a reference for land management departments. However, this study only studied the spatial variation of soil water and salt at the aqueduct scale after spring irrigation, and this information cannot be used to predict the future situation of soil salinization in the study area. In future studies, we will study soil salinization at the scale of canals by analyzing groundwater ion components, different irrigation and drainage systems, and different spatial configuration component models of cultivated wasteland. This will provide solid technical support for sustainable agricultural development [59,60].

5. Conclusions

The strong variability of soil water and salt content in the region was indicated by the coefficients of variation, which ranged from 0.1 to 1.0 for soil water content and from 0.77 to 1.75 for soil salt content. The index model provided a good fit for both soil water and salt, with a strong spatial correlation observed. This was supported by the nugget coefficients, all of which were less than 0.25. The spatial variation of soil water and salt content in the region was primarily influenced by structural factors.

A positive spatial correlation was observed between soil water and salt, with Moran's I values ranging from 0.09 to 0.48. The Z values were all greater than 0, indicating a significant relationship ($p < 0.05$). Furthermore, spatial interpolation revealed a well-distributed pattern of water and salt in the study area. Moreover, a strong exponential relationship between groundwater depth and soil salt was found. Consequently, it was determined that the levels of soil salinization were low in the study area.

A combination of geostatistics, ordinary Kriging, and spatial autocorrelation was used to reveal the spatial variation and distribution patterns of soil water and salt at the canal scale. The distribution map of soil water and salt was obtained via conventional Kriging interpolation and then divided according to a grid. Moran's I index was used to determine the accumulation behavior of water and salt in the soil samples at the aqueduct scale.

Author Contributions: Conceptualization, H.S.; Methodology, H.S.; Validation, H.S.; Formal analysis, Z.H. and Q.M.; Investigation, Z.H., Q.M., H.S., C.H., C.Y. and Y.M.; Resources, H.S.; Data curation, H.S.; Writing—original draft, Z.H.; Writing—review & editing, H.S. and W.F.; Visualization, Z.H., H.S. and W.F.; Supervision, H.S. and W.F.; Project administration, H.S. All authors have read and agreed to the published version of the manuscript.

Funding: The work was financially supported by the 14th Five-Year Plan National Key Research and Development Plan (2021YFC3201202-05), the Inner Mongolia Autonomous Region science and technology project (2023JBGS0003), the National Natural Science Foundation of China (52269014), and the Research Program of Science and Technology at Universities of Inner Mongolia Autonomous Region, China (2022YFHH0044).

Data Availability Statement: The data already exist in this manuscript.

Conflicts of Interest: The authors declare that they have no competing financial interests or personal relationships that could have influenced the work reported in this manuscript.

References

1. AbdelRahman, M.A.; Zakarya, Y.M.; Metwaly, M.M.; Koubouris, G. Deciphering soil spatial variability through geostatistics and interpolation techniques. *Sustainability* **2020**, *13*, 194. [CrossRef]
2. Kosma, C.; Triantafyllidis, V.; Zotos, A.; Pittaras, A.; Kouneli, V.; Karydogianni, S.; Mavroeidis, A.; Kakabouki, I.; Beslemes, D.; Tigka, E.L.; et al. Assessing spatial variability of soil properties in Mediterranean smallholder farming systems. *Land* **2022**, *11*, 557. [CrossRef]
3. Ke, Z.; Liu, X.; Ma, L.; Jiao, F.; Wang, Z. Spatial distribution of soil water and salt in a slightly salinized farmland. *Sustainability* **2023**, *15*, 6872. [CrossRef]
4. Pouryazdankhah, H.; Shahnazari, A.; Ahmadi, M.Z.; Khaledian, M.; Andersen, M.N. Rice yield estimation based on forecasting the future condition of groundwater salinity in the Caspian coastal strip of Guilan Province, Iran. *Environ. Monit. Assess.* **2019**, *191*, 492. [CrossRef]
5. Xu, X.; Huang, G.; Sun, C.; Pereira, L.S.; Ramos, T.B.; Huang, Q.; Hao, Y. Assessing the effects of water table depth on water use, soil salinity and wheat yield: Searching for a target depth for irrigated areas in the upper Yellow River basin. *Agric. Water Manag.* **2013**, *125*, 46–60. [CrossRef]
6. Nachshon, U. Cropland soil salinization and associated hydrology: Trends, processes and examples. *Water* **2018**, *10*, 1030.
7. Aeman, H.; Shu, H.; Abbas, S.; Aisha, H.; Usman, M. Sinking delta: Quantifying the impacts of saltwater intrusion in the Indus Delta of Pakistan. *Sci. Total Environ.* **2023**, *880*, 163356. [CrossRef]

8. Zheng, T.; Fang, Y.; Zhang, X.; Gao, S.; Li, P.; Zheng, X.; Liu, T.; Walther, M. Saltwater intrusion induced micro-scale mineral precipitation and evolution in porous media. *J. Hydrol.* **2023**, *625*, 129968. [CrossRef]
9. Shi, H.; Yang, S.; Li, R.; Li, X.; Li, W.; Yan, J.; Miao, Q.; Li, Z. Soil water and salt movement and soil salinization control in Hetao Irrigation District: Current state and future prospect. *J. Irrig. Drain* **2020**, *39*, 1–17.
10. Loescher, W.; Chan, Z.; Grumet, R. Options for developing salt-tolerant crops. *HortScience* **2011**, *46*, 1085–1092. [CrossRef]
11. Wang, X.; Gao, Q.; Lu, Q. Salt-water balance and dry drainage desalting in Hetao Irrigating Area, Inner Mongolia. *Sci. Geogr. Sin.* **2006**, *26*, 460.
12. Li, B.; Shi, H.; Yan, J.; Li, Z.; Zhang, J.; Zhou, J. Relation between groundwater depth and soil water and salt after water saving reform in salinization irrigation district. *J. Soil Water Conserv.* **2014**, *28*, 117–122.
13. Abd-Elaty, I.; Zelenakova, M. Saltwater intrusion management in shallow and deep coastal aquifers for high aridity regions. *J. Hydrol. Reg. Stud.* **2022**, *40*, 101026. [CrossRef]
14. Zhang, W.; Shi, H.; Li, X.; Li, Z.; Zhou, H.; Wang, W. Dynamic and transformational relationship between soil water and groundwater in typical areas of Hetao Irrigation District. *Trans. Chin. Soc. Agric. Mach.* **2022**, *53*, 352–362. (In Chinese)
15. Zhang, S.; He, Y.; Fang, H. Spatial variability of soil properties in the field based on GPS and GIS. *Nongye Gongcheng Xuebao (Trans. Chin. Soc. Agric. Eng.)* **2003**, *19*, 39–44.
16. Yao, R.; Yang, J.; Liu, G. Spatial variability of soil salinity and moisture and their estimations by co kriging method—A case study in characteristic field of Yellow River delta. *J. Soil Water Conserv.* **2006**, *20*, 133–138.
17. Qianji, H.; Shaoyuan, F.; Zhou, G. Intra-annual spatiotemporal variation in salt content in the plough layer in Hetao Irrigation District. *J. Irrig. Drain.* **2020**, *39*, 26–34.
18. Yan, A.; Jiang, P.; Sheng, J.; Wang, X.; Wang, Z. Spatial variability of surface soil salinity in Manas River basin. *Acta Pedofil. Sin* **2014**, *51*, 410–414.
19. Wu, C.; Lu, C.; Chen, J.; Liu, J.; Zhang, H.; Saidy, E.; Liu, B.; Shu, L. Spatio-temporal variability of hydraulic conductivity in the floodplain riverbank of a hyporheic zone. *CATENA* **2023**, *228*, 107172. [CrossRef]
20. Liu, J.; Ma, X.; Zhang, Z. Multifractal study on spatial variability of soil water and salt and its scale effect. *Trans. Chin. Soc. Agric. Eng.* **2010**, *26*, 81–86.
21. Guan, X.; Yang, P.; Lv, Y. Analysis on spatial variability of soil properties based on multifractal theory. *J. Basic Sci. Eng* **2011**, *19*, 712–720.
22. Mondal, P.; Walter, M.; Miller, J.; Epanchin-Niell, R.; Gedan, K.; Yawatkar, V.; Nguyen, E.; Tully, K.L. The spread and cost of saltwater intrusion in the US Mid-Atlantic. *Nat. Sustain.* **2023**, 1–11. [CrossRef]
23. Shi, H.; Wu, D.; Yan, J.; Li, X.; Zhu, K. Spatial temporal Variation of Soil Salinity after Water Saving Transformation in Salinized Irrigation District. *Trans. Chin. Soc. Agric. Mach.* **2020**, *51*, 318–331.
24. Zhang, F.; Li, Y.; Wang, D.; Wang, X.; Zhang, H.; Chen, L. Analysis of Distribution Patterns and Spatial Variability of Soil Salinity Affecting Factors in Topsoil Layer ofSalinized Soil in Jinghe Oasis. *J. Ecol. Rural Environ.* **2018**, *34*, 64–73. (In Chinese)
25. Chen, L.; Lü, Y.; Fu, B.; Wei, W. A framework on landscape pattern analysis and scale change by using pattern recognition approach. *Acta Ecol. Sin.* **2006**, *26*, 663–670.
26. Zhang, Q.; Pan, X.; Wang, H.; Liu, F.; Shi, X. Study on spatial distribution of soil quality and quantitative evaluation of soil fertility quality under middle spatial scale. *Chin. J. Soil Sci.* **2003**, *34*, 493–497.
27. Xu, Y.; Chen, Y.; Shi, H.; Wei, Z. Scale effect of spatial variability of soil water-salt. *Trans. Chin. Soc. Agric. Eng.* **2004**, *20*, 1–5.
28. Ren, D.; Xu, X.; Huang, G. Irrigation water use in typical irrigation and drainage system of Hetao Irrigation District. *Trans. Chin. Soc. Agric. Eng* **2019**, *35*, 98–105.
29. Rousseva, S. Data transformations between soil texture schemes. *Eur. J. Soil Sci.* **1997**, *48*, 749–758. [CrossRef]
30. Liu, W.; Ma, L.; Smanov, Z.; Samarkhanov, K.; Abuduwaili, J. Clarifying Soil Texture and Salinity Using Local Spatial Statistics (Getis-Ord Gi* and Moran's I) in Kazakh–Uzbekistan Border Area, Central Asia. *Agronomy* **2022**, *12*, 332. [CrossRef]
31. Li, L.; Yi, Y.; Ling, G.; Wang, S. Utilization of geostatistics in soil spatial variability. *Chin. J. Soil Sci.* **2005**, *36*, 265–268.
32. Franco, T.C.; Ferraz, G.A.; Carvalho, L.C.; Silva, F.M.; Alves, M.C.; Marin, D.B. Spatial variability of soil physical properties in longitudinal profiles. *An. Acad. Bras. Ciências* **2022**, *94*, e20200411. [CrossRef] [PubMed]
33. Yao, R.; Yang, J.; Liu, G.; Zou, P. Spatial variability of soil salinity in characteristic field of the Yellow River Delta. *Trans. CSAE* **2006**, *22*, 61–66.
34. Fang, L.; Li, Y.; Li, F.; Zhu, H. Analysis of spatial variation of soil moisture-salinity-nutrient in Ebinur Lake wetlands, China. *J. Agro-Environ. Sci.* **2019**, *38*, 157–167.
35. Li, Y.; Zhao, G. Spatial prediction of cultivated land soil nutrients in typical region of yellow river delta. *J. Nat. Resour.* **2018**, *33*, 489–503.
36. Liu, Y.; Sun, G.; Mao, W.; Cheng, X.; Zhu, Y.; Yang, J. Geostatistical Analysis of Spatial Variability of Soil Water and Salt in Hetao Irrigation District. *J. Irrig. Drain.* **2022**, *41*, 101–109. (In Chinese)
37. Guan, X.; Wang, S.; Gao, Z.; Lv, Y.; Fu, X. Spatio-temporal variability of soil salinity and its relationship with the depth to groundwater in salinization irrigation district. *Acta Ecol. Sin* **2012**, *32*, 1202–1210. [CrossRef]
38. Dou, X.; Shi, H.; Miao, Q.; Tian, F.; Yu, D.; Zhou, L.; Liang, Z. Temporal and spatial variability analysis of soil water and salt and the influence of groundwater depth on salt in saline irrigation area. *J. Soil Water Conserv.* **2019**, *33*, 246–253.
39. Bin, L.; Haibin, S.; Debao, T. Soil salinity profile characteristics and its spatial distribution before and after water saving. *Arid Zone Res.* **2015**, *32*, 663–673.

40. Paz-Ferreiro, J.; Vázquez, E.V.; Miranda, J.G.V. Assessing soil particle-size distribution on experimental plots with similar texture under different management systems using multifractal parameters. *Geoderma* **2010**, *160*, 47–56. [CrossRef]
41. Zhou, Z.; Zhou, Z. Investigating the hydrodynamic and biogeochemical evolutions of the hyporheic zone due to large-scale reservoir impoundment. *J. Hydrol.* **2023**, *620*, 129475. [CrossRef]
42. Mazzoncini, M.; Sapkota, T.B.; Barberi, P.; Antichi, D.; Risaliti, R. Long-term effect of tillage, nitrogen fertilization and cover crops on soil organic carbon and total nitrogen content. *Soil Tillage Res.* **2011**, *114*, 165–174. [CrossRef]
43. Xu, Y.; Ge, Z.; Wang, J.; Li, W.; Feng, S. Study on relationship between soil salinization and groundwater table depth based on indicator Kriging. *Trans. Chin. Soc. Agric. Eng.* **2019**, *35*, 123–130.
44. Xu, D.; Liu, C.; Cai, T.; Zhang, S. 3D spatial distribution characteristics of soil organic matter and total nitrogen in farmland. *Trans. Chin. Soc. Agric. Mach.* **2015**, *46*, 157–163.
45. Friedel, M.J. Estimation and scaling of hydrostratigraphic units: Application of unsupervised machine learning and multivariate statistical techniques to hydrogeophysical data. *Hydrogeol. J.* **2016**, *24*, 2103. [CrossRef]
46. Liu, G.; Wu, Y.; Yang, J.; Yu, S.; Wang, X. Regional 3-D soil salt spatial variability based on electromagnetic induction technology. *Nongye Jixie Xuebao = Trans. Chin. Soc. Agric. Mach.* **2013**, *44*, 78–82.
47. Laslett, G.; McBratney, A.; Pahl, P.J.; Hutchinson, M. Comparison of several spatial prediction methods for soil pH. *J. Soil Sci.* **1987**, *38*, 325–341. [CrossRef]
48. Hosseini, E.; Gallichand, J.; Marcotte, D. Theoretical and experimental performance of spatial interpolation methods for soil salinity analysis. *Trans. ASAE* **1994**, *37*, 1799–1807. [CrossRef]
49. Yao, R.; Yang, J.; Han, J. Comparative study of simulations of spatial variability of soil salinity in coastal polders. *Acta Pedol. Sin.* **2012**, *49*, 275–281.
50. Oliver, M.; Webster, R. A tutorial guide to geostatistics: Computing and modelling variograms and kriging. *Catena* **2014**, *113*, 56–69. [CrossRef]
51. Guo, X.; Fu, B.; Ma, K.; Chen, L. Spatial variability of soil nutrients based on geostatistics combined with GIS—A case study in Zunghua City of Hebei Province. *Chin. J. Appl. Ecol.* **2000**, *11*, 557–563. (In Chinese)
52. Wang, G.; Shi, H.; Li, X.; Zheng, Q.; Guo, J.; Wang, W. Analysis of water and salt transportation and balance during cultivated land, waste land and lake system in Hetao Irrigation Area. *J. Hydraul. Eng.* **2019**, *50*, 1518–1528.
53. Huang, Y.; Ma, Y.; Zhang, S.; Li, Z.; Huang, Y. Optimum allocation of salt discharge areas in land consolidation for irrigation districts by SahysMod. *Agric. Water Manag.* **2021**, *256*, 107060. [CrossRef]
54. Zhao, Y.; Feng, Q.; Yang, H. Soil salinity distribution and its relationship with soil particle size in the lower reaches of Heihe River, Northwestern China. *Environ. Earth Sci.* **2016**, *75*, 810. [CrossRef]
55. Xia, J.; Zhao, X.; Zhao, Z.; Chen, Y.; Liu, J. Migration characteristics of soil water and salt and their interaction under different groundwater levels. *Trans. Chin. Soc. Agric. Eng.* **2015**, *31*, 93–100.
56. Feng, W.; Yang, F.; Cen, R.; Liu, J.; Chen, H. Effects of straw biochar application on soil temperature, available nitrogen and growth of corn. *J. Environ. Manag.* **2021**, *277*, 111331. [CrossRef]
57. Feng, W.; Gao, J.; Cen, R.; Yang, F.; He, Z.; Wu, J.; Miao, Q.; Liao, H. Effects of polyacrylamide-based super absorbent polymer and corn straw biochar on the arid and semi-arid salinized soil. *Agriculture* **2020**, *10*, 519. [CrossRef]
58. Feng, W.; Wang, T.; Yang, F.; Cen, R.; Liao, H.; Qu, Z. Effects of biochar on soil evaporation and moisture content and the associated mechanisms. *Environ. Sci. Eur.* **2023**, *35*, 66. [CrossRef]
59. Inam, A.; Adamowski, J.; Prasher, S.; Albano, R. Parameter estimation and uncertainty analysis of the Spatial Agro Hydro Salinity Model (SAHYSMOD) in the semi-arid climate of Rechna Doab, Pakistan. *Environ. Model. Softw.* **2017**, *94*, 186–211. [CrossRef]
60. Inam, A.; Adamowski, J.; Prasher, S.; Halbe, J.; Malard, J.; Albano, R. Coupling of a distributed stakeholder-built system dynamics socio-economic model with SAHYSMOD for sustainable soil salinity management–Part 1: Model development. *J. Hydrol.* **2017**, *551*, 596–618. [CrossRef]

water

MDPI

Article

Characteristics, Source and Risk Assessment of Soil Polycyclic Aromatic Hydrocarbons around Oil Wells in the Yellow River Delta, China

Yuxuan Wu [1], Yilei Zhao [1], Yue Qi [1,*], Junsheng Li [2,*], Yuchen Hou [1], Haojing Hao [1], Nengwen Xiao [1] and Qiuying Zhi [1,3]

1 Chinese Research Academy of Environmental Sciences, Beijing 100012, China; wuyuxuan221@mails.ucas.ac.cn (Y.W.); 13309163638@163.com (Y.Z.); houych2020@lzu.edu.cn (Y.H.); whwh20@126.com (H.H.); xiaonw@163.com (N.X.); 220220932760@lzu.edu.cn (Q.Z.)
2 Command Center for Comprehensive Survey of Natural Resources, China Geological Survey Bureau, Beijing 100055, China
3 College of Ecology, Lanzhou University, Lanzhou 730000, China
* Correspondence: qiyue8351572@163.com (Y.Q.); lijunshen@cgs.gov.cn (J.L.)

Abstract: The Yellow River Delta (YRD) is the most complete wetland ecosystem in the warm temperate zone of China and is rich in oil resources. However, with petroleum extraction and the development of the economy, pollution of the YRD has been paid increasing attention, in particular, pollution via polycyclic aromatic hydrocarbons (PAHs), as they have caused great harm to human health and the ecosystem balance. Based on the investigations of a research group in 2009, this study re-collected samples according to the same sampling points and analyzed the concentration, composition, source, ecological risk and health risk of PAHs in 2021. The concentration of ΣPAH_{16} in the surface soil of YRD in 2009 ranged from 2.6 to 8275.46 ng/g, with an average of 1744.41 ng/g. The concentration of ΣPAH_{16} in 2021 ranged from 56.25 to 582.56 ng/g, with an average of 149.63 ng/g. Therefore, the pollution situation in the YRD in 2021 was significantly improved compared with 2009. The composition of PAHs in soil in 2009 and 2021 was similar, which was dominated by low-ring PAHs. The evaluation results of the toxicity equivalent factor method showed that there was no potential ecological risk in the soil in 2009 and 2021. The evaluation results of the lifetime cancer risk increment model showed that the incremental lifetime cancer risk models (ILCRs) of soil PAHs in 2009 and 2021 were lower than the safety threshold of 10^{-6}; therefore, there was no carcinogenic risk. The existing management measures for oil wells need to be further promoted to protect the regional ecological environment in the YRD.

Keywords: PAHs; the Yellow River Delta; source apportionment; carcinogenic risk; ecological risk assessment

Citation: Wu, Y.; Zhao, Y.; Qi, Y.; Li, J.; Hou, Y.; Hao, H.; Xiao, N.; Zhi, Q. Characteristics, Source and Risk Assessment of Soil Polycyclic Aromatic Hydrocarbons around Oil Wells in the Yellow River Delta, China. *Water* **2023**, *15*, 3324. https://doi.org/10.3390/w15183324

Academic Editor: Catherine N. Mulligan

Received: 29 July 2023
Revised: 11 September 2023
Accepted: 15 September 2023
Published: 21 September 2023

1. Introduction

The Yellow River Delta (YRD) is the most primitive, youngest and most complete wetland ecosystem in the warm temperate zone of China and even the world [1]. There are abundant mineral resources dominated by oil in the Yellow River Delta. With the in-depth development of oil resources and the production and construction of other industries, the ecological environment of the Yellow River Delta has also undergone great changes [2]. The discharge of a large number of pollutants seriously threatens the safety of the ecosystem in the region [3,4], among which the problem of oil pollution is particularly prominent. Previous studies have shown that PAH pollution in surface water and groundwater was low, and the soil and sediment pollution was moderate. The former pollution was mainly from oil input and oil combustion, and the latter was mainly from wood, coal combustion and oil input [5].

The pollution of polycyclic aromatic hydrocarbons (PAHs) in the YRD has attracted the attention of scholars. PAHs are a class of persistent organic pollutants that are widely present in various environments [6]. The sources of PAHs in the environment are mainly divided into natural sources and anthropogenic sources [7]. The main cause of PAH pollution at present is anthropogenic sources, including petroleum and pyrolysis sources. The large quantity of PAHs produced by human activities usually enter the atmospheric environment first and then eventually converge into the soil, sediment and water environment through surface runoff, dry and wet deposition, etc. [8]. PAHs can also be bioenriched, accumulating and transferring into the food chain, thus increasing the risk level [9]. PAHs are not only carcinogenic, teratogenic and mutagenic but also phototoxic [10,11], causing great harm to human health.

Previous pollution investigations in soil and sediment have been performed in the YRD [10,12–14]. For example, the concentration of PAHs in 61 surface soil samples from the YRD has been studied. Results have shown that the total concentration of PAHs ranged from 27 to 735 ng/g, with the highest concentration occurring in the central and southern YRD, which was related to oil exploration. The PAHs in the surface soil of the YRD were mainly from coal and biomass burning, oil leakage and/or vehicle emissions (Yuan et al. 2014). In addition, it was found that the concentration of PAHs in soil around oil wells at different ages increased with the increase in mining time. From the source point of view, the soil near the new oil wells was mainly polluted by oil sources, while the old oil wells were mostly related to combustion sources [15]. The distribution of coastal sediments in the YRD is different. The concentration of PAHs in surface sediments near the Yellow River estuary is lower than that in the northern and southern coastal wetlands of the YRD [16]. Other studies have found that the PAH pollution level in the Yellow River Delta Nature Reserve is relatively low, and the potential ecological risk level is extremely low, but there is a potential carcinogenic risk. In addition, the spatial distribution characteristics of PAH concentration and PAH health risk value are similar, which indicates that the carcinogenic risk of PAHs is proportional to its content [17]. These research results have important theoretical significance for the protection of the ecological environment and the management of the Yellow River Delta.

At present, the study of oil pollution in the YRD is mainly the result of single sampling observations, which renders it difficult to reflect the long-term change characteristics of pollution in the YRD. There are few studies on the distribution of PAHs and their changes in different years in this area. Based on the research group's survey in the Yellow River Delta in 2009, this paper collected samples from the same sampling points to analyze the concentration, composition, source, ecological risk and health risk of PAHs and to understand the change in PAH pollution in the region in the recent 10 years.

2. Materials and Methods

2.1. Study Area

The Yellow River Delta is located in East China (118°07′–119°18′ E, 36°55′–38°12′ N) (Figure 1). About 96% of its area is distributed in Dongying City, Shandong Province. It is one of the most important wetland protection sites in the world. It is the breeding habitat and migration transit site of rare birds in Northeast Asia and the Western Pacific. The research area of this paper is mainly located in Xianhe Town, Gudao Town and Huanghekou Town of Dongying City, with a total area of 2152.7 km^2. Affected by the Eurasian continent and the Western Pacific, the study area belongs to the warm temperate semi-humid continental monsoon climate, with four distinct seasons and a hot summer and cold winter. The annual average temperature is between 11.7 and 12.8 °C. The average annual precipitation is about 530~630 mm, of which the summer rainfall accounts for more than half of the year. The rain and heat are in the same period, which is conducive to the growth of plants and crops and can easily cause flood disasters [18,19]. According to the statistics of soil census data, the study area can be divided into five soil types: cinnamon soil, saline soil, fluvo-aquic soil, mortar black soil and paddy soil, among which fluvo-aquic soil and saline soil account for 95% of the total soil area. The YRD is rich in natural resources, with a large number of brine, oil and gas, geothermal, mineral and other resources. It is an important energy base in China [20].

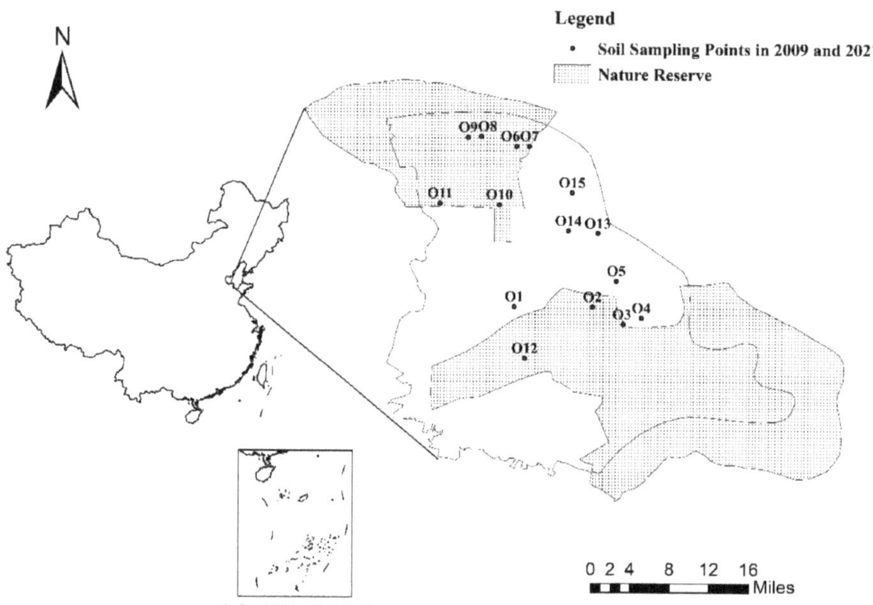

Figure 1. Locations of sampling sites in the study area of the Yellow River Delta.

2.2. Sample Collection

The soil samples in this study were collected in July 2021, and the collection areas were mainly concentrated in Xianhe Town, Gudao Town and Huanghekou Town of Dongying City. The sampling points were collected in accordance with the sampling points of the field survey conducted in 2009 around the wells. At each sample point, soil samples of 0–20 cm of surface layer were taken according to the plum sampling method. The basic physical and chemical properties such as pH and temperature were measured. The samples were mixed evenly; then, about 1kg of the evenly mixed samples were taken by the quartering method and put into a Ziplock bag for marking. A total of 15 soil samples were collected and transported back to the laboratory for freezing (−20 °C) preservation at low temperature and in darkness. The notations O1–O15 are soil points in 2009 and 2021 (Figure 1).

2.3. Laboratory Analysis

2.3.1. Determination of PAHs

According to the HJ805-2016 standard, polycyclic aromatic hydrocarbons in soil were determined by gas chromatography–mass spectrometry. The determination process included sample preparation, extraction, concentration, purification, elution and constant volume determination as follows. After removing impurities such as gravel and roots in the sample, 20 g was weighed and an appropriate amount of anhydrous sodium sulfate was added to dehydrate. The sample was ground into fine particles. About 5 g of samples was weighed and extracted by the Soxhlet extraction method. An amount of 100 mL acetone-n-hexane mixed solvent was added to the sample and extracted for 16~18 h, refluxing 4~6 times per hour. The extract was concentrated to about 2 mL using a parallel concentrator. Then, the samples were purified by a silica gel chromatography column. An amount of 25 mL dichloromethane-pentane mixed solvent was added for elution and the eluent was collected. After the eluent was concentrated again, an appropriate amount of acetone-n-hexane mixed solvent was added, and the volume was fixed to 1 mL for determination. Gas chromatography–mass spectrometry with a quartz column (DB-5MS, 30 m × 0.25 mm × 0.25 μm) was used to quantify the PAH concentrations. The carrier gas type is high-purity helium. The sample was injected without a shunt. The injection volume was 1 μL and the inlet temperature was 280 °C. The oven temperature program was as follows: the initial temperature was 80 °C, held for 2 min, then heated at 20 °C/min to 180 °C, then heated at 10 °C/min to 290 °C, and held for 5 min.

2.3.2. Quality Control

In order to ensure the accuracy and reliability of the data, strict quality control was carried out throughout the experiment. In the determination of PAHs in soil and sediment, not only was o-fluorobiphenyl added to each sample for recovery detection but blank experiments and parallel sample tests were also carried out. The spiked recoveries of 16 PAHs were between 40% and 150%. The blank experiment was performed once every 20 samples, and the relative standard deviations of parallel samples were all within 20%. The external standard method was used to quantitatively analyze PAHs, and the correlation coefficients of the standard curves were all above 0.995.

2.4. PAH Source Apportionment Method

2.4.1. Positive Definite Matrix Factorization Model (PMF)

The most critical step in the source apportionment of the PMF model is the determination of the number of factors. In the robust model, the number of factors is set to 3~6, and 20 iterations are carried out, respectively. Considering the factors such as the residual and fitting effect, the 4-factor model was used to analyze the source of PAHs across the 21 years, and the 6-factor model was used to analyze the source of PAHs in the soil in 2009.

2.4.2. Diagnostic Ratios

Diagnostic ratios are commonly used in the source apportionment of PAHs [21]. Based on the characteristics of similar kinetic mass transfer coefficients and thermodynamic distribution coefficients between isomeric PAHs, the heat source and petroleum source can be distinguished by their concentration ratio [22]. To differentiate the PAHs at our sampling sites with respect to petroleum vs. combustion sources, we applied two isomeric ratios.

2.5. Ecological Risk Assessment Method of PAHs

2.5.1. Toxic Equivalent Factor Method (TEQ)

Among the 16 US EPA priority PAHs, Benzo[a]pyrene (BaP) is used as a marker for estimating the carcinogenic potency of PAHs [23]. To evaluate and quantify the individual PAH and $\sum PAH_{7c}$ carcinogenic toxic potency, the toxicity equivalency factors (TEFs), which define the carcinogenic potency of the individual PAH relative to Benzo[a]pyrene (BaP), were used to estimate the toxic equivalent of PAH monomer relative to BaP (TEQ_{BaP}) in this study [24]. This evaluation approach pays attention to the estimation of carcinogenic PAHs. The toxic equivalent factor (TEF) of BaP was set to 1, and the remaining PAHs were assigned to the corresponding TEF according to the difference in toxicity (Table 1). The TEFs listed in Table 1 are the most reasonable values for the relative titers of PAHs currently available and using them would significantly reduce the uncertainty in risk assessments involving PAHs [24]. The calculation formula is as follows:

$$TEQ_{BaP} = \sum C_i \times TEF_i \quad (1)$$

where C_i is the exposure concentration of the ith PAHs; TEF_i is the toxic equivalent factor corresponding to the ith PAHs. The greater the toxic equivalent factor of monomer PAHs, the greater the toxicity.

Table 1. Toxicity equivalent factors of individual PAHs.

Monomer PAHs	Ring Number	Toxic Equivalent Factor	Monomer PAHs	Ring Number	Toxic Equivalent Factor
Naphthalence (Nap)	2	0.001	Pyrene (Pyr)	4	0.001
Anthracence (Ant)	3	0.01	Fluoranthene (Fla)	4	0.001
Phenanthrene (Phe)	3	0.001	Dibenzo[a,h]anthracene (DahA) *	5	1
Fluorene (Flu)	3	0.001	Benzo[a]pyrene (BaP) *	5	1
Acenapthene (Ace)	3	0.001	Benzo[k]fluoranthene (BkF) *	5	0.1
Acenaphthylene (Acpy)	3	0.001	Benzo[b]fluoranthene (BbF) *	5	0.1
Chrysene (Chr) *	4	0.01	Benzo[g,h,i]perylene (BghiP)	6	0.01
Benzo[a]anthracene (BaA) *	4	0.1	Indenol [1,2,3-cd]pyrene (IcdP) *	6	0.1

Note: * means carcinogenic compound.

2.5.2. Effect Interval Low Median Method (ERL/ERM)

The evaluation criteria proposed by Long et al. [25] are an effective tool to identify the potential ecological risks of PAHs [26,27]. The tool uses the effects range low (ERL) and effects range median (ERM) as thresholds; if the PAH concentration is less than ERL, the probability of adverse toxic effects of PAHs on organisms is less than 10%, and ecological risks rarely occur. ERL greater than or equal to ERM indicates that PAHs have more than 50% probability of causing negative effects on the ecological environment, and risks may occur frequently. Between the two, the PAH concentration will occasionally harm the ecological environment; that is, the concentration has potential ecological risks.

In order to quantitatively predict the comprehensive ecological toxicity of various PAHs, the ecological risk can be further evaluated by the mean quotient of ERM (M-ERM-Q). The calculation formula is:

$$\mathrm{M-ERM-Q} = \sum\nolimits_{i=1}^{n} \frac{C_i}{\mathrm{ERM}_i} / n \quad (2)$$

where C_i is the exposure concentration of individual PAHs; ERM_i is the median effect range (ERM) corresponding to individual PAHs; and n is the type of PAHs. When M-ERM-Q < 0.1, the possibility of ecological risk is low; when 0.1 < M-ERM-Q < 0.5, there is a 30% probability of toxicity; when 0.5 < M-ERM-Q < 1.5, the probability of medium and high toxicity is about 50%; when M-ERM-Q > 1.5, the ecological risk is high, and the probability of high toxicity is about 75%.

2.6. Health Risk Assessment Methods of PAHs

Incremental Lifetime Cancer Risk Models (ILCRs)

The incremental lifetime cancer risks (ILCRs) model was used in this paper to quantitatively evaluate the health risk of PAHs in soil, that is, the incidence of cancer caused by a certain dose of carcinogens at a certain time. There are three main ways for humans to be directly exposed to soil PAHs: ingestion ($\mathrm{ILCR_{ingestion}}$), inhalation ($\mathrm{ILCR_{inhalation}}$) and dermal contact ($\mathrm{ILCR_{dermal}}$) [28]. The cancer risk calculation formula for each pathway is as follows:

$$\mathrm{CS} = \sum(\mathrm{PAH_i} \times \mathrm{TEF_i}) \quad (3)$$

$$\mathrm{ILCR_{ingestion}} = \mathrm{CS} \times \left(\mathrm{CSF_{ingestion}} \times \sqrt[3]{\frac{\mathrm{BW}}{70}}\right) \times \mathrm{IR_{soil}} \times \mathrm{EF} \times \mathrm{ED/BW} \times \mathrm{AT} \times 10^6 \quad (4)$$

$$\mathrm{ILCR_{inhalation}} = \mathrm{CS} \times \left(\mathrm{CSF_{inhalation}} \times \sqrt[3]{\frac{\mathrm{BW}}{70}}\right) \mathrm{IR_{air}} \times \mathrm{EF} \times \mathrm{ED/BW} \times \mathrm{AT} \times \mathrm{PEF} \quad (5)$$

$$\mathrm{ILCR_{dermal}} = \mathrm{CS} \times \left(\mathrm{CSF_{dermal}} \times \sqrt[3]{\frac{\mathrm{BW}}{70}}\right) \times \mathrm{SA} \times \mathrm{AF} \times \mathrm{ABS} \times \mathrm{EF} \times \mathrm{ED/BW} \times \mathrm{AT} \times 10^6 \quad (6)$$

$$ILCR_s = ILCR_{ingestion} + ILCR_{inhalation} + ILCR_{dermal} \quad (7)$$

where CS is the toxic equivalent concentration (mg/kg) of monomer PAHs; PAH_i is the exposure concentration of the i th PAHs; TEF_i is the toxic equivalent factor of monomer PAHs (Table 2); CSF is the carcinogenic slope factor (mg/(kg·d)), which was determined by the carcinogenic ability of BaP. The values of $CSF_{ingestion}$, $CSF_{inhalation}$ and CSF_{dermal} were 7.3, 3.85 and 25, respectively. BW is body weight, IR_{soil} is soil inhalation rate, EF is exposure frequency, ED is exposure time, AT is the average life of carcinogens, IR_{air} is inhalation rate, PEF is soil particle emission factor, SA is skin surface exposure area, AF is skin adhesion factor, and ABS is skin adsorption parameter. The specific parameter values are shown in Table 2. ILCRs are the sum of the carcinogenic risks of the three exposure pathways: when ILCRs $< 10^{-6}$, the carcinogenic risk is negligible; when ILCRs $> 10^{-4}$, this indicates a high carcinogenic risk; between the two indicates a potential carcinogenic risk.

Table 2. Parameters of incremental lifetime cancer risks.

Parameter	Unit	Adult	Child
BW	Kg	61.5	15
IR_{soil}	$mg \cdot d^{-1}$	100	200
EF	$d \cdot a^{-1}$	350	350
ED	a	24	6
AT	d	25,550	25,550
IR_{air}	$m^3 \cdot d^{-1}$	20	10
PEF	$m^3 \cdot kg^{-1}$	1.36×10^9	1.36×10^9
SA	$cm^2 \cdot d^{-1}$	5700	2800
AF	$mg \cdot cm^{-2}$	0.07	0.2
ABS	zero dimension	0.13	0.13

3. Results and Discussion

3.1. PAH Concentration Characteristics

In 2009, the concentration of PAHs in the samples varied greatly, ranging from 2.6 to 8275.46 ng/g, with an average of 1744.41 ng/g, according to the analysis of surface soil samples collected by the research group in the YRD (Table 3). Compared with ΣPAH_{16} concentrations in the YRD studied in other years, such as Yuan et al. (27–753 ng/g, with an average of 118 ± 132 ng/g) [29]; Yuan et al. (79.2–311 ng/g, with an average of 119 ng/g) [30]; and Fu et al. (278.7–733.5 ng/g, with an average of 382.5 ± 128.4 ng/g) [15], PAH concentrations in soil samples investigated in 2009 were higher on average. Among soil samples, four samples exceeded the mean value, accounting for 26.67% of total soil samples, and eleven samples were below the mean value, accounting for 73.33% of total soil samples. The concentration of seven carcinogenic PAHs ranged from ND to 481.35 ng/g, with an average of 59.42 ng/g, accounting for 3.41% of ΣPAH_{16}. The BaP concentration ranged from ND to 22.71 ng/g, with an average of 2.51 ng/g, respectively, accounting for 4.23% and 0.14% of ΣPAH_7 and ΣPAH_{16}. The other PAH monomers were detected to varying degrees, except for Ace and DahA, and the detection rate was between 7% and 100%. Flu had the highest detection rate, followed by Nap and Phe, and Acpy had the lowest detection rate (Table 3). According to Maliszewska-Kordybach's [31] definition of soil PAH pollution, only one sample had a concentration below 200 ng/g, accounting for 6.67% of total soil samples. There were two samples with a concentration range of 200–600 ng/g, two samples with a concentration range of 600–1000 ng/g, and ten samples with a concentration higher than 1000 ng/g, respectively, accounting for 13.33%, 13.33% and 66.67% of total soil samples. The above results show that more than half of the soil samples in the study area reached the level of heavy pollution, which suggests the pollution situation is serious and corresponding measures need to be taken to repair and control the situation.

Table 3. Descriptive statistics of PAHs in soils in 2009 (ng/g).

Monomer PAH	Ring Number	Range	Mean Value	Standard Deviation	Coefficient of Variation	Detection Rate
Nap	2	ND~218.67	86.82	58.33	0.67	93
Ant	3	ND~188.81	25.22	46.65	1.85	73
Phe	3	ND~1002.96	117.00	243.43	2.08	93
Flu	3	2.6~7651.99	1409.20	1794.76	1.27	100
Ace	3	ND	ND	ND	ND	ND
Acpy	3	ND~2.16	0.14	0.54	3.74	7
Chr	4	ND~230.93	29.71	58.16	1.96	67
BaA	4	ND~227.56	19.40	56.38	2.91	47
Pyr	4	ND~137.97	24.59	44.82	1.82	80
Fla	4	ND~52.52	16.20	14.82	0.91	80
DahA	5	ND	ND	ND	ND	ND
BaP	5	ND~22.71	2.51	5.94	2.36	40
BkF	5	ND~14.78	2.03	4.80	2.36	20
BbF	5	ND~19.93	4.44	6.67	1.50	60
BghiP	6	ND~86.77	5.82	21.64	3.72	20
IcdP	6	ND~10.77	1.32	3.37	2.56	13
ΣPAH_7		ND~481.35	59.42	118.88	2.00	73
ΣPAH_{16}		2.6~8275.46	1744.41	1941.10	1.11	100

Note: ND means "Not Detected".

In 2021, the concentration of PAHs in soil samples ranged from 56.25 to 582.56 ng/g, with an average of 149.63 ng/g, collected from the YRD corresponding to the location information in 2009 (Table 4). According to the study of PAH pollution in the soil around oil wells in the Yellow River Delta in other years, the PAH concentration decreased with the increase in years, which indicates that the PAH pollution has been effectively managed [15,32] (Table 5). From the perspective of spatial distribution, the PAH concentration in the soil of the northern oil field area of the Yellow River Delta is higher than that in the southern area of the Yellow River, the Yellow River estuary and the former route [29,30]. Among them, four samples exceeded the mean value, accounting for 26.67% of total soil samples, and eleven samples were below the mean value, accounting for 73.33% of total soil samples. The concentration of seven carcinogenic PAHs ranged from ND to 291.28 ng/g, with an average of 28.3 ng/g, accounting for 18.91% of ΣPAH_{16}. The BaP content ranged from ND to 107.88 ng/g, with an average of 13.19 ng/g, respectively, accounting for 46.61% and 8.82% of ΣPAH_7 and ΣPAH_{16}. The other PAH monomers, except for Ace, Acpy, DahA, BkF and BghiP, were detected to varying degrees, and the detection rate was between 7% and 100%. The detection rate of Nap was the highest, followed by BaP. The detection rate of Chr was the lowest (Table 4). In previous reports, Nap and Phe dominated ΣPAH_{16} in the soil of the YRD [33], and Chr, Phe and Pyr dominated ΣPAH_{16} in the soil of the YRD [15]. Among the 15 samples, only two samples were slightly polluted, accounting for 13.33% of total soil samples, and the remaining samples were not polluted. Previous research showed that 60.1% of the soil samples were slightly polluted by ΣPAH_{16}, and 18.1% of the soil samples were moderately polluted [15]. Combined with our research results, the pollution situation in the study area was significantly improved compared with 2009 (Figure 2).

Table 4. Descriptive statistics of PAHs in soils in 2021 (ng/g).

Monomer PAH	Ring Number	Range	Mean Value	Standard Deviation	Coefficient of Variation	Detection Rate
Nap	2	45~291.28	112.79	59.64	0.53	100
Ant	3	ND~10.2	1.30	3.32	2.55	13
Phe	3	ND~10.2	1.30	3.32	2.55	13
Flu	3	ND~8.48	1.63	3.26	2.00	20
Ace	3	ND	ND	ND	ND	ND
Acpy	3	ND	ND	ND	ND	ND
Chr	4	ND~21.58	1.44	5.38	3.74	7
BaA	4	ND~32.36	2.89	8.34	2.88	13
Pyr	4	ND~11.25	1.37	3.51	2.56	13
Fla	4	ND~12.63	2.94	4.91	1.67	27
DahA	5	ND	ND	ND	ND	ND
BaP	5	ND~107.88	13.19	26.18	1.98	53
BkF	5	ND	ND	ND	ND	ND
BbF	5	ND~53.94	5.04	13.58	2.69	20
BghiP	6	ND	ND	ND	ND	ND
IcdP	6	ND~75.52	5.74	18.83	3.28	13
ΣPAH_7		ND~291.28	28.30	71.04	2.51	53
ΣPAH_{16}		56.25~582.56	149.63	122.33	0.82	100

Note: ND means "Not Detected".

Table 5. Concentrations (ng/g) of PAHs in sediments or soils from other research on the YRD.

Sampling Points	Sampling Year	∑PAHs, ng/g	Concentration Mean, ng/g	Reference
61 Stations within the YRD	2006–2008	27–753	118 ± 132	[29]
The Experimental, Buffer and Core area of the YRDNR	2013	79.2–311	119	[30]
Around Oil Wells with Different Extraction Histories in the YRD	2015	278.7–733.5	382.5 ± 128.4	[15]
The Oil Field Soil and the Estuary Soil of the YRD	2018	157.8–481.7	274	[32]
Around the wells in Xianhe Town, Gudao Town and Huanghekou Town of Dongying City	2009 2021	2.6–8275.46 56.25–582.56	1744.41 149.63	This study This study

Note: YRDNR means "Yellow River Delta Natural Reserve".

The main reasons may be related to the improvement in production conditions, the enhancement of staff's awareness of environmental protection and the improvements in technology. According to the research, the water source of the YRD was mostly surface runoff, which caused serious soil and water loss [34,35]. Severe soil erosion may also carry away some PAH pollutants from the soil into rivers and oceans [20,36,37]. This may also be part of the reason for the drop in pollution levels. In addition, there were also studies on oil pollution and soil microbial remediation. Two strains selected by previous researchers from oil-contaminated soil can deal with phenanthrene pollution in a salinity environment, among which strained FM6-1 metabolizes phenanthrene through the "phthalic acid" pathway and strained FM8-1 metabolizes phenanthrene through the "naphthalene" pathway [38]. This may also explain a decrease in the concentration of PAHs in the soil near the well in 2021. However, the proportion of seven carcinogenic PAHs and the most carcinogenic BaP were increasing. The management of PAHs in this area was still worthy of attention.

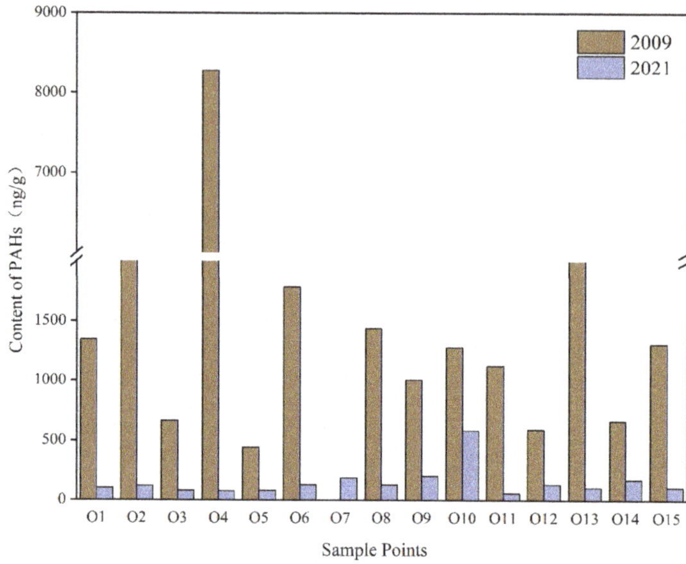

Figure 2. The concentrations of PAHs at each sampling site in 2009 and 2021.

3.2. PAH Component Characteristics

The composition of ΣPAH_{16} in 2009 and 2021 is shown in Figure 3. In 2009, the soil samples were dominated by low-ring PAHs (2–3 rings). The proportion of each ring in 2009 was 3 rings (88.94%) > 4 rings (5.15%) > 2 rings (4.98%) > 5 rings (0.52%) > 6 rings (0.41%). Among the low-ring PAHs, Flu contributed significantly, accounting for 86.01%. The proportion of each ring in the soil in 2021 was similar to that in 2009, and the low-ring PAHs were dominant. The proportion of each ring in 2021 was 2 rings (75.38%) > 5 rings (12.19%) > 4 rings (5.78%) > 6 rings (3.83%) > 3 rings (2.82%). The difference is that Nap contributed the most to 96.39% of the low-ring PAHs in the soil in 2021, and Flu accounted for only 1.39%. Low-ring PAHs mainly exist in the environment in a non-persistent gaseous form [39,40]. In contrast, high-ring PAHs are relatively resistant to microorganisms and photolysis [41,42]. In addition, the overall decrease in low cyclic PAHs was 15.72%. With the change in time, the proportion of low-ring PAHs gradually decreased [15,39,43]. The proportion of high-ring PAHs gradually increased from 6.08% to 21.8%. It is speculated that this may be related to the rapid economic development in the past ten years. With the increasing number of chemical factory enterprises, vehicles and the increasing intensity of human activities, oil pollution is still the long-term and main source of PAHs in this area.

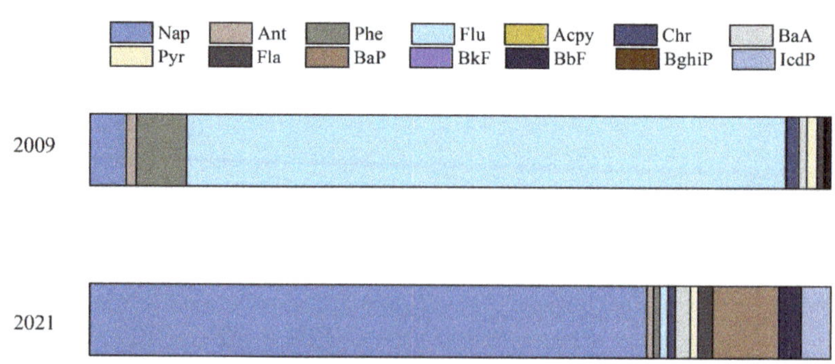

Figure 3. Composition of PAHs in 2009 and 2021.

3.3. PAH Source Apportionment

Now, the main cause of PAH pollution is anthropogenic sources, including oil sources and pyrolysis sources. Specifically, oil source refers to oil spill events that occur in the process of oil exploration and development, transportation and storage and offshore ship operations; pyrolysis source refers to the production activities of chemical enterprises such as petroleum processing, coking gas, steel making and iron making, the incomplete combustion of fuel oil, coal, gas, wood and other biomass in daily life, as well as waste incineration, smoking, cooking and frying, and the emission of various vehicle exhaust [44,45]. In general, an individual PAH ratio is used to identify the source of PAHs in the soil [33,46], different PAH molecular weights or different aromatic ring numbers are used to identify different PAH sources [47,48].

Figure 4 shows the source composition of PAHs in soil in 2009 and 2021. The results show that soil PAHs in 2009 were mainly affected by six sources (Figure 4a). The contribution rate of factor 1 to high-ring PAHs is higher, such as BaA (77.7%), Chr (56.36%) and BbF (35.51%). Gasoline and diesel produce BaA and Chr when they are not completely burned [49]. BbF is a typical substance emitted by diesel vehicles [50], so this factor can be identified as traffic pollution source 1. The main load of factor 2 is Ant, which represents the biomass combustion source [51]. Factor 3 has a large load on 2~3 ring PAHs, including NaP, Flu, Ant and Phe. These low molecular weight components are closely related to oil pollution [52–54], so factor 3 is considered to be an oil pollution source. BghiP (87.88%), IcdP (67.49%) and Pyr (48.34%) were found to be the higher-loading compounds in factor 4. BghiP is a typical substance of gasoline combustion emissions [55], so factor 4 is a mixed source of traffic pollution and fossil fuel combustion. The load of factor 5 was mainly distributed on BkF (77.98%) and BaP (76.31%), followed by BbF (37.59%). These high molecular weight PAHs represent traffic pollution source 2. The highest contribution rate of factor 6 to Fla was 66.32%. Phe and Flu were medium loads. So, factor 6 could be considered as the combustion of coal and biomass [56].

In 2021, soil PAHs were mainly affected by four sources (Figure 4b). Factor 1 not only has a higher load on low-ring PAHs such as NaP and Flu but also has a greater contribution to the high-ring components BaA, BaP and BkF. Therefore, factor 1 can be identified as a mixed source of oil pollution and traffic pollution [57–59]. Factor 2 is dominated by Pyr load, and the contribution rate reaches 98.92%, which is much higher than other compounds, reflecting the source characteristics of fossil fuel combustion [52,60]. Factor 3 has the highest contribution rate to Fla, which is 92.41%, representing the source of coal combustion [52,53,59]. The main loads of factor 4 are Phe (99.38%) and Ant (98.78%), which can be attributed to the combustion of coal and firewood. Therefore, factor 4 can be identified as the combustion source of coal and biomass [51].

Figure 5 shows the contribution rate of each pollution source of soil PAHs in different periods via PMF model analysis. The positive matrix factorization model (PMF) is recommended by the United States Environmental Protection Agency for pollution. A factorization model for source apportionment was first proposed by Paatero et al. [61]. In recent years, it has been widely used in soil, sediment, atmospheric particles and so on. The model can avoid the negative value of factor loading and scores in the application process and further optimize the data by using uncertainty for the model output results to achieve strong reliability.

Common isomeric ratios such as Fla/(Fla + Pyr), Ant/(Ant + Phe), BaA/(BaA + Chr) and Fla/(Fla + Pyr) were used to analyze the possible sources of PAHs in the soil around oil wells in the Yellow River Delta. According to the ratio analysis of Fla/(Fla + Pyr) and Ant/(Ant + Phe) [62,63], the main source of PAHs in the soil of samples A9, A11, A12, A13 and A15 in 2009 was biomass and coal combustion, and the main source of PAHs in the soil of sample A2 in 2009 was petroleum combustion. The PAHs in the soil of other plots were mainly derived from petroleum (Figure 6a). According to the ratio analysis of BaA/(BaA + Chr) and Fla/(Fla + Pyr) [64,65], the main source of PAHs in the soil of samples A11 and A13 in 2009 and B12 in 2021 was biomass and coal combustion, and the main source of PAHs in the soil of plots A2 in 2009 was petroleum combustion. The main source of PAHs in other soil samples was petroleum pollution (Figure 6b).

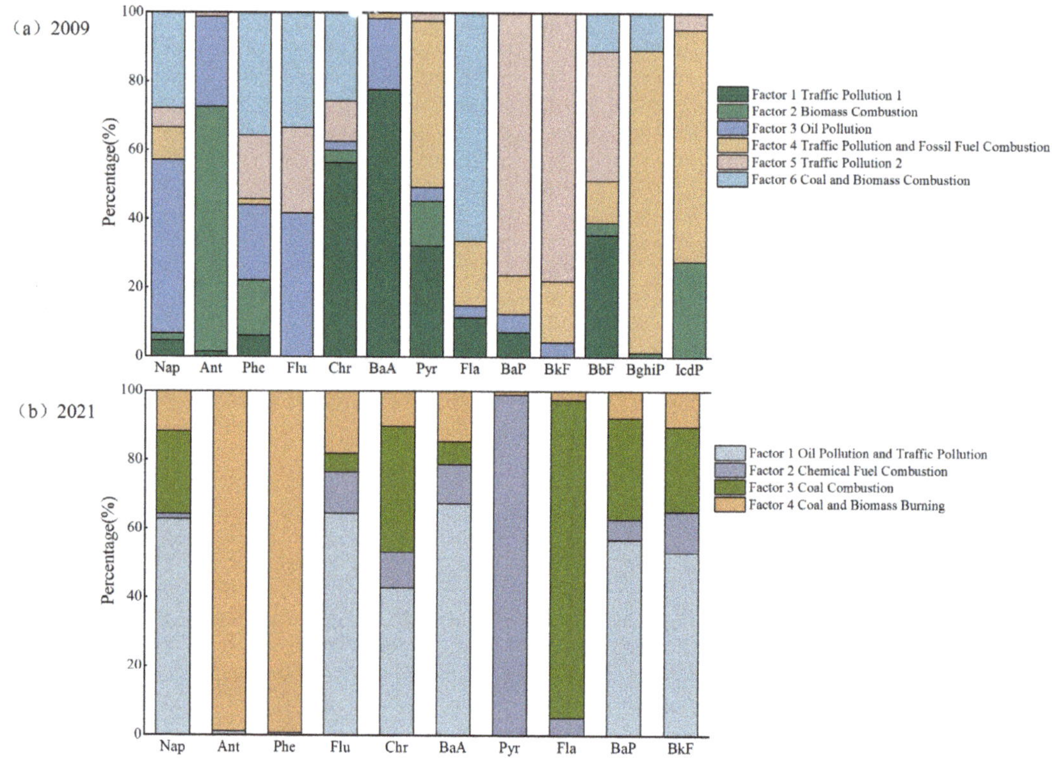

Figure 4. Sources of soil PAHs in 2009 and 2021 via PMF model analysis.

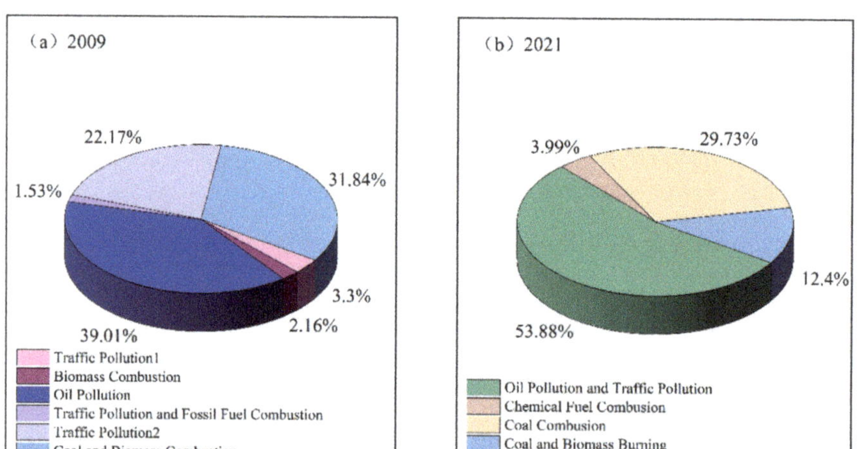

Figure 5. Contribution rate of soil PAHs in 2009 and 2021.

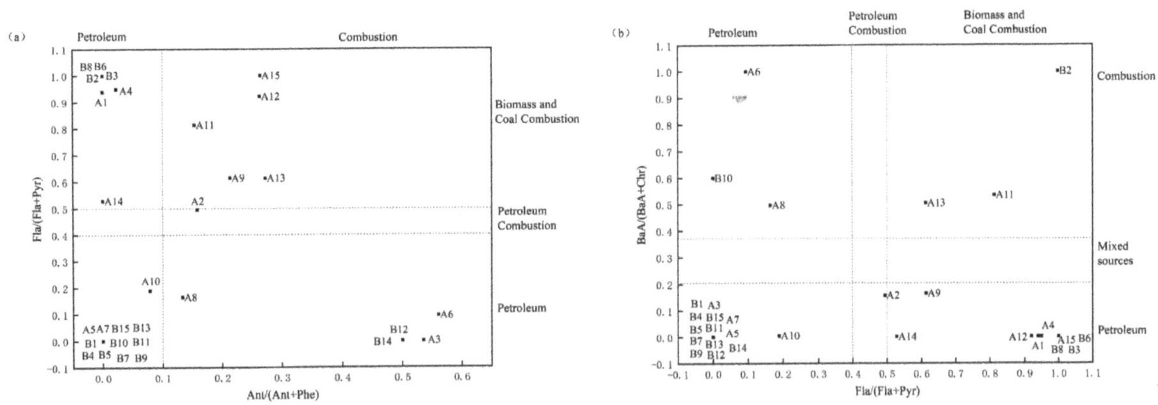

Figure 6. Diagnostic ratios of PAH sources at different sampling sites in 2009 and 2021. A1–A15 represent soil samples from 15 plots in 2009, and B1–B15 represent soil samples from 15 plots in 2021.

On the whole, oil pollution, traffic pollution, coal and biomass combustion are the long-term and main sources of PAHs in this area. This is consistent with previous research reports [12,66]. The source contribution rates of PAHs in soil in 2009 and 2021 are slightly different (Figure 5). In 2009, the contribution rate of oil pollution sources was the highest, reaching 39.01%, followed by coal and biomass combustion sources, and traffic pollution and fossil fuel combustion sources (1.53%) contributed the least. In 2021, oil pollution and traffic pollution were the main sources. The other three pollution sources and their contribution rates were coal combustion (29.73%), coal and biomass combustion (2.4%) and fossil fuel combustion (3.99%). Zhao found that PAHs in the surface sediments of the YRD were mainly derived from oil leakage and fossil fuel combustion and had a moderate risk to the local ecosystem [12]. Xu found that PAHs in sediments were mainly derived from oil leakage and fossil fuel combustion and had low to moderate risks to local ecosystems [66]. Studying the sources of PAHs in different periods can not only effectively prevent and control pollution but also provide a theoretical basis for relevant departments to formulate corresponding laws and regulations.

3.4. Ecological Risk Assessment of PAHs

3.4.1. Toxic Equivalent Factor Method (TEQ)

The ecological risk assessment was performed via the toxic equivalent of PAH monomer relative to BaP (TEQ_{BaP}) [24]. The TEQ concentrations of ΣPAH_{16} in soil ranged from 0.003 to 34.76 ng/g in 2009, with an average of 7.49 ng/g. The TEQ of seven carcinogenic PAHs accounted for 73.79% of the total TEQ. The monomer with the largest contribution to the total TEQ was BaP (33.53%), followed by BaA (25.89%). The TEQ concentrations of ΣPAH_{16} in soil in 2021 ranged from 0.06 to 124.57 ng/g, with an average of 14.71 ng/g, which was about twice the TEQ of 2009. The TEQ of seven carcinogenic PAHs accounted for 99.1% of the total TEQ, an increase of 25.31% from 2009. The carcinogenic risk was also increasing. In general, the TEQ values of different years showed a trend of high-ring PAHs being greater than low-ring PAHs, indicating that high-ring PAHs had high toxicity. Compared with the reference value (33 ng/g), there was basically no potential ecological risk in 2009 and 2021. It is worth noting that although the pollution level in 2009 was much more serious than that in 2021, the ecological risk did not show the same results. On the contrary, the TEQ value in 2021 was larger, which was mainly due to the dominant position of the Flu monomer in soil PAHs in 2009, but its toxic equivalent factor was smaller. In 2021, BaP in the soil was not only the most abundant but also the most toxic, accounting for 89.7% of the total TEQ. Therefore, the TEQ was larger, especially in the O10 plot, which exceeded the target value by about 3.8 times (Figure 7), and may have been affected by PAHs, with potential ecological risks. Based on the analysis of sampling points and human activities, the sampling point is located beside the road. The emissions of motor vehicle exhausts and the lifestyles of the surrounding residents are conducive to the increase in PAH toxicity in the soil. Therefore, the traffic flow should be controlled to reduce the ecological risk. The ecological risk of the O4 plot has been significantly reduced (Figure 7), which may be related to the reduction in the use of fuels such as coal and biomass, which effectively prevents and controls Flu pollution.

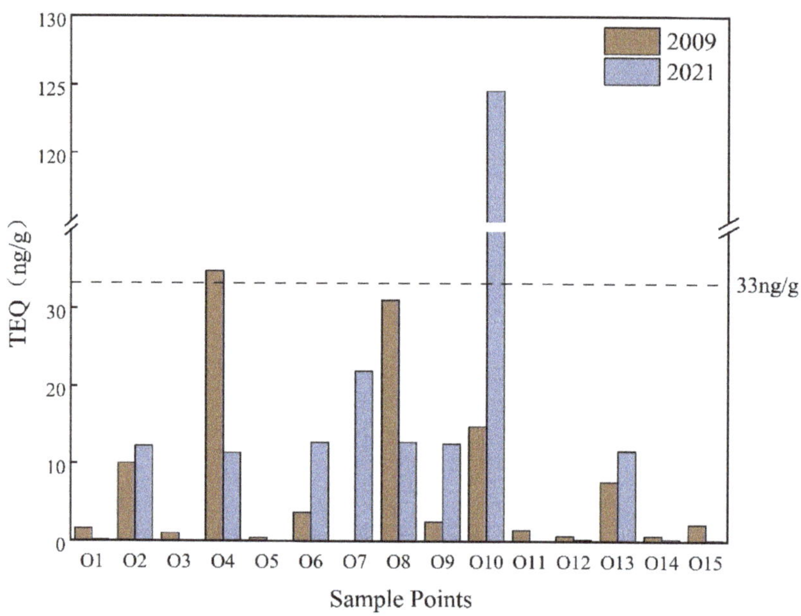

Figure 7. TEQ values of PAHs at each sampling site in 2009 and 2021.

3.4.2. Effect Interval Low Median Method (ERL/ERM)

Figures 8 and 9 show the ecological risk assessment results of individual PAH, low-ring PAHs, high-ring PAHs and ΣPAH_{16} in soil in 2009 and 2021. According to Figure 8, Flu concentrations detected in 12 of the 15 soil samples in 2009 were higher than the ERM value, indicating that 80% of the samples had frequent ecological risks. In addition, the concentrations of Nap in two sites (13%), Ant in one site (7%), Phe in two sites (13%) and Flu in two sites (13%) were between ERM and ERL, indicating that these monomeric PAHs occasionally produced ecological risks in their corresponding sampling points. At the same time, IcdP was also detected at two sampling points, which was harmful to the ecological environment. The ecological risk of other PAHs was small. In 2021, the ecological risk of each monomer PAH in the soil was significantly improved, and only three sampling points (20%) had Nap concentrations between ERM and ERL, with potential ecological risks. The concentrations of other PAHs in all sampling points were lower than ERL, and there was no negative ecological effect.

Compared with the low-ring PAHs and ΣPAH_{16} in some samples in 2009 that will cause harm to the ecological environment, the concentrations of low-ring PAHs, high-ring PAHs and ΣPAH_{16} in soil samples in 2021 will not only decrease but also be less than the ERL values, and there is basically no ecological risk (Figure 9). It shows that the ecological environment quality of the study area is developing well, and the environmental protection system and policy formulated on the prevention and control of PAH pollution have achieved initial success. The future protection in the YRD will continue to be aimed at regulating and reducing PAH pollution and promoting ecological protection and high-quality development in the Yellow River basin.

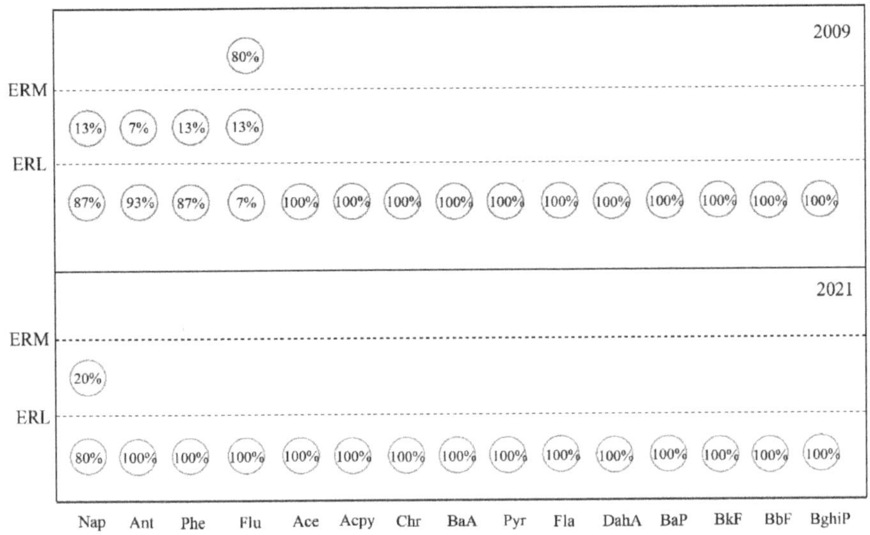

Figure 8. Probability of ecological risk of individual PAH concentration at each sampling site in 2009 and 2021.

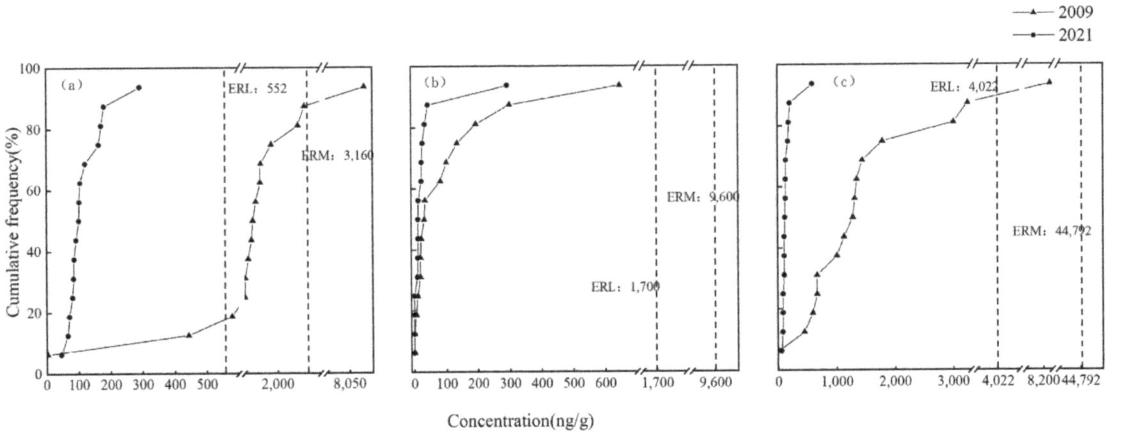

Figure 9. Accumulative probability distributions of LMW (**a**), HMW PAHs (**b**) andΣPAH_{16} concentrations (**c**) in 2009 and 2021.

3.5. Health Risk Assessment of PAHs

Domestic and foreign researchers mainly use the incremental lifetime cancer risk model to assess the health risks of PAHs in soil. Halfadji et al. studied the health risks of PAHs in soil in northwestern Algeria and found that adults and children had the highest potential cancer risk and adolescents had the lowest risk [67]; Zheng et al. combined the PMF model with the lifetime cancer risk increment model to analyze the source of cancer risk and found that biomass burning was the main factor causing the carcinogenicity of soil PAHs in the Chengdu Economic Zone [68]. Table 6 shows the carcinogenic risk values of soil PAHs under three exposure pathways in 2009 and 2021.

Table 6. ILCR values of PAHs in soils under three exposure pathways in 2009 and 2021.

Exposure Pathways	2009		2021	
	Adult	Child	Adult	Child
ILCRingestion	8.56×10^{-9}	4.28×10^{-9}	1.68×10^{-8}	8.4×10^{-9}
ILCRinhalation	6.64×10^{-13}	4.74×10^{-20}	1.3×10^{-12}	9.31×10^{-20}
ILCRdermal	1.52×10^{-8}	5.34×10^{-9}	2.99×10^{-8}	1.05×10^{-8}
ILCRs	2.38×10^{-8}	9.62×10^{-9}	4.67×10^{-8}	1.89×10^{-8}

In 2009, the carcinogenic risks of soil PAHs to adults and children through three exposure pathways of ingestion of soil, respiratory intake and skin contact were, respectively, 8.25×10^{-12}~1.1×10^{-7} and 3.34×10^{-12}~4.46×10^{-8}, and the average values were, respectively, 2.38×10^{-8} and 9.62×10^{-9} (Table 6). The carcinogenic risks of PAHs in soil to adults and children in 2021 were 1.78×10^{-10}~3.95×10^{-7} and 7.22×10^{-11} and 1.6×10^{-7}, respectively, with an average of 4.67×10^{-8} and 1.89×10^{-8}, respectively (Table 6). In terms of time, ILCRs in both years were lower than the safety threshold of 10^{-6}, and there was no carcinogenic risk. Compared with other studies on health risk assessment of soil PAHs in the YRD [14,17], the health risks of soil PAHs in this study were lower in 2009 and 2021 on average (Table 7). Consistent with the trend of ecological risk, the health risk in 2021 was slightly higher than that in 2009 mainly because the toxic equivalent factor of BaP was large. Despite this, the average ΣPAH_{16} content in the surface soil of the Yellow River Delta in 2009 was ten times that of the average ΣPAH_{16} content in 2021. The soil PAH status in 2021 was at the unpolluted level. Therefore, compared with 2009, the soil pollution situation in 2021 is significantly better. It is worth noting that in many organizations, a lifetime cancer risk in the vicinity of 10^{-6} and 10^{-4} connotes a potential hazard. For instance, Tarafdar et al. (2018) found that fly ash PAH ILCR values of adults and children near oil refineries in India were, respectively, 1.823×10^{-5} and 1.854×10^{-5} [69]. The simulated risk exceeded the negligible risk of 10^{-6} but fell below the acceptable risk benchmark of 10^{-4}, which meant that people from both age groups have a plausible threat of cancer occurring [69].

Table 7. ILCR values of PAHs in soil samples reported in the previous literature.

Study Area	Study Year	Mean of ILCRs (Adult)	Mean of ILCRs (Child)	Reference
Yellow River Delta Natural Reserve	2017	1.14×10^{-6}	1.23×10^{-6}	[17]
Farmland soil in the YRD	2020	9.00×10^{-6}	3.60×10^{-5}	[14]
Xianhe Town, Gudao Town and Huanghekou Town of Dongying City in the YRD	2009	2.38×10^{-8}	9.62×10^{-9}	This Study
Farmland soil in the YRD	2021	4.67×10^{-8}	1.89×10^{-8}	This Study

Skin contact and ingestion of soil are the main means of soil PAH exposure, which was consistent with previous studies [13,14,70]. Zheng et al. used the lifetime cancer risk increment (ILCR) to evaluate the polycyclic aromatic hydrocarbons in farmland soil in Ningde. The results showed that the ILCR value ranged from 7.1×10^{-4} to 1.1×10^{-3}, which mainly caused moderate to high cancer risk to human health through soil intake and skin contact [13]. Xie found that among the three exposure pathways, skin contact was the most important exposure mode of soil PAHs, followed by oral ingestion, and inhalation exposure was minimal [14]. Studies have shown that long-term exposure to high concentrations of PAHs in the environment through skin contact, oral and nasal intake would cause damage to the respiratory system, nervous system, liver and kidneys, causing lung cancer, nasopharyngeal carcinoma, skin cancer, gastric cancer and other diseases, which would cause great harm to health [71–74]. Therefore, it is essential to clarify the concentration, content and health risk of ΣPAH_{16} [43,75] and actively study and improve the pollution status of PAHs. From the perspective of the population, the health risk of soil PAH pollution in adults is higher than that in children, because $ILCR_{dermal}$ contributes the most to ILCRs, which was inconsistent with previous studies [14,17,67,69]. The specific causes need to be further studied, so caution should be taken when referring to the management measures of the threat of PAHS pollution to adults and children in the YRD.

4. Conclusions

Based on the investigation of the research group in 2009, this study re-collected samples according to the same sampling points and analyzed the concentration, composition, source, ecological risk and health risk of PAHs. The concentration of ΣPAH_{16} in the surface soil of the Yellow River Delta in 2009 ranged from 2.6 to 8275.46 ng/g, with an average of 1744.41 ng/g, which reached the level of severe pollution. The concentration of ΣPAH 16 in 2021 ranged from 56.25 to 582.56 ng/g, with an average of 149.63 ng/g, which was unpolluted. The composition and distribution characteristics of PAHs in the soil in 2009 and 2021 were similar, and low-ring PAHs are dominant. The main source of soil PAHs in 2009 was oil pollution. In 2021, soil PAHs were mainly affected by oil pollution and traffic pollution, which further confirmed the analysis results of the proportion of rings. The evaluation results of the toxicity equivalent factor method showed that there was no potential ecological risk in the soil in 2009 and 2021, but the TEQ value in 2021 was large. It was mainly because the Flu monomer in the soil PAHs in 2009 was dominant, but its toxicity equivalent factor was small. The ecological risk of each monomer of PAH also improved significantly. Only 20% of the NaP samples had potential ecological risks, and the remaining PAHs had no negative ecological effects. The evaluation results of the lifetime cancer risk increment model showed that the ILCRs of soil PAHs in 2009 and 2021 were lower than the safety threshold of 10^{-6}, and there was no carcinogenic risk. From the perspective of exposure pathways, the highest carcinogenic risk was skin contact, followed by ingestion of soil, and the lowest carcinogenic risk was respiratory intake.

Author Contributions: Conceptualization, Y.W. and Y.Z.; methodology, Y.W., Y.Z. and Y.Q.; software, Y.W., Y.Z., Q.Z. and Y.H.; validation, Y.W. and Y.Z.; formal analysis, Y.W., Y.Z., Y.H. and H.H.; investigation, Y.W., Y.Z., Y.H., H.H. and Q.Z.; resources, Li Junsheng, Y.Q. and N.X.; data curation, Y.W. and Y.Z.; writing—original draft preparation, Y.W. and Y.Z.; writing—review and editing, Y.W. and Y.Z.; visualization, Y.W., Y.Z. and Y.H.; supervision and project administration, J.L., Y.Q. and N.X.; funding acquisition, J.L. and Y.Q. All authors have read and agreed to the published version of the manuscript.

Funding: This research was funded by the Budget Surplus of Central Financial Science and Technology Plan: 2021-JY-05.

Data Availability Statement: The data presented in this study are available on request from the corresponding authors.

Conflicts of Interest: The authors declare no conflict of interest.

References

1. Wang, C.; Du, J.; Gao, X.; Duan, Y.; Sheng, Y.J. Chemical characterization of naturally weathered oil residues in the sediment from Yellow River Delta, China. *Mar. Pollut. Bull.* **2011**, *62*, 2469–2475. [CrossRef]
2. Li, X.; Hou, X.; Song, Y.; Shan, K.; Zhu, S.; Yu, X.; Mo, X. Assessing changes of habitat quality for shorebirds in stopover sites: A case study in Yellow River Delta, China. *Wetlands* **2018**, *39*, 66–77. [CrossRef]
3. Feng, W.; Wang, T.; Zhu, Y.; Sun, F.; Giesy, J.P.; Wu, F. Chemical composition, sources, and ecological effect of organic phosphorus in water ecosystems: A review. *Carbon Res.* **2023**, *2*, 12. [CrossRef]
4. Kuenzer, C.; Ottinger, M.; Liu, G.; Sun, B.; Baumhauer, R.; Dech, S. Earth Observation-based Coastal Zone Monitoring of the Yellow River Delta: Dynamics in China's Second Largest Oil Producing Region observed over four Decades. *Appl. Geogr.* **2014**, *55*, 92–107. [CrossRef]
5. Li, J.; Li, F.; Liu, Q.J.C. PAHs behavior in surface water and groundwater of the Yellow River estuary: Evidence from isotopes and hydrochemistry. *Chemosphere* **2017**, *178*, 143–153. [CrossRef] [PubMed]

6. Menzie, C.A.; Potocki, B.B.; Santodonato, J.J. Exposure to carcinogenic PAHs in the environment. *Environ. Sci. Technol.* **1992**, *26*, 1278–1284. [CrossRef]
7. Deng, W.; Li, X.G.; Li, S.Y.; Ma, Y.Y.; Zhang, D.H. Source apportionment of polycyclic aromatic hydrocarbons in surface sediment of mud areas in the East China Sea using diagnostic ratios and factor analysis. *Mar. Pollut. Bull.* **2013**, *70*, 266–273. [CrossRef]
8. Huang, W.; Wang, Z.; Wen, Y. Distribution and sources of polycyclic aromatic hydrocarbons (PAHs) in sediments from Zhanjiang Bay and Leizhou Bay, South China. *Mar. Pollut. Bull.* **2012**, *64*, 1962–1969. [CrossRef]
9. Gu, Y.G.; Lin, Q.; Lu, T.T.; Ke, C.L.; Sun, R.X.; Du, F.Y. Levels, composition profiles and sources of polycyclic aromatic hydrocarbons in surface sediments from Nan'ao Island, a representative mariculture base in South China. *Mar. Pollut. Bull.* **2013**, *75*, 310–316. [CrossRef]
10. Zheng, B.; Wang, L.; Lei, K.; Nan, B.J.C. Distribution and ecological risk assessment of polycyclic aromatic hydrocarbons in water, suspended particulate matter and sediment from Daliao River estuary and the adjacent area, China. *Chemosphere* **2016**, *149*, 91–100. [CrossRef]
11. Cetin, B.; Yurdakul, S.; Gungormus, E.; Ozturk, F.; Sofuoglu, S.C. Source apportionment and carcinogenic risk assessment of passive air sampler-derived PAHs and PCBs in a heavily industrialized region. *Sci. Total Environ.* **2018**, *633*, 30–41. [CrossRef] [PubMed]
12. Zhao, M.; Wang, W.; Liu, Y.; Dong, L.; Jiao, L.; Hu, L.; Fan, D.J. Distribution and sources of polycyclic aromatic hydrocarbons in surface sediments from the Bering Sea and western Arctic Ocean. *Mar. Pollut. Bull.* **2016**, *104*, 379–385. [CrossRef] [PubMed]
13. Zheng, H.; Qu, C.; Zhang, J.; Talpur, S.A.; Ding, Y.; Xing, X.; Qi, S. Polycyclic aromatic hydrocarbons (PAHs) in agricultural soils from Ningde, China: Levels, sources, and human health risk assessment. *Environ. Geochem. Health* **2019**, *41*, 907–919. [CrossRef] [PubMed]
14. Xie, X.; Liu, Y.; Qiu, H.; Zhang, M.; Wang, W.; Yang, X.; Du, Z.; Zhang, R. Distribution characteristics and risk assessment of polycyclic aromatic hydrocarbons in farmland soil-corn system from oil mining area of Yellow River Delta. *Environ. Geochem. Health* **2021**, *44*, 987–997.
15. Fu, X.W.; Li, T.Y.; Ji, L.; Wang, L.L.; Zheng, L.W.; Wang, J.N.; Zhang, Q. Occurrence, sources and health risk of polycyclic aromatic hydrocarbons in soils around oil wells in the border regions between oil fields and suburbs. *Ecotoxicol. Environ. Saf.* **2018**, *157*, 276–284. [CrossRef]
16. Qi, Y.; Li, J.; Ma, Y.; He, J.; Fu, G.; Shen, Q.; Zhao, C.; Cao, M. Distribution and Risk Assessment of Heavy Metals of Surface Sediments in Intertidal Flats of the Yellow River Delta, China. *Res. Environ. Sci.* **2020**, *33*, 1488–1496.
17. Zhang, K.; Liu, Z.; Wang, Y.; Wang, J.; Cui, D.; Liu, X. Risk assessment and spatial characteristics of PAHs in soils in the Yellow River Delta Nature Reserve. *Ecol. Environ. Sci.* **2022**, *31*, 2198–2205.
18. Qi, Y.; Zhao, Y.; Fu, G.; Li, J.; Zhao, C.; Guan, X.; Zhu, S. The Nutrient and Heavy Metal Contents in Water of Tidal Creek of the Yellow River Delta, China: Spatial Variations, Pollution Statuses, and Ecological Risks. *Water* **2022**, *14*, 713. [CrossRef]
19. Zhao, Y.; Li, J.; Qi, Y.; Guan, X.; Zhao, C.; Wang, H.; Zhu, S.; Fu, G.; Zhu, J.; He, J. Distribution, sources, and ecological risk assessment of polycyclic aromatic hydrocarbons (PAHs) in the tidal creek water of coastal tidal flats in the Yellow River Delta, China. *Mar. Pollut. Bull.* **2021**, *173*, 113110. [CrossRef]
20. Zheng, M. The Environmental Effects of Oil Pollution in Yellow River Delta. Master's thesis, Jinan University, Guangzhou, China, 2013; p. 56.
21. Soclo, H.H.; Garrigues, P.H.; Ewald, M. Origin of polycyclic aromatic hydrocarbons (PAHs) in coastal marine sediments: Case studies in Cotonou (Benin) and Aquitaine (France) areas. *Mar. Pollut. Bull* **2000**, *40*, 387–396. [CrossRef]
22. Che, L.N.; Liu, S.; Yu, Y.; Wan, L.H. Ecological risk assessment of polycyclic aromatic hydrocarbons pollution in snowmelt runoff in Harbin. *Acta Sci. Circumst.* **2019**, *39*, 3508–3515.
23. Nyarko, E.; Botwe, B.O.; Klubi, E.J. Polycyclic Aromatic Hydrocarbons (PAHs) Levels in Two Commercially Important Fish Species from the Coastal Waters of Ghana and their Carcinogenic Health Risks. *West Afr. J. Appl. Ecol.* **2011**, *19*, 53–66.
24. Nisbet, I.C.; Lagoy, P.K. Toxic equivalency factors (TEFs) for polycyclic aromatic hydrocarbons (PAHs). *Regul. Toxicol. Pharmacol.* **1992**, *16*, 290–300. [CrossRef] [PubMed]
25. Long, E.R.; Macdonald, D.D.; Smith, S.L.; Calder, F.D. Incidence of adverse biological effects within ranges of chemical concentrations in marine and estuarine sediments. *Environ. Manag.* **1995**, *19*, 81–97. [CrossRef]
26. Liu, G.; Niu, J.; Guo, W.; An, X.; Zhao, C. Ecological and health risk-based characterization of agricultural soils contaminated with polycyclic aromatic hydrocarbons in the vicinity of a chemical plant in China. *Chemosphere* **2016**, *163*, 461–470. [CrossRef]
27. Zhang, J.; Li, B.; Bi, E.; Hao, Z.; Huang, J. Pollutant characteristics and risk assessment of polycyclic aromatic hydrocarbons in the sediment of the North Canal Basin (Beijing section). *Res. Environ. Sci.* **2019**, *32*, 9.
28. Yu, X.; Ding, Y. Research progress of environmental distribution and bioremediation of polycyclic aromatic hydrocarbons. *J. Dalian Marit. Univ.* **2004**, *30*, 55–59.
29. Yuan, H.; Li, T.; Ding, X.; Zhao, G.; Ye, S. Distribution, sources and potential toxicological significance of polycyclic aromatic hydrocarbons (PAHs) in surface soils of the Yellow River Delta, China. *Mater. Proc.* **2014**, *83*, 258–264. [CrossRef]
30. Yuan, Z.; Liu, G.; Da, C.; Wang, J.; Liu, H. Occurrence, Sources, and Potential Toxicity of Polycyclic Aromatic Hydrocarbons in Surface Soils from the Yellow River Delta Natural Reserve, China. *Arch. Environ. Contam. Toxicol.* **2015**, *68*, 330–341. [CrossRef]
31. Maliszewska-Kordybach, B.J. Polycyclic aromatic hydrocarbons in agricultural soils in Poland: Preliminary proposals for criteria to evaluate the level of soil contamination. *Appl. Geochem.* **1996**, *11*, 121–127. [CrossRef]

32. Cao, J.; Liu, Y.; Yu, S. The concentrations and sources of PAHs and PCBs in soil from an oil field and estuary in the Yellow River Delta, China. *Toxicol. Pollut. Environ.* **2022**, *10*, 1028299.
33. Xie, W.; Chen, A.; Li, J.; Liu, Q.; Lu, Z. County-scale distribution of polycyclic aromatic hydrocarbons in topsoil of the Yellow River Delta Region. *Environ. Lett.* **2012**, *47*, 1419–1427. [CrossRef] [PubMed]
34. Chen, M.; Mu, N.; Wang, Y.; Liu, G.; Zhao, Z. Quantitative Analysis of the Adsorded Nitrogen Pollution of the Yellow River Delta in the Past Three Decades. *Chin. J. Ecol.* **2022**, *41*, 1324–1333.
35. Gao, Z. Experimental Stduy on Remediation of Groundwater/Soil Polluted by Petroleum Hydrocarbon in Dongying City. *China Univ. Geosci.* **2012**, *8*, 147.
36. Gao, Y.F.; Zhang, D.; Zhao, B.; Yin, B.; Zhang, J.; Li, J.; Cao, X. Dynmamic Changes of Soil and Water Loss in the Yellow River Basin from 1990 to 2019. *Soil Water Conserv. China* **2020**, *10*, 64–67.
37. Wang, Y. Analysis on the Water and Sediment Variation in the Yellow River and its Influence on the Change of the Yellow River Delta. Master's Thesis, University of Chinese Academy of Sciences (Research Center for Soil and Water Conservation and Ecological Environment, Ministry of Education, Chinese Academy of Sciences), Beijing, China, 2018; p. 67.
38. Xu, X.; Liu, W.; Wang, W.; Tian, S.; Jiang, P.; Qi, Q.; Li, F.; Li, H.; Wang, Q.; Li, H. Potential biodegradation of phenanthrene by isolated halotolerant bacterial strains from petroleum oil polluted soil in Yellow River Delta. *Sci. Total Environ.* **2019**, *664*, 1030–1038. [CrossRef]
39. Agarwal, T.; Khillare, P.S.; Shridhar, V.; Ray, S. Pattern, sources and toxic potential of PAHs in the agricultural soils of Delhi, India. *J. Hazard. Mater.* **2009**, *163*, 1033–1039. [CrossRef]
40. Oyo-Ita, O.E.; Oyo-Ita, I.O. PAH depositional history sources in recent sediment core from Ukwa Ibom Lake, S.E. Nigeria. *Environ. Geochem. Health* **2013**, *35*, 185–199. [CrossRef]
41. Jiang, Y.F.; Wang, X.T.; Wang, F.; Jia, Y.; Wu, M.H.; Sheng, G.Y.; Fu, J.M. Levels, composition profiles and sources of polycyclic aromatic hydrocarbons in urban soil of Shanghai, China. *Chemosphere* **2009**, *75*, 1112–1118. [CrossRef]
42. Liu, S.; Xia, X.; Yang, L.; Shen, M.; Liu, R. Polycyclic aromatic hydrocarbons in urban soils of different land uses in Beijing, China: Distribution, sources and their correlation with the city's urbanization history. *Environ. Res.* **2010**, *177*, 1085–1092. [CrossRef]
43. Wang, X.C.; Sun, S.; Ma, H.Q.; Liu, Y. Sources and distribution of aliphatic and polyaromatic hydrocarbons in sediments of Jiaozhou Bay, Qingdao, China. *Mar. Pollut. Bull.* **2006**, *52*, 129–138. [CrossRef] [PubMed]
44. Abdel-Shafy, H.I.; Mansour, M.S.M. A review on polycyclic aromatic hydrocarbons: Source, environmental impact, effect on human health and remediation. *Egypt. J. Pet.* **2016**, *25*, 107–123. [CrossRef]
45. Khairy, M.A.; Lohmann, R.J.C. Source apportionment and risk assessment of polycyclic aromatic hydrocarbons in the atmospheric environment of Alexandria, Egypt. *Atmos. Environ.* **2013**, *91*, 895–903. [CrossRef]
46. Liu, Y.; Chen, L.; Zhao, J.; Huang, Q.; Zhu, Z.; Gao, H. Distribution and sources of polycyclic aromatic hydrocarbons in surface sediments of rivers and an estuary in Shanghai, China. *Environ. Pollut.* **2008**, *154*, 298–305. [CrossRef] [PubMed]
47. Wang, X.T.; Miao, Y.; Zhang, Y.; Li, Y.C.; Wu, M.H.; Yu, G. Polycyclic aromatic hydrocarbons (PAHs) in urban soils of the megacity Shanghai: Occurrence, source apportionment and potential human health risk. *Sci. Total Environ.* **2013**, *447*, 80–89. [CrossRef]
48. Ping, L.F.; Luo, Y.M.; Zhang, H.B.; Li, Q.B.; Wu, L. Distribution of polycyclic aromatic hydrocarbons in thirty typical soil profiles in the Yangtze River Delta region, east China. *Environ. Pollut.* **2007**, *147*, 358–365. [CrossRef]
49. Bi, X.; Luo, W.; Gao, J.; Xu, L.; Guo, J.; Zhang, Q.; Romesh, K.Y.; Giesy, J.P.; Kang, S.; de Boer, J. Polycyclic aromatic hydrocarbons in soils from the Central-Himalaya region: Distribution, sources, and risks to humans and wildlife. *Sci. Total Environ.* **2016**, *556*, 12–22. [CrossRef]
50. Hamid, N.; Syed, J.H.; Junaid, M.; Mahmood, A.; Li, J.; Zhang, G.; Malik, R.N. Elucidating the urban levels, sources and health risks of polycyclic aromatic hydrocarbons (PAHs) in Pakistan: Implications for changing energy demand. *Sci. Total Environ.* **2017**, *619*, 165–175. [CrossRef]
51. Yu, G.G.; Wang, T.G.; Wu, D.P. Study on fingerprints of PAHs from the combustion of bavin and coal. *Ecol. Environ.* **2007**, *16*, 285–289.
52. Khalili, N.R.; Scheff, P.A.; Holsen, T.M. PAH source fingerprints for coke ovens, diesel and, gasoline engines, highway tunnels, and wood combustion emissions. *Atmos. Environ.* **1995**, *29*, 533–542. [CrossRef]
53. Simcik, M.F.; Eisenreich, S.J.; Lioy, P.J. Source apportionment and source/sink relationships of PAHs in the coastal atmosphere of Chicago and Lake Michigan. *Atmos. Environ.* **1999**, *33*, 5071–5079. [CrossRef]
54. Yunker, M.B.; Macdonald, R.W.; Vingazan, R.; Mitchell, R.H.; Goyette, D.; Sylvestre, S.J.O.G. PAHs in the Fraser River basin: A critical appraisal of PAH ratios as indicators of PAH source and composition. *Org. Geochem.* **2002**, *33*, 489–515. [CrossRef]
55. Wang, C.; Wu, S.; Zhou, S.; Wang, H.; Li, B.; Chen, H.; Yu, Y.; Shi, Y. Polycyclic aromatic hydrocarbons in soils from urban to rural areas in Nanjing: Concentration, source, spatial distribution, and potential human health risk. *Sci. Total Environ.* **2015**, *527–528*, 375–383. [CrossRef] [PubMed]
56. Yang, B.; Zhou, L.; Xue, N.; Li, F.; Li, Y.; Vogt, R.D.; Cong, X.; Yan, Y.; Liu, B.J. Source apportionment of polycyclic aromatic hydrocarbons in soils of Huanghuai Plain, China: Comparison of three receptor models. *Sci. Total Environ.* **2013**, *443*, 31–39. [CrossRef] [PubMed]
57. Yang, W.; Lang, Y.; Li, G.; Liu, W. Distribution and health risk assessment of polycyclic aromatic hydrocarbons in Jiaozhou Bay wetland soil. *Environ. Chem.* **2013**, *32*, 1989–1990.

58. Kannan, K.; Johnson-Restrepo, B.; Yohn, S.S.; Giesy, J.P.; Long, D. Spatial and temporal distribution of polycyclic aromatic hydrocarbons in sediments from Michigan inland lakes. *Environ. Sci. Technol.* **2005**, *39*, 4700–4706. [CrossRef]
59. Ravindra, K.; Sokhi, R.; Van Grieken, R. Atmospheric polycyclic aromatic hydrocarbons: Source attribution, emission factors and regulation—ScienceDirect. *Atmos. Environ.* **2008**, *42*, 2895–2921. [CrossRef]
60. Larsen, R.K.; Baker, J.E. Source Apportionment of Polycyclic Aromatic Hydrocarbons in the Urban Atmosphere: A Comparison of Three Methods. *Environ. Sci. Technol.* **2003**, *37*, 1873–1881. [CrossRef]
61. Paatero, P.; Tapper, U.J.E. Positive matrix factorization: A non-negative factor model with optimal utilization of error estimates of data values. *Environmetrics* **1994**, *5*, 111–126. [CrossRef]
62. Qi, H.X.; Chen, X.L.; Du, Y.E.; Niu, X.; Guo, F.; Li, W. Cancer risk assessment of soils contaminated by polycyclic aromatic hydrocarbons in Shanxi, China. *Ecotoxicol. Environ. Saf.* **2019**, *182*, 109381. [CrossRef]
63. Tarafdar, A.; Sinha, A. Health risk assessment and source study of PAHs from roadside soil dust of a heavy mining area in India. *Arch. Environ. Occup. Health* **2019**, *74*, 252–262. [CrossRef] [PubMed]
64. Li, C.; Rong, Q.Y.; Zhu, C.M.; Han, J.; Li, P. Distribution, sources, and risk assessment of polycyclic aromatic hydrocarbons in the estuary of Hongze Lake, China. *Environments* **2019**, *6*, 92. [CrossRef]
65. Liang, X.X.; Junaid, M.; Wang, Z.F.; Li, T.; Xu, N. Spatiotemporal distribution, source apportionment and ecological risk assessment of PBDEs and PAHs in the Guanlan River from rapidly urbanizing areas of Shenzhen, China. *Environ. Pollut.* **2019**, *250*, 695–707. [CrossRef] [PubMed]
66. Xu, J.; Yu, Y.; Wang, P.; Guo, W.; Dai, S.; Sun, H. Polycyclic aromatic hydrocarbons in the surface sediments from Yellow River, China. *Chemosphere* **2007**, *67*, 1408–1414. [CrossRef] [PubMed]
67. Halfadji, A.; Naous, M.; Bettiche, F.; Touabet, A. Human Health Assessment of Sixteen Priority Polycyclic Aromatic Hydrocarbons in Contaminated Soils of Northwestern Algeria. *J. Health Pollut.* **2021**, *11*, 210914. [CrossRef]
68. Zheng, H.; Xing, X.; Hu, T.; Zhang, Y.; Zhang, J.; Zhu, G.; Li, Y.; Qi, S. Biomass burning contributed most to the human cancer risk exposed to the soil-bound PAHs from Chengdu Economic Region, western China. *Ecotoxicol. Environ. Saf.* **2018**, *159*, 63–70. [CrossRef]
69. Abhrajyoti, T.; Alok, S. Public health risk assessment with bioaccessibility considerations for soil PAHs at oil refinery vicinity areas in India. *Sci. Total Environ.* **2018**, *616–617*, 1477–1484.
70. Cai, C.Y.; Li, J.Y.; Wu, D.; Wang, X.; Tsang, D.; Li, X.D.; Sun, J.T.; Zhu, L.Z.; Shen, H.Z.; Shu, T. Spatial distribution, emission source and health risk of parent PAHs and derivatives in surface soils from the Yangtze River Delta, eastern China. *Chemosphere* **2017**, *178*, 301. [CrossRef]
71. Huang, X.D.; Zeiler, L.F.; Dixon, D.G.; Greenberg, B.M. Photoinduced toxicity of PAHs to the foliar regions of Brassica napus (canola) and Cucumbis sativus (cucumber) in simulated solar radiation. *Ecotoxicol. Environ. Saf.* **1996**, *35*, 190–197. [CrossRef]
72. Lu, D.; Zhang, F. Researh progress on occupational hazards of polycyclic aromatic hydroarbons, China. *Occup. Med.* **2017**, *44*, 4.
73. Sheng, S.; Cao, X.; Song, Y.; Liu, Y.; Sheng, Z.; Qin, L. Properties and hazards of polycyclic aromatic hydrocarbons. *Guizhou Chem. Industy* **2008**, *33*, 61–63.
74. Toyooka, T.; Ibuki, Y. DNA damage induced by coexposure to PAHs and light. *Environ. Toxicol. Pharmacol.* **2007**, *23*, 256–263. [CrossRef] [PubMed]
75. Wang, C.; Wang, W.; He, S.; Du, J.; Sun, Z.J. Sources and distribution of aliphatic and polycyclic aromatic hydrocarbons in Yellow River Delta Nature Reserve, China. *Appl. Geochem.* **2011**, *26*, 1330–1336. [CrossRef]

water

MDPI

Article

Study on Water and Salt Transport under Different Subsurface Pipe Arrangement Conditions in Severe Saline–Alkali Land in Hetao Irrigation District with DRAINMOD Model

Feng Tian [1,2,†], Qingfeng Miao [1,2,†], Haibin Shi [1,2,*], Ruiping Li [1,2], Xu Dou [1,2], Jie Duan [1,2], Jing Liu [3] and Weiying Feng [4,*]

1 College of Water Conservancy and Civil Engineering, Inner Mongolia Agricultural University, Hohhot 010010, China; tianfeng@emails.imau.edu.cn (F.T.); 15049121836@126.com (Q.M.); nmglrp@163.com (R.L.); nmgdouxu@163.com (X.D.); nmgduanjie@163.com (J.D.)

2 High Efficiency Water-Saving Technology and Equipment and Soil and Water Environment Effect in Engineering Research Center of Inner Mongolia Autonomous Region, Hohhot 010011, China

3 Environment Research Institute, Shandong University, Qingdao 266237, China; liu_jing@email.sdu.edu.cn

4 School of Materials Science and Engineering, Beihang University, Beijing 100191, China

* Correspondence: shb@imau.edu.cn (H.S.); fengweiying@buaa.edu.cn (W.F.)

† These authors contributed equally to this work.

Citation: Tian, F.; Miao, Q.; Shi, H.; Li, R.; Dou, X.; Duan, J.; Liu, J.; Feng, W. Study on Water and Salt Transport under Different Subsurface Pipe Arrangement Conditions in Severe Saline–Alkali Land in Hetao Irrigation District with DRAINMOD Model. *Water* **2023**, *15*, 3001. https://doi.org/10.3390/w15163001

Academic Editor: William Frederick Ritter

Received: 14 July 2023
Revised: 13 August 2023
Accepted: 18 August 2023
Published: 20 August 2023

Abstract: As an effective method to improve saline–alkali land, the drainage from subsurface pipes has been extensively studied in typical arid and semi-arid agricultural areas (Hetao Irrigation District). However, there are few studies on the improvement of subsurface pipe layout and the long-term soil salinization control in the process of leaching and soil amendment with subsurface pipes in this area. This study investigated the water and salt migration in the process of amending the heavy saline soil. Field experiments growing sunflowers and numerical model calculation were combined in this research. It was found in the field experiment that the salt concentration in the surface pipe drainage was positively correlated with the salt content in the soil and the depth of the pipe, while it was negatively correlated with the amount of irrigation water and the spacing of crops. Thus, the soil desalting rate (N) and salt control rate (SCR) were positively correlated with the depth of the pipe, and they were negatively correlated with the spacing. The leaching effect of irrigation would decrease when the soil salt content decreased. On the basis of field experiments, the DRAINMOD model and drainmod equation were used to calculate the water and salt migration in 38 different field plots during 2019 and 2020. When N was the same, the soil salinity in several plots with large burial depth could be controlled below the salt tolerance threshold of sunflowers during the growth period in the second year. The quantitative relationship between N and SCR, soil salt content before leaching, water amount of leaching, pipe spacing and buried depth was already established. These results can help develop strategies for desalination and salt control in the soil in the arid and semi-arid areas with the optimal layout of subsurface pipes.

Keywords: Hetao Irrigation District; saline–alkali soil; subsurface drainage system; DRAINMOD; water and salt transport

1. Introduction

As a global agricultural issue, soil salinization has been widely concerned because it reduces the utilization efficiency of fertilizers, increases the loss of nutrients in the soil, and affects the growth, yield and quality of crops [1–5]. Soil salinization has affected more than 30% of arid and semi-arid areas worldwide [6,7]. As a typical arid and semi-arid irrigated area, 63.8% of the soil in the Hetao Irrigated District of Inner Mongolia has been affected by salinization [8]. Subsurface pipe drainage has been widely recognized as an effective measure for amending saline farmland [9], due to its advantages of small land occupancy and high land use efficiency for improving soil conditions, reducing soil salinity and increasing crop yield [10–12]. There have been many studies on the displacement, drainage quality and the influence on groundwater level of the subsurface pipe with different layout parameters [13,14]. However, the proposed layout parameters in those studies may not be applicable to the area for desalting and long-term salt control in the arid and semi-arid areas.

The key points of the research on the subsurface pipe mainly include the drainage, salt and nutrient transport, the response of crop yield, and the optimal layout parameters of subsurface pipes under different soil conditions [15–17]. Many studies have carried out quantitative analysis of various farmland indicators with different layout parameters of subsurface pipes. Laboratory and field experiments showed that the depth of subsurface pipe positively affected the quantity and quality of discharged water, while the spacing of subsurface pipe presented negative influence on these parameters [18,19].

The migration of water and salt under different pipe layout conditions cannot be fully investigated due to the limitations of laboratory and field tests. The numerical model is an effective and complementary method to solve this issue. Some of the effective and widely accepted models include but are not limited to the HYDRUS [20], SWAT [21] and DRAINMOD [22]. Dou et al. used HYDRUS-2D to simulate the subsurface pipe drainage in the farmland compared the drainage depth of the subsurface pipe at different growth stages of sunflowers, and they determined that the optimal drainage depth in medium salinized soil was 50 cm [23]. Addab and Bailey simulated the migration of eight major salt ions in the soil with subsurface pipe drainage in the salinized area, and they developed effective land and water management strategies with the SWAT model [24,25]. DRAINMOD, based on water balance, is widely used to simulate farmland drainage and crop yield because of its simplicity, accuracy and practicability [26,27]. The model was extended to simulate the macropore seepage [28]. Meanwhile, this model well simulated the recycling of subsurface pipe drainage water and the related economic benefits [29].

Hetao Irrigation District is a typical salinization area, and many studies have been conducted on the improvement of soil quality in this region [30,31]. However, few studies focused on the subsurface pipe drainage and the water and salt migration in the soil with subsurface pipes in long time series. The groundwater level is high, and soil permeability is poor in the heavily salinized area. As a result, the salinity of the soil can only be reduced in a short period of time even if subsurface drainage is used. This study mainly investigated the water and salt migration in the salinized farmland growing sunflowers with buried drainage pipes in Hetao Irrigation District. Based on this field experiment, simulation was conducted with DRAINMOD model for water migration and calculate the soil salt migration. The index of salt control rate (*SCR*) was proposed to describe the effect of subsurface pipe on the salinity of main root layer soil of sunflowers, and the best arrangement of the subsurface pipe under different desalting and salt control requirements with planting sunflower was determined to provide technical support for the improvement of subsurface pipe drainage in the arid and semi-arid areas.

2. Materials and Methods

2.1. Study Site

From May 2019 to November 2020, a trial of leaching was conducted in a field with an area of approximately 5.4 ha. The field is located in the Comprehensive Improvement Test Base of Saline–alkali Land with Subsurface Pipe Drainage in the Wulat irrigation area, Hetao Irrigation District, Inner Mongolia, northwest China, with an altitude of 1018.88 m at 40°45′28″ N and 108°38′16″ E [32]. The region has a typical temperate continental climate, dry and windy, with an annual evaporation of 2173 mm [33]. The groundwater is relatively shallow with a depth of 1.6 m, and the salinity of groundwater is 29.66 dS/m. Flood irrigation is mainly used for irrigation with water from the Yellow River with a total dissolved solids content of 0.67 g/L. The drainage method is mostly open ditch, which is mostly used for drainage in this area. The large amount of irrigation water use, untimely drainage, high groundwater level and intense surface evaporation lead to prominent soil salinization and seriously restrict the development of local agriculture.

2.2. Field Test Design

Seven test fields were constructed. The Dutch INTERDRAIN all-in-one pipe laying machine was used to complete the ditching and laying of the subsurface pipes, which were single-wall bellows pipes with a diameter of 8 cm, a length of 200 m and a buried slope of 1‰. The pipes were arranged in the short direction of the field, and then they were wrapped with geotextiles, covered with sand of 10 cm, backfilled with original soil to the initial elevation, and finally processed via laser leveling.

The leaching water was discharged from the subsurface pipe to the drainage ditch excavated along one side of the field. Referring to previous studies [34], the drainage depth of the pipes was set at 1.1 m, 1.2 m and 1.3 m, and the drainage spacing was set at 10 m, 20 m and 30 m. Seven field experiments were arranged, as shown in Figure 1a. Due to the limitation of field tests, the drain depth ranged from 1.1 m to 1.3 m, and the drain spacing ranged from 10 m to 30 m. DRAINMOD model was used to simulate the desalination and salt control effects of different layout parameters of the pipes under the same initial soil conditions. There were a total of 38 prediction scenarios, denoted as F_1–F_{38}, with 9 drain depths (1 m, 1.1 m, 1.2 m, 1.3 m, 1.4 m, 1.5 m, 1.6 m,1.8 m and 2 m) and 5 drain spacings (10 m, 15 m, 20 m, 25 m and 30 m). Field and prediction layout of subsurface pipes are shown in Table 1.

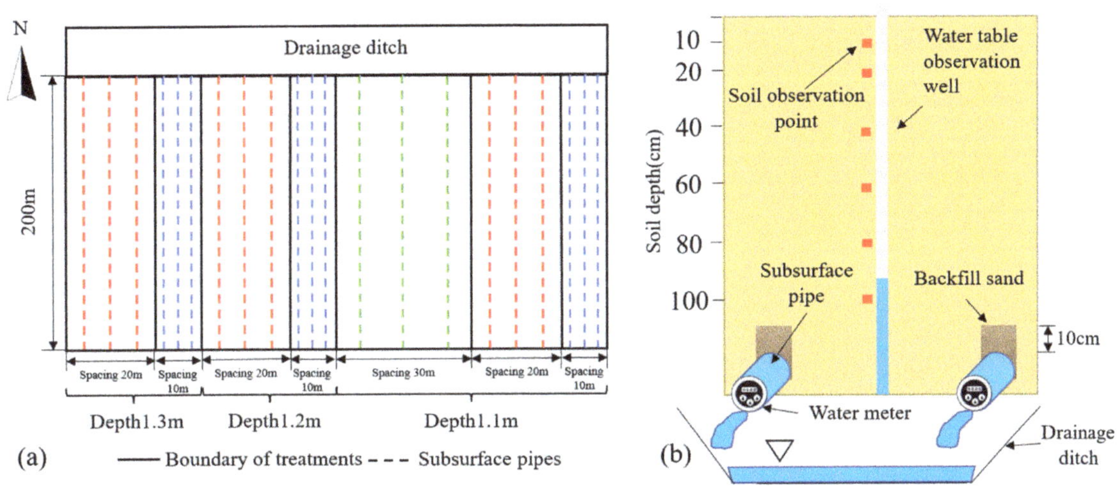

Figure 1. Schematic diagram of (**a**) subsurface drainage pipe layout of each treatment and (**b**) drainage structure of subsurface pipe and observation point of water and soil samples (The green line is a subsurface pipe with a spacing of 30 m. The red line is a subsurface pipe with a spacing of 20 m. The blue line is a subsurface pipe with a spacing of 10 m).

2.3. Sampling, Measurement and Calculation

2.3.1. Meteorological Data

The weather input files required by the model included precipitation, solar radiation, maximum and minimum temperatures, wind speed, etc. The hourly data were obtained from field weather stations (HOBO-U30) used at the test field.

Table 1. Layout parameters of the 7 experiment (T) and 38 prediction (F) scenarios.

Depth (cm)	Spacing (cm)				
	1000	1500	2000	2500	3000
100	F_1	F_2	F_3	F_4	F_5
110	T_1	F_6	T_4	F_7	T_7
120	T_2	F_8	T_5	F_9	F_{10}
130	T_3	F_{11}	T_6	F_{12}	F_{13}
140	F_{14}	F_{15}	F_{16}	F_{17}	F_{18}
150	F_{19}	F_{20}	F_{21}	F_{22}	F_{23}
160	F_{24}	F_{25}	F_{26}	F_{27}	F_{28}
180	F_{29}	F_{30}	F_{31}	F_{32}	F_{33}
200	F_{34}	F_{35}	F_{36}	F_{37}	F_{38}

2.3.2. Irrigation Time and Water Volume

Flood irrigation was used in the study. The quantity of irrigation water was controlled by installing a circular inlet gate of each plot for measuring the inlet water quantity. The time and frequency of irrigation mainly depend on the local Yellow River inflow time and irrigation management system. During the experiment, seven times of irrigation with different water amounts were carried out, respectively, on 15 May 2019 (300 mm), 26 June 2019 (150 mm), 17 July 2019 (150 mm), 25 October 2019 (300 mm), 3 May 2020 (350 mm), 28 June 2020 (250 mm) and 19 October 2020 (350 mm).

2.3.3. Soil Sampling

In order to monitor the desalination effect of subsurface pipes, soil sampling was conducted one day before each drainage test and after the end of drainage. The soil sampling points were arranged at the intermediate points between the middle subsurface pipe and the adjacent subsurface pipe in each test area. The surface soil was sampled from 0–10 cm and 10–20 cm, and then the soil was sampled every 20 cm below the surface to a depth of 100 cm (Figure 1b).

2.3.4. Water Sampling

A water meter was connected to the subsurface pipe near the drain. In order to facilitate water meter disassembly and data collection, A manhole was built above water meter. The daily subsurface pipe displacement and drainage quality were monitored during the irrigation and drainage stage.

2.3.5. Water Table

Groundwater level observation wells were arranged near the soil sampling points in each test plot. Meter ruler and plumb hammer were used to measure the groundwater level in the initial stage of the test, and a groundwater level monitor DATA-6216 was placed in the observation wells to monitor the real-time groundwater level during the test.

2.3.6. Soil Moisture Content, Salt Content, Desalting Rate and Salt Control Rate

Some soil samples were weighed to determine the wet weight and then dried in a hot blast oven at 105 °C for 24 h to measure the dry weight and calculate the moisture content. Next, 10 g of some soil sample was well mixed in 50 mL distilled water. After standing for 15 min, the conductivity of the solution (*EC*) was measured using a lightning magnetic DDS-307A conductivity meter. The soil salt content was obtained using Equation (1) [35]. The soil desalting rate was calculated using Equation (2) [23].

$$S = 3.7657EC_{(1:5)} - 0.2405 \quad (1)$$

where S is soil salt content (g/kg), and $EC_{1:5}$ is the EC of the soil extract solution made by mixing the soil with water at a soil:water ratio of 1:5, (mS/cm).

$$N = \frac{S_1 - S_2}{S_1} \times 100\% \tag{2}$$

In the equation, N is the soil desalting rate (%), S_1 is the initial value of soil salinity before irrigation (g/kg), and S_2 is the soil salinity after irrigation (g/kg).

The index of salt control rate (denoted as SCR), proposed in this study, represents the percentage of days when the salinity in the soil was lower than the threshold of crop salinity in the total growth period, which was used to reflect the control ability of subsurface pipe drainage on soil salinity, so as to select the suitable subsurface pipe layout.

2.4. DRAINMOD Model

2.4.1. Model Description

DRAINMOD is a two-dimensional model to obtain information of soil hydrological processes using a water balance method and empirical relationships with data about meteorology, soil properties, crop growth and field irrigation/drainage [22]. An hourly or daily simulation of the water balance at the midpoint of two parallel drainage pipes, including evapotranspiration, surface runoff, subsurface drainage, infiltration and groundwater level. The salt module of the model is rarely used because the calculation process is not rigorous, resulting in a large difference between the simulation results and the actual situation. This paper only uses it for hydrological simulation.

Water balance in this model is classified as two compartments: In the surface balance, the model can be applied to compute the permeability, irrigation and runoff using the following Equation (3). In the subsurface balance, DRAINMOD calculates the water balance for a section of the soil in the middle of two drains according to the following Equation (4) [22].

$$\Delta W = P + I - F - RO \tag{3}$$

$$\Delta Va = D + ET + VLS - F \tag{4}$$

In the equation, ΔW is the change in surface water storage (mm); P is the precipitation (mm); I is the irrigation amount (mm); F is the infiltration amount (mm); RO is the surface runoff (mm); ΔV_a is the change in water content of anhydrous pore space in soil profile (mm); D is the drainage depth (mm); ET is the evapotranspiration (mm); and VLS is the vertical and lateral flow (mm).

The flow from spawn to sink to drain was assumed instantaneous in the DRAINMOD model. The amount of irrigation water was measured in the field. The amount of infiltration was estimated with Green–Ampt equation with the input of soil physical properties. In the equation, surface runoff was not included since all the irrigation water was discharged from the field through subsurface drainage due to the 1 m high ridge surrounding the field. The efficiency of subsurface drainage depends on the depth of groundwater, soil hydraulic conductivity and the layout of buried pipes. When the underground water level is below the surface, Hooghoudt steady flow equationis are used to calculate the underground pipe displacement; when the underground water level rises to completely submerge the ground, the Kirkham equation is used to estimate the underground drainage flow [36]. Vertical and lateral seepage were estimated using a simple method based on the Darcy equation and Dupuit-Forchheimer hypothesis [22,37].

2.4.2. Model Input

The selected field was an abandoned farmland. A comprehensive soil survey was carried out before burying the pipes (Table 2). The drainage depth and spacing of the field drainage system depended on the buried subsurface pipe parameters in each test area. The depth from the surface to the impermeable layer was set to 2 m, the initial depth of the groundwater was set to the actual measured value (160 cm), and the depth at which water flows to the subsurface pipe (Kirkham depth) was set to 0.3 cm. The actual irrigation time and irrigation amount are integrated into the meteorological data and then input into the model. The model calculates the potential evapotranspiration (*PET*) based on the input meteorological data and the Penman–Monteith formula, and then it distributes the calculated *PET* to each hour of the simulation period. In the calculation process, the model will self-check whether evapotranspiration is limited by soil moisture conditions. If it is not affected by soil water supply capacity, the actual evapotranspiration (*ET*) is equal to *PET*. Otherwise *ET* is equal to soil water supply capacity (upward flux of diving).

Table 2. Soil physical and chemical characteristics of experimental site.

Soil Layer (cm)	Particle Composition/%			Bulk Density ($g{\cdot}cm^{-3}$)	Soil Salt Content ($g{\cdot}kg^{-1}$)	pH	Field Capacity ($cm^3\ cm^{-3}$)
	Sand	Clay	Silt				
0–10	36.6	4.22	59.18	1.44	25.23	7.45	32.35
10–20	32.32	3.86	63.82	1.45	21.79	7.60	33.29
20–40	23.98	2.20	73.82	1.47	18.22	7.64	36.61
40–60	8.27	2.47	89.26	1.48	15.53	7.72	35.08
60–80	3.36	5.70	90.94	1.49	11.79	7.51	36.33
80–100	12.91	5.02	82.07	1.49	10.60	7.62	36.52

As one of the main planting cash crops in Hetao Irrigation Distcict, sunflowers have high salt tolerance (18.28 g/kg) and are generally sown in early June and harvested in early or middle September, with a growth period of about 105 d. The irrigation water in this study was from the Yellow River, and the time of each irrigation test was consistent with the local irrigation schedule. The potential evapotranspiration of bare land is PET_1. In order to simulate the soil water balance when sunflowers were planted, the dual crop coefficient [38,39] method was used to calculate crop transpiration (T_c) and soil evaporation (E_s), respectively, to estimate potential evapotranspiration (PET_2) during sunflower planting.

$$T_c = K_{cb} ET_o \quad (5)$$

$$E_s = ET_o \quad (6)$$

Here, K_{cb} is the basal crop coefficient; K_e is soil evaporation coefficient; and ET_o is the reference crop evapotranspiration.

Basal crop coefficient refers to the ratio of crop evapotranspiration to reference evapotranspiration at a given potential rate without soil evaporation. Miao et al. provided the K_{cb} reference values in the early (0.10), middle (1.15) and late (0.25) growth periods of sunflower after calibration [40]. The evaporation coefficient K_e is used to describe the portion of soil evaporation in the actual evapotranspiration, which depends on the amount of water available for evaporation in the upper soil and on the amount of energy available at the soil surface in conjunction with the energy consumed by transpiration [41].

$$K_e = K_r \left(K_{c\,max} - K_{cb}\right) \le f_{ew}\, K_{c\,max} \quad (7)$$

In the equation, K_r is the evaporation reduction coefficient, depending on the cumulative depth of surface soil water loss, and $K_{c\,max}$ is the maximum value of K_c after rainfall or irrigation. The calculation of K_r is divided into two stages. When the accumulated evaporation loss from soil surface (D_e, mm) does not exceed the amount of evaporable water (*REW*, mm), K_r is equal to 1. When D_e exceeds *REW*, K_r decreases, and when the evaporation from surface soil is 0, $K_r = 0$. Equation (8) was used to calculate the K_r of the second stage.

$$K_r = \frac{TEW - D_{e,i-1}}{TEW - REW} \quad (8)$$

In the equation, *TEW* is the maximum depth of evaporation of surface soil after wetting (mm); *REW* is the amount of water easily evaporated without water limitation (the cumulative evaporation depth at the end of the first stage); $D_{e,i-1}$ is the cumulative depth of soil surface evaporation water at the end of day $i-1$(mm).

f_{ew} is the fraction of soil that is exposed to radiation and moistened by water, and it is related to the percentage of the ground covered by crops.

$$f_{ew} = \min(1 - f_c, f_w) \tag{9}$$

In the equation, f_c is the fraction of the surface soil area covered by plants. f_w is the fraction of soil surface wetting by rainfall or irrigation.

2.5. Water and Salt Balance in Soil Profile

In the calculation of water balance in the soil profile of the test site including shallow groundwater utilization, the influence of soil pore distribution difference on water migration is ignored for the convenience of the study, and the piston flow is considered. When calculating the water amount of precipitation, irrigation, underground drainage, shallow groundwater utilization, evapotranspiration and infiltration, the salt carried in each part of water is defined to clarify the water–salt balance of the profile. The calculation formula is as follows:

Moisture:

$$\Delta V_a = P + I + G - D - ET - L \tag{10}$$

Salt:

$$S = S_O + S_I + S_G - S_D - S_L \tag{11}$$

In the equation, G is the utilization of shallow groundwater (cm); L is the deep percolation (cm); S_O is the initial salt content of soil profile (g); S_I is the salt brought by irrigation water (g); S_G is the rising salinity by groundwater (g); S_D is the salt discharged by underground drainage (g); S_L is the salt introduced into groundwater (g).

When the water table rises after rain or irrigation, the water quantity L of deep seepage can be calculated using the following equation.

$$L = \mu \Delta H \tag{12}$$

Here, μ is the water supply degree. According to experience, the value of μ in the test site ranges from 0.04 to 0.06 [42], calibrated via moisture calculation, and the value in this paper is 0.05. ΔH is the change in depth of groundwater after rainfall or irrigation (cm).

The salt brought into groundwater can be calculated using Equation (13):

$$S_L = \frac{S_O + S_I}{V_a + L} \tag{13}$$

where V_a is the moisture content of soil before rainfall or irrigation.

2.6. Shallow Groundwater Utilization and Salt Accumulation Rate

The simulation was carried out under the two conditions, i.e., the bare land and land with crops. Under the two conditions, the corresponding underground displacements were D_1 and D_2, evapotranspirations were ET_1 and ET_2, infiltrations were F_1 and F_2, deep percolation were L_1 and L_2, and groundwater levels were h_1 and h_2 (usually $h_1 < h_2$). The difference in the profile water quantity between bare land and land with planted crops was mainly caused by the utilization of groundwater by crops. Therefore, by calculating the difference in water balance between the two conditions, the amount of shallow groundwater used by crops can be obtained using Equation (14).

$$G = (D_1 - D_2) + (ET_1 - ET_2) + (F_2 - F_1) + (L_2 - L_1) + (h_2 - h_1)\theta_s \tag{14}$$

In the equation, θ_s is the saturated water content ($cm^3 cm^{-3}$). Equation (15) can be used to obtain the total amount of salt carried by crops using groundwater, and Equation

(16) can be used to calculate the salt accumulation rate of soil, and the salt accumulation degree (soil profile at a certain time and the rate of increase in soil salt content compared to the previous period) of 0–100 cm soil layer in each test plot is analyzed without being affected by the initial salt content.

$$S_G = GC_G \tag{15}$$

Here, C_G is the average salt concentration of groundwater.

$$t = \frac{S_i - S_{i-1}}{S_i} \times 100\% \tag{16}$$

In the equation, t is soil salt accumulation rate (%); S_i is the soil salt content in period i (g/kg); and S_{i-1} is the soil salt content in period $i-1$ (g/kg).

2.7. Calibration and Verification of the Models and Fitting Equations

The model was evaluated using graphical and statistical methods. The mean absolute error (MAE), root mean square error ($RMSE$) and *Nash–Sutcliffe* efficiency coefficient (NSE) were used to evaluate the consistency between the measured and the predicted data. When MAE and $RMSE$ are close to 0, NSE is close to 1, indicating that the simulation is accurate. When daily $NSE > 0.4$ or monthly $NSE > 0.5$, the simulation is considered acceptable; when daily $NSE > 0.6$ or monthly $NSE > 0.7$, the simulation is considered good; when daily $NSE > 0.75$ or monthly $NSE > 0.8$, the simulation is considered excellent [22].

$$MAE = \frac{\sum_{i=1}^{n}|A_i - O_i|}{n} \tag{17}$$

$$RMSE = \sqrt{\frac{1}{n}\sum_{i=1}^{n}(A_i - O_i)^2} \tag{18}$$

$$NSE = 1 - \frac{\sum_{i=1}^{n}(A_i - O_i)^2}{\sum_{i=1}^{n}(A_i - \overline{O})^2} \tag{19}$$

The coefficient of determination (R^2) was used to evaluate the regression equation.

$$R^2 = \frac{\left(\sum_{i=1}^{n}(O_i - \overline{O})(A_i - \overline{A})\right)^2}{\sum_{i=1}^{n}(O_i - \overline{O})^2 \sum_{i=1}^{n}(A_i - \overline{A})^2} \tag{20}$$

where A_i is the simulated value; $\overline{A}$ is the average observed value; O_i is the average simulated value; $\overline{O}$ is the average observed value; and n is the total number of the observations. The R^2 value is between 0 and 1, and it is acceptable when the value is larger than 0.5.

3. Results and Analysis

3.1. Changes in Soil Water Content and Salinity

The measured soil moisture content and soil salt content of the profile before and after each irrigation test of seven field-arranged experimental treatment plots during the period 2019–2020 were, respectively, provided in Figures 2 and 3. At the beginning of the experiment, there was little difference in the initial water content of each soil layer in each plot. The average soil water content of 0–100 cm soil layer was 36.85–37.89 cm^3cm^{-3}, and the soil water content of the profile gradually increased with the soil depth. After irrigation, the water content of the profile increased, and the water distribution from the top to the bottom tended to be uniform. With the increase in the buried depth or the decrease in the distance between the buried pipes, the water discharge rate from the soil accelerated gradually. With the same spacing, the average soil moisture content of 0–100 cm soil layer in the test plot after leaching decreased by 0.54 cm^3cm^{-3} for every 0.1 m increase in the burial depth. The average soil moisture content of 0–100 cm soil in the test plot increased by 0.08 cm^3cm^{-3} when the burial depth remained unchanged, and the spacing increased by 10 m. Before the irrigation test, the average salt content in 0–100 cm soil layer of each test plot was 16.93–17.56 g/kg, which gradually decreased with the burial depth, and the spatial variability of soil salt content was large in the same profile (Figure 3). After each leaching test, the soil salt content in each plot showed similar decreasing trends, but the spatial variability was still large. During the interval of two leaching tests, the salt content of the profile increased under the joint influence of surface evaporation and soil capillarity, and the most significant increase occurred in the 0–40 cm soil layer. From 4 June to 25 June 2019, the average salt content increased by 0.53–1.01 g/kg in the 0–100 cm soil layer in each plot, and the average salt content increased by 0.81–2.79 g/kg in the 0–40 cm soil layer. No leaching was carried out from November 2019 to May 2020, but the overall salt content of the soil decreased. This was because the water in the soil melted and carried part of the salt out of the soil, and the salt in the 0–40 cm layer of the surface layer significantly decreased. The soil moisture content and salt content in T_3 plot were significantly lower than those in other plots during the experiment, while those in T_7 plot were higher than those in other plots (Figures 2 and 3). Water discharge led to a more obvious reduction in soil salinity with a deeper burial depth and smaller spacing of the subsurface pipes. Moreover, the influence of subsurface pipe on the surface 0–40 cm soil layer was more significant compared to other layers, and the difference between the water content and salt content of deep soil in the subsurface pipe is small among different plots.

3.2. The Salt Content and Soil Desalting Rate under Experimental Conditions and the Influencing Factors

Linear regression was carried out between the salt concentration of the drainage in the subsurface pipe (C_D) and the soil salt content (S_o), irrigation amount (I), spacing (s) and buried depth (d) as shown in Equation (21).

$$C_D = 5.057S_o - 0.086I - 0.016s + 0.531d - 39.848 \tag{21}$$

The statistical parameters of the regression model are shown in Table 3. The soil salt content before leaching and the buried depth of pipes had a significant positive influence on the salt concentration of drainage, and the irrigation amount and spacing of the pipes had a significant negative influence on the salt concentration of drainage. After each leaching, the soil salt content varied greatly across the test plots. To compare the changes in the soil salt content, Equation (2) was used to calculate and list the N of each plot after each leaching. N was the desalting capacity of subsurface pipes under different layout conditions without considering leaching (Table 4). After the arrangement of the subsurface pipe, the soil was desalted by irrigation in each plot, and the N range was 10.83–37.94%. N decreased with the increase in pipe spacing when the burial depth of pipes was the same. With the same spacing, N increased with the increase in burial depth. However, the change in soil salt content did not show a similar trend with N. After the irrigation test on 3 May 2020, the highest N was 37.94% in T_3 plot, but the change in soil salt content was 2.36 g/kg, which was lower than that in other plots. The maximum change in soil salt content in T_4 plot was 2.65 g/kg, and N was only 23.31%. With the progress of the experiment, the salt content of the soil in the test plot decreased as a whole, it was more difficult to discharge the salt from the soil, and the leaching effect of the subsurface pipe on the soil also decreased.

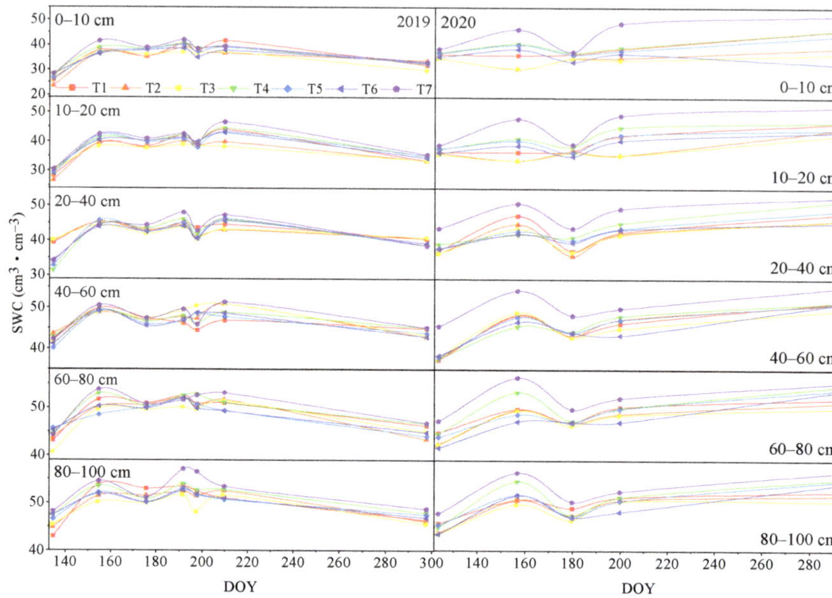

Figure 2. Changes in soil water content (SWCs) of different soil layers in each test plot in 2019 and 2020.

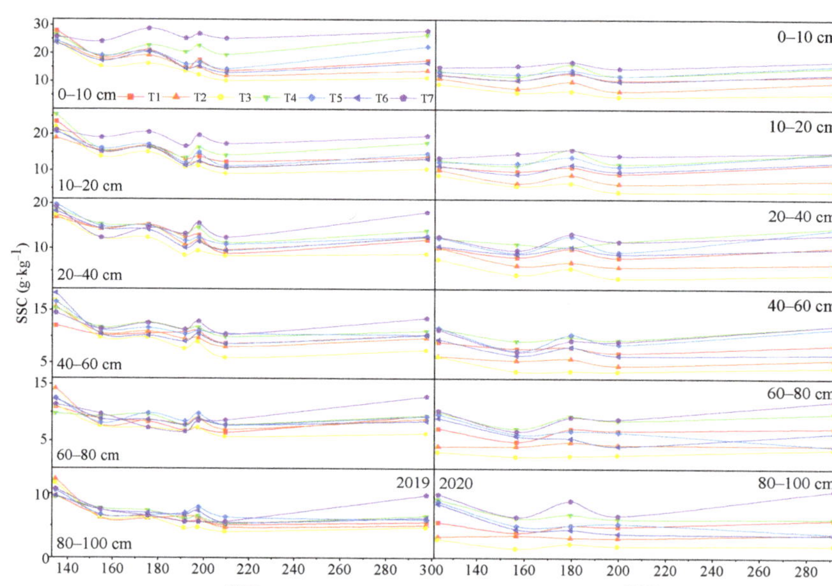

Figure 3. Changes in soil salt content (SSCs) in different soil layers in each test plot in 2019 and 2020.

Table 3. Multiple linear regression model test and parameter statistics for salt concentration in subsurface pipe drainage.

Parameter	Denormalization Coefficient	Standardization Coefficient	*VIF*	*p*	R^2	F
	B	Beta				
Constant	−39.848	-	-	0.018 *		
S_o	5.057	0.811	1.307	0.000 **		
I	−0.086	−0.344	1.070	0.000 **	0.902	F = 101.123 p = 0.000 **
s	−0.016	−0.577	1.298	0.000 **		
d	0.531	0.224	1.125	0.000 **		

Note: *, $p < 0.05$; **, $p < 0.01$.

Table 4. Change in soil salt content and desalination rate in each test plot after leaching in 2019–2020.

Treatment	Time	Soil Salt Content Change (g·kg^{-1})	Soil Desalinization Rate (%)	Time	Soil Salt Content Change (g·kg^{-1})	Soil Desalinization Rate (%)	Time	Soil Salt Content Change (g·kg^{-1})	Soil Desalinization Rate (%)
T_1		4.64	27.38		2.45	18.75		2.19	19.25
T_2		4.98	29.23		2.49	19.51		2.08	19.40
T_3		6.36	36.93		2.55	22.71		1.78	19.77
T_4	15 May 2019	4.42	25.19	26 June 2019	2.02	14.39	17 July 2019	2.14	15.86
T_5		4.65	26.77		2.55	18.89		1.94	15.95
T_6		4.85	28.32		2.89	22.58		1.85	16.74
T_7		3.05	17.88		1.93	12.81		1.61	10.83

Treatment	Time	Soil salt content change(g·kg^{-1})	Soil desalinization rate (%)	Time	Soil salt content change (g·kg^{-1})	Soil desalinization rate (%)
T_1		1.69	18.30		1.36	14.92
T_2		1.92	25.59		1.49	22.38
T_3		2.36	37.94		1.15	25.43
T_4	3 May 2020	2.09	18.04	28 June 2020	1.54	13.38
T_5		2.65	23.31		1.67	15.92
T_6		2.59	25.59		1.75	19.63
T_7		2.06	17.03		1.41	11.47

3.3. Calibration and Verification of Model Parameters

DRAINMOD model was calibrated and verified with the predicted and field measurements. For example, the groundwater level data collected in 2019 was used for model calibration, and the model parameters were adjusted within a reasonable range to minimize the difference between the predicted and measured values. In the calibration process, lateral saturated hydraulic conductivity, hydrodynamic dispersion coefficient, maximum depth of surface water, the actual distance from the surface to impermeable layer, drainage coefficient, initial groundwater depth and *PET* correction coefficient were mainly adjusted. After that, the measured groundwater level data in 2020 were used to verify the calibrated model. The results of statistical evaluation of DRAINMOD for simulating groundwater level of 7 test plots in 2019–2020 are presented in Table 5. *MAE*, *RMSE* and *NES* were, respectively, in the range of 3.51–5.83 cm, 13.78–41.64 cm and 0.57–0.81. The simulated results of groundwater level by the 2019–2020 model were analyzed; it was found that some output items were significantly different from the measured values, and they were mainly concentrated in the time period of irrigation or rainfall. This is because the DRAINMOD model is sensitive to soil water inputs. And the model assumes that the process from leaching to production and drain into the pipes is immediate, but in practice, the process from leaching to drain took more than a few hours. In addition, the soil composition of the field was complex and presented spatial variability. The basic input data of soil profile for the model could not fully reflect the actual soil condition of the whole plot, which resulted in the failure to promptly and accurately reflect the change in the groundwater level in individual periods. However, the model performed well on the trend of the groundwater level in annual time series. In a word, the whole simulation results were overall acceptable. The water-related outputs were used to calculate the salinity of the plot profile.

Table 5. Evaluation of model simulation accuracy.

Treatment	2019 *MAE* (cm)	2019 *RMSE* (cm)	2019 *NSE*	2020 *MAE* (cm)	2020 *RMSE* (cm)	2020 *NSE*
T_1	4.11	21.65	0.76	3.82	16.01	0.81
T_2	4.37	20.35	0.71	3.51	13.78	0.70
T_3	5.07	28.29	0.69	4.54	22.98	0.78
T_4	4.74	24.61	0.62	3.85	16.37	0.80
T_5	5.25	29.51	0.64	4.16	18.98	0.69
T_6	5.83	38.32	0.57	4.77	24.97	0.75
T_7	6.15	41.64	0.60	5.01	36.90	0.69

Note: *NSE*, *Nash–Sutcliffe* efficiency coefficient; *MAE*, mean absolute error; *RMSE*, root mean square error.

3.4. Soil Salts Simulation with DRAINMOD Model

The salt content of soil in the 0–100 cm layer in 2019 and 2020 was simulated and compared with the measured results (Figure 4 and Table 6). The *MAE*, *RMSE* and *NSE* of soil salt content in 2019 were 0.32–1.36 g/kg, 0.40–1.41 g/kg and 0.41–0.89, respectively. The *MAE*, *RMSE* and *NSE* of soil salt content in 2020 were 0.24–1.03 g/kg, 0.30–1.11 g/kg and 0.40–0.92, respectively. From the results of statistical evaluation, it can be seen that the model is more accurate to calculate the salt of the existing test plot, and it was feasible to use this method to simulate the change in the salt profile under different layouts of the subsurface pipe, which can be used to improve the layout of the subsurface pipe. According to the simulation, the salt accumulated during each leaching interval. Since there was no utilization of shallow groundwater, the salt increase only caused via surface evaporation. In addition, the *ET* results for each plot obtained via the model during the same period were not significantly different, so the accumulated salt values of each plot had a small difference. After two years of leaching, the soil salt content decreased significantly. The leaching effect of T_3 plot was the best with an *N* of 77.12%, while the leaching effect of T_7 plot was the worst with an *N* of 36.63%. In 2019, the desalting effect was more significant than that in 2020. After four cycles of leaching in 2019, the *N* value of each plot was 33.28–67.00% from the beginning of the experiment to the end of the year, and after three cycles of leaching from the beginning of 2020 to the end of the year, the *N* of each plot was 3.48–30.70%. With the decrease in soil salt content, the *N* of each plot was also decreasing gradually. When the soil salt content decreased to a certain extent, the leaching effect of the plots with shallow burial depth and large spacing of the subsurface pipe arrangement was no longer obvious, and the *N* of T_4 and T_7 plots in the second year was only 3.48% and 5.03%. This provides a direction for the determination of buried pipe parameters and leaching scheme.

Table 6. Evaluation of simulation accuracy of soil salt content.

Treatment	2019 *MAE* ($g \cdot kg^{-1}$)	2019 *RMSE* ($g \cdot kg^{-1}$)	2019 *NSE*	2020 *MAE* ($g \cdot kg^{-1}$)	2020 *RMSE* ($g \cdot kg^{-1}$)	2020 *NSE*
T_1	0.32	0.41	0.89	1.03	1.11	0.46
T_2	0.78	0.82	0.70	0.29	0.42	0.84
T_3	1.36	1.41	0.49	0.85	1.03	0.40
T_4	0.34	0.40	0.86	0.24	0.30	0.92
T_5	0.81	0.84	0.56	0.28	0.35	0.87
T_6	1.06	1.09	0.41	0.63	0.73	0.61
T_7	0.70	0.84	0.47	0.51	0.59	0.66

Note: *NSE*, *Nash–Sutcliffe* efficiency coefficient; *MAE*, mean absolute error; *RMSE*, root mean square error.

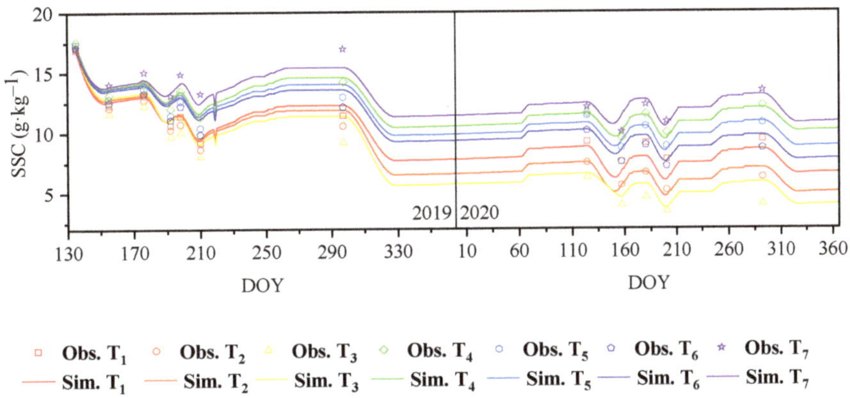

Figure 4. Simulated and measured average soil salt content (SSCs) in the 0–100 cm soil layer in each test plot in 2019 and 2020.

3.5. Soil Salt Content Simulation in Existing Plot and Predicted Plot after Sunflower Planting

The salt content changes in the 0–100 cm soil profile in the subsurface pipe control area after sunflowers planting in the existing and predicted test plots during the period 2019–2020 were simulated, respectively, and the simulation results are shown in Figure 5. The simulation showed that under the influence of shallow groundwater utilization and evapotranspiration, during the growth period of sunflowers over two years, each plot experienced a period when the soil salt content was higher than its threshold. In 2019, the days of soil salt content being higher than the threshold of salt tolerance accounted for 13.33–83.81% of the total growth period, and in 2020, the proportion was 0–93.33%. In the first year of leaching, because of the higher soil salt content, all the plots could reduce the yield of sunflower or fail to meet the growing conditions of sunflower. In the second year of leaching, the soil salt content of most plots decreased, and the time below the salt tolerance threshold of sunflowers increased. The partial test plot kept the soil salt content below the threshold of sunflower salt tolerance during the growth period, and it had no influence on the yield of sunflower. However, the phenomenon of salt accumulation in the soil in individual plots became increasingly serious, and it was difficult to meet the requirement of sunflower growth, especially in F_5 plots; the ratio of time when the salt content in the soil was above the threshold of salt tolerance to the total growth period of sunflowers increased from 83.81% to 93.33%, and sunflowers could not grow well from the seedling stage. On the one hand, these phenomena are related to the desalting capacity of each plot during leaching. On the other hand, they are related to the groundwater level caused by different layouts pipes, and the groundwater level of each plot also changed with different layouts of pipes. The deeper the subsurface pipe is, the smaller the spacing is, and the deeper the groundwater level is during the growth of sunflower. During the irrigation period, the utilization of groundwater by sunflower decreased, and the salt above the underground water entering the subsurface pipe also decreased. Therefore, the soil salt content in the experimental plot was kept below the salt tolerance threshold during the growth of sunflower. In conclusion, the normal growth conditions of sunflowers could not be satisfied after only one year of leaching. In the second year, sunflower could be successfully planted in some plots without reducing production. In the shallower and more interspaced plots, when sunflower used groundwater, the soil salt content did not decrease, and serious salt accumulation occurred. Therefore, in the improvement of the heavily saline–alkali land in Hetao Irrigation District, the layout parameters of the subsurface pipe with deeper burial depth and smaller spacing should be selected, which can not only achieve good salt leaching effect in the irrigation process, but also have a significant inhibition effect on soil salt accumulation during the non-leaching period, so that crops can grow normally, and the yield will not be affected.

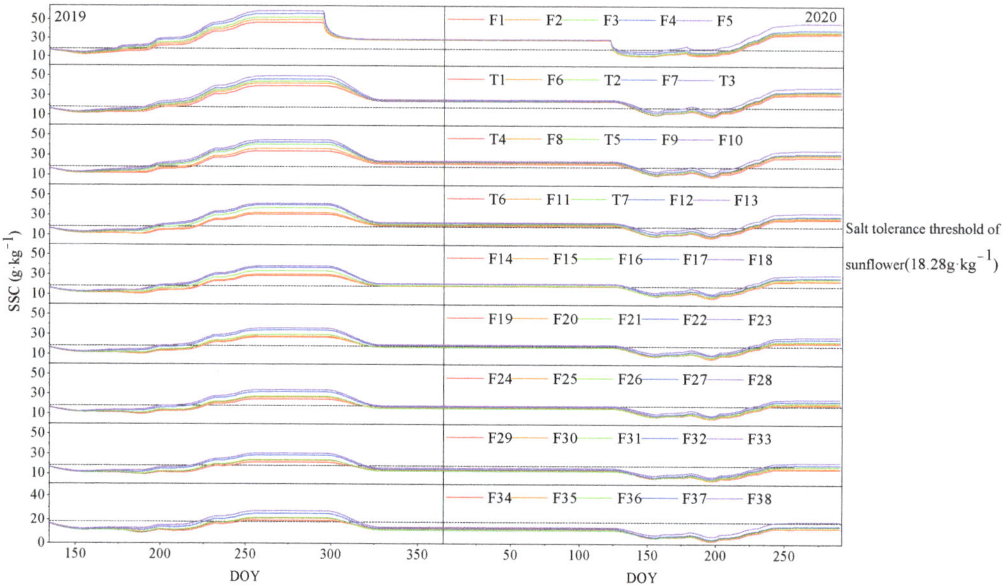

Figure 5. Simulation of average soil salt content (SSCs) in 0–100 cm soil layer after planting sunflower in existing and predicted test plots in 2019 and 2020.

3.6. Analysis of Rational Layouts of Pipes

With the data of the measured and predicted drainage and salt discharge by the subsurface pipe, *N* was correlated with S_o, *s*, *d* and *I*, *SCR* was correlated with *s*, *d* and annual soil desalting rate (N_t) of the subsurface pipe, and a clear quantitative relationship was established (Table 7 and Figure 6).

Table 7. Multiple linear regression model test and parameter statistics of soil desalination rate under sunflower planting conditions.

Parameter	Denormalization Coefficient	Standardization Coefficient	*VIF*	*p*	R^2	*F*
	B	Beta				
Constant	−61.548	-	-	0.000 **	0.911	*F* = 678.203 *p* = 0.000 **
S_o	0.390	0.157	1.481	0.000 **		
I	0.254	0.868	1.228	0.000 **		
s	−0.004	−0.121	1.041	0.000 **		
d	0.151	0.209	1.212	0.000 **		

Note: **, $p < 0.01$.

According to the above research results, the buried depth of the subsurface pipe should be as deep as possible, so the buried depth of the subsurface pipe is set as the maximum excavation depth of the machine used, namely 210 cm, which is also in the recommended critical depth range for groundwater control in arid areas [34]. In order to meet different requirements of salt drainage and salt control, the optimal buried space and desalting water volume of the pipes can be obtained using the fitting Equations (22) and (23), based on the initial salt content of this experiment (16.93 g/kg) and the buried depth of the subsurface pipe set to 210 cm, in order to keep 100% *SCR* of the soil in the control area during the sunflower growth period. The aim is to remove 60%, 70%, 80%, 90% and 100% soil salt from 100 cm of soil. According to the calculation, the buried spacing of the subsurface pipe should be set as 562 cm, 902 cm, 1241 cm, 1580 cm and 1919 cm, respectively, and the corresponding optimal irrigation water volume is 33.65 cm, 38.13 cm, 42.60 cm, 47.07 cm and 51.54 cm. When the target *N* is 100%, the required water flow is 1.53 times of that when the target *N* is 60%, but the number of subsurface pipes buried is only nearly 1/3 of that when the target *N* is 60%. In conclusion, when soil desalting target (*N*) is low, less water is required for irrigation, the pipe spacing can be increased, and the number of subsurface pipes buried to meet the salt control requirements in the growth period of sunflowers can be increased. When soil desalting target (*N*) is high, the amount of irrigation water increases, the number of buried pipes decreases, and the buried cost of subsurface pipes in the test area decreases.

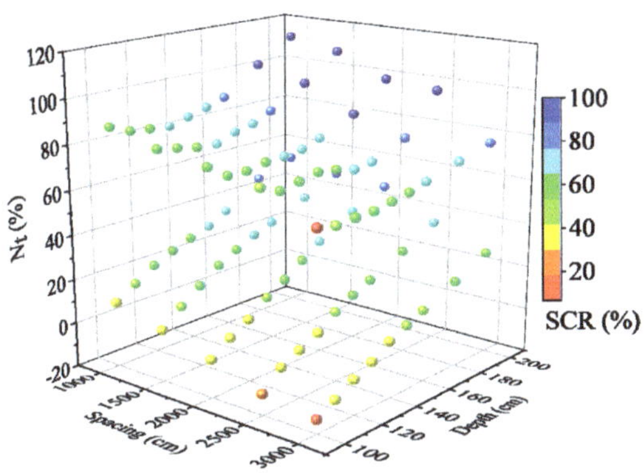

Figure 6. Relationship between salt control rate and burial depth, spacing and sum of soil desalination rate during the year.

According to the results, deep-buried subsurface pipes are beneficial to improve *N*. On the premise of satisfying the desalting effect, the construction cost can be reduced, and the groundwater level in the control area of pipe can be reduced. Thus, soil salinity in the root layer of the subsurface pipe control area was kept at a relatively low level, which could meet the soil conditions for crop growth. According to the initial soil salt content and construction conditions in the improved area, the optimal subsurface pipe layout parameters suitable for this area were obtained using Equations (22) and (23), Equation (23) R^2 is 0.906.

$$N = 0.390S_o + 0.254I - 0.004s + 0.151d - 61.548 \quad (22)$$

$$SCR = 0.375d - 0.007s + 0.240N_t + 10.790 \quad (23)$$

where *N* is the soil desalting rate (%); S_O is the soil salt content before leaching (g), *I* is the irrigation quantity (cm); *s* is the spacing of subsurface pipes (cm); *d* is the buried depth of subsurface pipes (cm); *SCR* is the salt control rate (%); and N_t is the total soil desalting rate in the year (%).

4. Discussion

4.1. Effects of Subsurface Pipe Drainage on Soil Moisture and Salinity

As an important drainage management measure in farmland, most studies showed that buried subsurface drainage has a great influence on the migration of soil water and salt [13,18,43]. With the increase in buried depth or the decrease in spacing, the water discharge rate of soil accelerated. Compared with the soil water content before irrigation, the soil water content after irrigation was distributed more evenly from top to bottom, and the average soil water content decreased by 0.75 cm^3cm^{-3} compared with that before irrigation. This is consistent with the results of Wiskow and van der Ploeg [44]. With the increase in the water displacement, the amount of salt discharged from the subsurface pipe also increases [45], and the salt concentration and N of the water discharged from the subsurface pipe are negatively correlated with the spacing, but positively correlated with the buried depth. The influence of irrigation on the 0–40 cm soil layer of different plots is greater, but the influence on the water content and salt content of the deep soil is less.

In arid and semi-arid areas where evaporation is large and groundwater level is shallow, the salt in the bottom layer will accumulate to the surface with water due to evaporation and crop transpiration after irrigation, and there is still a risk of salinization even after irrigation [46]. However, there are relatively few studies on salt accumulation in subsurface pipe test plots after irrigation. This study found that plots with good performance in irrigation can also play an effective role in inhibiting salt during non-irrigation, especially plots with the same spacing, and a large burial depth can control soil salt below the threshold of crop salt during the crop growth period. This indicates that the buried depth of the subsurface pipe has more obvious influence on the water and salt transport in soil profile than the spacing in saline–alkali soil improvement.

4.2. The Influence of Field Conditions on the Layout of Pipes

Under different soil and water environment of farmland and research objectives, the recommended layout parameters of concealed pipe will be different. When the precipitation of the study area is large and the degree of salinization is light, a shallow buried depth of the subsurface pipe is generally recommended [47,48]. In the arid and semi-arid irrigated area where this study is located, rainfall is low, and soil salinization degree is serious. In order to achieve effective salt leaching effect, it is urgent to increase the drainage and drainage amount of subsurface pipe to reduce soil salt. The results show that the buried depth of the subsurface pipe with better desalting effect in the test plot is above 1.8 m, and the soil salt content in the plot with a buried depth of 1 m increases rather than decreases during the two-year irrigation test. Therefore, based on the results of this study, it is recommended to bury the buried pipe with a depth greater than 1.8 m in saline–alkali soil improvement under current conditions. This is consistent with the findings of Qian et al., whose research showed that N is close to 0 when the buried depth of the pipe is less than 1 m [34].

Along with the increase in buried depth of subsurface pipes, it also leads to the loss of water and nutrients in soil to a certain extent, which will lead to eutrophication of drainage zone [49,50]. In order to solve this problem, on the basis of the single-layer drainage arrangement, multiple drainage methods such as double-layer drainage, controlled drainage and the combination of concealed pipe and shaft drainage were developed, which not only improved irrigation efficiency and alleviated drought stress, but also reduced nutrient loss [23,51]. However, there are few related research results of the above layout of subsurface pipe drainage, so there are certain limitations in the promotion. There are many research results on the arrangement of single-layer subsurface pipes, and the arrangement of single-layer subsurface pipe has better applicability. In this study, soil salinization can be successfully reduced, and crops can grow normally after two years of irrigation under the single-layer pipe arrangement. In order to avoid the negative effects of drought threat and nutrient loss caused by deep-buried pipes, based on the current research results, various other improvement methods can be considered, such as adding organic fertilizer and straw biochar in the field [32] or laying low permeability mulching film [52]. The improvement on the basis of not changing the layout of subsurface pipe provides a solution for the farmland that has laid subsurface pipes and encountered the above problems. The selection of the layout parameters of the subsurface pipe should be the target of the next research.

5. Conclusions

In the arid and semi-arid region of Hetao irrigation district, Inner Mongolia, the effects of subsurface pipe layout on soil salinity were studied via field experiment and numerical simulation. The smaller the spacing and the deeper the buried pipe, the more beneficial the drainage of water and salt in soil. The influence of buried pipes on the water content and salt content of 0–40 cm surface soil is significant, and the difference of deep soil leaching in different plots is small. As the salt content of soil decreases, the more difficult it is for the salt to discharge from the soil, and the leaching effect also decreases. Based on the consideration of salt drainage and salt control, if the field conditions allow, the buried pipe should be the main buried pipe in the improvement of arid saline–alkali land. To make the results more convincing, the DRAINMOD model and drainmod equation were used to simulate the water and salt migration in soil. The results showed that *MAE*, *RMSE* and *NSE* were 0.24–1.36 g/kg, 0.30–1.41 g/kg and 0.40–0.92, respectively. The proposed method could accurately simulate the dynamic changes in soil water and salt content in the period 2019–2020. This is very beneficial to the improvement of saline–alkali land in arid and semi-arid areas. By using the calibrated model combined with the calculation of the water and salt profile, we can obtain the most suitable local concealed pipe layout parameters and the best irrigation water amount under different salt discharge and salt control objectives, which reduces the cost and greatly improves the work efficiency of the concealed pipe improvement of saline–alkali soil. The calculated pipe layout parameters can control the soil salt content below the threshold of crop salt tolerance during the crop growth period, which provides technical support for the sustainable development of farmland.

Author Contributions: F.T.: Formal analysis, Investigation, Data Curation, Formal Analysis, Visualization and Writing—Original Draft; Q.M.: Formal analysis and Investigation; H.S.: Conceptualization, Methodology, Validation, Formal analysis, Resources, Data Curation, Writing—Review and Editing, Visualization, Supervision and Project administration; R.L.: Investigation; X.D.: Investigation; J.D.: Investigation; J.L.: Writing—Review and Editing; W.F.: Writing—Review and Editing, Visualization, Supervision and funding. All authors have read and agreed to the published version of the manuscript.

Funding: The work was financially supported by the Science and Technology Major Projects of Inner Mongolia (zdzx2018059), the National Natural Science Foundation of China (52269014 and 52009056), and the Research Program of Science and Technology at Universities of lnner Mongolia Autonomous Region, China (2022YFHH0044).

Data Availability Statement: The data already exist in the manuscript.

Conflicts of Interest: The authors declare that they have no known competing financial interests or personal relationships that could have appeared to influence the work reported in this paper.

References

1. Liu, Y.; Ao, C.; Zeng, W.; Kumar Srivastava, A.; Gaiser, T.; Wu, J. Simulating water and salt transport in subsurface pipe drainage systems with HYDRUS-2D. *J. Hydrol.* **2021**, *592*, 125823. [CrossRef]
2. Dregne, H.E. Land degradation in the drylands. *Arid Land Res. Manag.* **2002**, *16*, 99–132. [CrossRef]
3. Singh, A. Soil salinization and waterlogging: A threat to environment and agricultural sustainability. *Ecol. Indic.* **2015**, *57*, 128–130. [CrossRef]
4. Li, P.Y.; Wu, J.H.; Qian, H. Regulation of secondary soil salinization in semi-arid regions: A simulation research in the Nanshantaizi area along the Silk Road, northwest China. *Environ. Earth Sci.* **2016**, *75*, 698. [CrossRef]
5. Feng, W.Y.; Wang, T.K.; Zhu, Y.R.; Sun, F.H.; Giesy, J.P.; Wu, F.C. Chemical composition, sources, and ecological effect of organic phosphorus in water ecosystems: A review. *Carbon Res.* **2023**, *2*, 12. [CrossRef]
6. Asfaw, E.; Suryabhagavan, K.V.; Argaw, M. Soil salinity modeling and mapping using remote sensing and GIS: The case of Wonji sugar cane irrigation farm, Ethiopia. *J. Saudi Soc. Agric. Sci.* **2018**, *17*, 250–258. [CrossRef]
7. Li, J.; Pu, L.; Han, M.; Zhu, M.; Zhang, R.; Xiang, Y. Soil salinization research in China: Advances and prospects. *J. Geogr. Sci.* **2014**, *24*, 943–960. [CrossRef]
8. Dou, X.; Shi, H.B.; Li, R.P.; Miao, Q.F.; Tian, F.; Yu, D.D. Distribution characteristics of salinity and nutrients in salinized soil profile and estimation of salt migration. *Trans. Chin. Soc. Agric.* **2022**, *53*, 279–290+330.
9. Wang, Z. *Salt Movement Trends in Cotton Fields with Long-Term Drip Irrigation under Mulch in Typical Oasis and Irrigation Management*; China Agricultural University: Beijing, China, 2014.
10. Haj-Amor, Z.; Bouri, S. Subsurface drainage system performance, soil salinization risk, and shallow groundwater dynamic under irrigation practice in an arid land. *Arabian J. Sci. Eng.* **2019**, *44*, 467–477. [CrossRef]

11. Ritzema, H.; Chultz, B. Optimizing subsurface drainage practices in irrigated agriculture in the semi-arid and arid regions: Experience from Egypt, India AND Pakistan. *Irrig. Drain.* **2011**, *60*, 360–369. [CrossRef]
12. Sharma, D.P.; Gupta, S.K. Subsurface drainage for reversing degradation of waterlogged saline lands. *Land Degrad. Dev.* **2010**, *17*, 605–614. [CrossRef]
13. He, X.L.; Liu, H.G.; Ye, J.Y.; Yang, G.; Li, M.S.; Gong, P. Comparative investigation on soil salinity leaching under subsurface drainage and ditch drainage in Xinjiang arid region. *Int. J. Agric. Biol. Eng.* **2016**, *9*, 109–118.
14. Hornbuckle, J.W.; Christen, E.W.; Faulkner, R.D. Evaluating a multi-level subsurface drainage system for improved drainage water quality. *Agric. Water Manag.* **2007**, *89*, 208–216. [CrossRef]
15. Ceuppens, J.; Wopereis, M.; Miézan, K.M. Soil salinization processes in rice irrigation schemes in the Senegal river delta. *Soil Sci. Soc. Am. J.* **1997**, *61*, 1122–1130. [CrossRef]
16. Salo, H.; Mellin, I.; Sikkilä, M.; Nurminen, J.; Äijö, H.; Paasonen-Kivekäs, M. Performance of subsurface drainage implemented with trencher and trenchless machineries. *Agric. Water Manag.* **2019**, *213*, 957–967. [CrossRef]
17. Buckland, G.D.; Bennett, D.R.; Mikalson, D.E.; Jong, E.D.; Chang, C. Soil salinization and sodication from alternate irrigations with saline-sodic water and simulated rain. *Can. J. Soil Sci.* **2002**, *82*, 297–309. [CrossRef]
18. Skaggs, R.W.; Brevé, M.A.; Gilliam, J.W. Hydrologic and water quality impacts of agricultural drainage. *Crit. Rev. Environ. Sci. Technol.* **1994**, *24*, 1–32.
19. Nozari, H.; Azadi, S.; Zali, A. Experimental study of the temporal variation of drain water salinity at different drain depths and spacing in the presence of saline groundwater. *Sustain. Water Resour. Manag.* **2018**, *4*, 887–895.
20. Šimůnek, J.; van Genuchten, M.T.; Šejna, M. Recent developments and applications of the HYDRUS computer software packages. *Vadose Zone J.* **2016**, *15*, 25. [CrossRef]
21. Bailey, R.T.; Tavakoli-Kivi, S.; Wei, X. A salinity module for SWAT to simulate salt ion fate and transport at the watershed scale. *Hydrol. Earth Syst. Sci.* **2019**, *23*, 3155–3174. [CrossRef]
22. Skaggs, R.W.; Youssef, M.A.; Chescheir, G.M. DRAINMOD: Model use, calibration, and validation. *Trans. ASABE* **2012**, *55*, 1509–1522. [CrossRef]
23. Dou, X.; Shi, H.B.; Li, R.P.; Miao, Q.F.; Yan, J.W.; Tian, F. Simulation and evaluation of soil water and salt transport under controlled subsurface drainage using HYDRUS-2D model. *Agric. Water Manag.* **2023**, *273*, 107899. [CrossRef]
24. Addab, H.; Bailey, R.T. Simulating the effect of subsurface tile drainage on watershed salinity using SWAT. *Agric. Water Manag.* **2022**, 262. [CrossRef]
25. Hosseini, P.; Bailey, R.T. Investigating the controlling factors on salinity in soil, groundwater, and river water in a semi-arid agricultural watershed using SWAT-Salt. *Sci. Total Environ.* **2022**, *810*, 152293.
26. Luo, W.; Skaggs, R.; Madani, A.; Cizikci, S.; Mavi, A. Predicting field hydrology in cold conditions with drainmod. Transactions of the ASAE. *Trans. ASAE Soc. Agric.Eng.* **2001**, *44*, 825–834.
27. Youssef, M.A.; Skaggs, R.W.; Chescheir, G.M.; Gilliam, J.W. The nitrogen simulation model, DRAINMOD-N II. *Trans. ASABE* **2005**, *48*, 611–626. [CrossRef]
28. Askar, M.H.; Youssef, M.A.; Chescheir, G.M.; Negm, L.M.; King, K.W.; Hesterberg, D.L. DRAINMOD Simulation of macropore flow at subsurface drained agricultural fields: Model modification and field testing. *Agric. Water Manag.* **2020**, *242*, 106401. [CrossRef]
29. Moursi, H.; Youssef, M.A.; Chescheir, G.M. Development and application of DRAINMOD model for simulating crop yield and water conservation benefits of drainage water recycling. *Agric. Water Manag.* **2022**, *266*, 107592.
30. Ren, D.; Huang, G.; Huang, Q.; Ramos, T.B.; Xu, X.; Huo, Z. Modeling and assessing the function and sustainability of natural patches in salt-affected agro-ecosystems: Application to tamarisk (Tamarix chinensis lour.) in Hetao, upper Yellow River basin. *J. Hydrol.* **2017**, *552*, 490–504. [CrossRef]
31. Cao, Z.D.; Zhu, T.J.; Cai, X.M. Hydro-agro-economic optimization for irrigated farming in an arid region: The Hetao Irrigation District, Inner Mongolia. *Agric. Water Manag.* **2023**, *277*, 108095. [CrossRef]
32. Feng, W.Y.; Yang, F.; Cen, R.; Liu, J.; Qu, Z.Y.; Miao, Q.F. Effects of straw biochar application on soil temperature, available nitrogen and growth of corn. *J. Environ. Manag.* **2021**, *277*, 111331. [CrossRef]
33. Feng, Z.Z.; Miao, Q.F.; Shi, H.B.; Feng, W.Y.; Li, X.Y.; Yan, J.W. Simulation of water balance and irrigation strategy of typical sand-layered farmland in the Hetao Irrigation District, China. *Agric. Water Manag.* **2023**, *280*, 108236. [CrossRef]
34. Qian, Y.Z.; Zhu, Y.; Ye, M.; Huang, J.S.; and Wu, J.W. Experiment and numerical simulation for designing layout parameters of subsurface drainage pipes in arid agricultural areas. *Agric. Water Manag.* **2021**, *243*, 106455. [CrossRef]
35. Tong, W.; Chen, X.; Wen, X.; Chen, F.; Zhang, H.; Chu, Q. Applying a salinity response function and zoning saline land for three field crops: A case study in the Hetao Irrigation District, Inner Mongolia, China. *J. Integr. Agric.* **2015**, *14*, 178–189. [CrossRef]
36. Bouwer, H.; van Schilfgaarde, J. Simplified method of predicting fall of water table in drained land. *J. Water Resour. Prot.* **1963**, *11*, 0288–0291.
37. Kirkham, D. The Ponded Water Case. In *Drainage of Agricultural Lands*; Luthin, J.N., Ed.; American Society of Agronomy: Madison, WI, USA, 1957; pp. 139–181.
38. Minhas, P.S.; Ramos, T.B.; Ben-Gal, A.; Pereira, L.S. Coping with salinity in irrigated agriculture: Crop evapotranspiration and water management issues. *Agric. Water Manag.* **2020**, *227*, 105832.

39. Liu, M.H.; Shi, H.B.; Paredes, P.; Ramos, T.B.; Dai, L.P.; Feng, Z.Z.; Pereira, L.S. Estimating and partitioning maize evapotranspiration as affected by salinity using weighing lysimeters and the SIMDualKc model. region. *Agric. Water Manag.* **2022**, *261*, 107362. [CrossRef]
40. Miao, Q.F.; Rosa, R.D.; Shi, H.B.; Paredes, P.; Zhu, L.; Dai, J.X. Modeling water use, transpiration and soil evaporation of spring wheat–maize and spring wheat–sunflower relay intercropping using the dual crop coefficient approach. *Agric. Water Manag.* **2016**, *165*, 211–229. [CrossRef]
41. Rosa, R.D.; Paredes, P.; Rodrigues, G.C.; Alves, I.; Fernando, R.M.; Pereira, L.S.; Allen, R.G. Implementing the dual crop coefficient approach in interactive software. 1. Backgr. *Comput. Strategy Agric. Water Manag.* **2012**, *103*, 8–24. [CrossRef]
42. Zhang, W.Z.; Zhang, Y.F. The specific yield and pore ratio of soils. *J. Irrig. Drain.* **1983**, *2*, 1–16.
43. Luo, W.; Sands, G.R.; Youssef, M.; Strock, J.S.; Song, I.; Canelon, D. Modeling the impact of alternative drainage practices in the northern Corn-belt with DRAINMOD-NII. *Agric. Water Manag.* **2010**, *97*, 389–398. [CrossRef]
44. Wiskow, E.; van der Ploeg, R.R. Calculation of drain spacings for optimal rainstorm flood control. *J. Hydrol.* **2003**, *272*, 163–174. [CrossRef]
45. Jafari-Talukolaee, M.; Shahnazari, A.; Ahmadi, Z.M.; Darzi-Naftchali, A. Drain discharge and salt load in response to subsurface drain depth and spacing in paddy fields. *J. Irrig. Drain. Eng.* **2015**, *141*, 1–6. [CrossRef]
46. Geng, X.; Boufadel, M.C. Numerical modeling of water flow and salt transport in bare saline soil subjected to evaporation. *J. Hydrol.* **2015**, *524*, 427–438. [CrossRef]
47. Ghane, E.; Askar, M.H. Predicting the effect of drain depth on profitability and hydrology of subsurface drainage systems across the eastern USA. *Agric. Water Manag.* **2021**, *258*, 107072. [CrossRef]
48. Darzi-Naftchali, A.; Motevali, A.; Keikha, M. The life cycle assessment of subsurface drainage performance under rice-canola cropping system. *Agric. Water Manag.* **2022**, *266*, 107579. [CrossRef]
49. Sands, G.R.; Song, I.; Busman, L.M.; Hansen, B.J. The effects of subsurface drainage depth and intensity on nitrate loads in the northern cornbelt. *Trans. ASABE* **2008**, *51*, 937–946. [CrossRef]
50. Krejcová, J.; Vicentini, F.; Flynn, T.; Mudrák, O.; Frouz, J. Biodiversity loss caused by subsurface pipe drainage is difficult to restore. *Ecol. Eng.* **2021**, *170*, 106336. [CrossRef]
51. Schott, L.; Lagzdins, A.; Daigh, A.L.M.; Craft, K.; Pederson, C.; Brenneman, G. Drainage water management effects over five years on water tables, drainage, and yields in southeast Iowa. *J. Soil Water Conserv.* **2017**, *72*, 251–259. [CrossRef]
52. Zhang, J.X.; Werner, A.D.; Lu, C.H. Improving salt leaching efficiency of subsurface drainage systems using low-permeability surface mulch. *Adv. Water Resour.* **2022**, *162*, 104147. [CrossRef]

water

MDPI

Communication

Synergistic Piezo-Catalytic Inactivation of Bacteria by Dual-Frequency Ultrasound (120 + 1700 kHz) Using Persulfate and ZnO Nano- and Microparticles

Irina Tsenter [1], Elena Kobunova [1,2], Galina Matafonova [1,*] and Valeriy Batoev [1]

[1] Laboratory of Engineering Ecology, Baikal Institute of Nature Management SB RAS, 670047 Ulan-Ude, Russia
[2] Chemistry Faculty, Buryat State University, 670000 Ulan-Ude, Russia
* Correspondence: ngal@binm.ru

Abstract: Dual-frequency ultrasound (DFUS) coupled with sonocatalysts has emerged to be an advanced tool for antimicrobial applications in medicine but remains scarcely studied for water disinfection. In the present work, we first integrated high-frequency DFUS (120 + 1700 kHz), persulfate ($S_2O_8^{2-}$) and ZnO nano- (50 nm) and microparticles (1 μm) for eradicating *Escherichia coli* and *Enterococcus faecalis* in synthetic water. For *E. coli*, the efficiency of DFUS-based processes can be ranked as follows: DFUS < DFUS/ZnO < DFUS/$S_2O_8^{2-}$ < DFUS/ZnO/$S_2O_8^{2-}$. A similar efficiency of the DFUS/$S_2O_8^{2-}$ and DFUS/ZnO/$S_2O_8^{2-}$ processes was found for more resistant *E. faecalis*. In the absence of persulfate, the performance of 1 μm ZnO was higher than that observed with 50 nm for inactivating *E. coli* via the DFUS/ZnO and 1700 kHz/ZnO processes. A synergy of DFUS in terms of 5-log (total) reduction was found in the $S_2O_8^{2-}$/ZnO-based systems, being higher for *E. faecalis* (synergistic coefficient = 1.8–3.0). The synergistic effect was proposed to be driven by the boosted generation of reactive oxygen species and sonoporation. This study opens prospects for the development of novel DFUS-based piezo-catalytic systems for efficient water disinfection.

Keywords: dual-frequency ultrasound; 120 kHz; 1700 kHz; synergistic effect; zinc oxide; persulfate; piezo-catalysis; microbial inactivation

Citation: Tsenter, I.; Kobunova, E.; Matafonova, G.; Batoev, V. Synergistic Piezo-Catalytic Inactivation of Bacteria by Dual-Frequency Ultrasound (120 + 1700 kHz) Using Persulfate and ZnO Nano- and Microparticles. *Water* **2023**, *15*, 2937. https://doi.org/10.3390/w15162937

Academic Editors: Andrea G. Capodaglio, Weiying Feng, Fang Yang and Jing Liu

Received: 7 July 2023
Revised: 4 August 2023
Accepted: 10 August 2023
Published: 15 August 2023

1. Introduction

Environmentally safe, effective and rapid water disinfection remains an up-to-date problem in view of limited drinking water supply in many parts of the world. Ultrasonication is known as an efficient green method for inactivating pathogenic microorganisms in water using reactive oxygen species (ROS), such as hydroxyl radicals (•OH). These are produced from the collapse of microbubbles during acoustic cavitation and the sonoactivated catalysts/oxidants if used. The latter refers to ultrasound-based advanced oxidation processes (US-AOPs), such as semiconductor sonocatalysis and sonophotocatalysis, which are regarded as powerful tools for water treatment and disinfection due to boosted ROS generation and hence shortened exposure times and reduced energy requirements [1–5]. It is important to emphasize that previous research on microbial inactivation in aqueous media by US-AOPs dealt mainly with single-frequency ultrasound.

The other way of improving the performance of ultrasonication is simultaneous exposure to two or more frequencies. This may lead to synergy in terms of increased inactivation efficiency (rate), which is higher than the sum of that obtained under single frequencies. In our review, dual-frequency ultrasonication coupled with catalysts/oxidants, i.e., DFUS-AOPs, has been identified as a promising strategy for water disinfection [6]. Unlike the degradation of organic contaminants, the inactivation of microbial pathogens in water by DFUS-AOPs is still poorly studied. So far, DFUS at low frequencies (≤100 kHz) was employed for water disinfection in pairs of 16 + 20 kHz [7], 17 + 33 and 70 + 100 kHz combined with NaClO [8,9]. Nonetheless, DFUS at high frequencies (>100 kHz up to 2.4 MHz) was successfully used over the last decade in medical research on sonodynamic therapy (SDT) and sonoantimicrobial chemotherapy [10–14]. Briefly, SDT induces the apoptosis of target (tumor) cells by ROS, which are generated upon the ultrasonic excitation of organic and inorganic sonosensitizers. Historically, organic sonosensitizers, such as porphyrins and xanthones, were derived from SDT. Inorganic sonosensitizers (piezo-catalysts), primarily modified TiO_2 and ZnO and their composites, have wider applications and also showed a high efficiency for microbial inactivation via single-frequency sonocatalysis in the low-frequency range of 20–48 kHz [15–17]. Meanwhile, recent single-frequency SDT studies have confirmed the high efficiency of modified TiO_2- and ZnO-based catalysts upon excitation with a high frequency of 1 MHz [18–22]. Regarding dual-frequency SDT research, only Ninomiya et al. (2014) [23] reported enhanced •OH production by the simultaneous application of DFUS (0.5 + 1 MHz) and TiO_2 nanoparticles modified with targeting protein. Given the promising results of SDT, we consider integrating high-frequency DFUS and inorganic sonocatalysts as a novel hybrid DFUS-AOP for improved water disinfection.

ZnO represents a bio- and eco-friendly, stable and low-cost product with numerous applications. It is known that ZnO exhibits high piezoelectric properties due to its noncentrosymmetric crystal structure [24]. This means that under the action of mechanical forces, the electric charge (electron and holes that move in opposite directions) appears on the crystals [25]; therefore, under exposure to ultrasonic waves in aqueous media, the occurred electric charge changes with the same frequency. Electrons and holes react with O_2 and H_2O with the generation of ROS, such as $•O_2^-$, •OH and H_2O_2, via a known mechanism for piezoelectric catalysts [26]. The generated ROS facilitate lipid peroxidation that affects membrane structure and fluidity and leads to significant damage to membrane proteins [27]. As such, piezo-catalytic disinfection over the last few years has emerged as a highly efficient and inexpensive technology [28]. Recently, Li et al. (2023) [29] comprehensively surveyed the mechanisms of piezoelectric effect-mediated catalysis with different piezoelectric materials and their applications in environmental pollution remediation, including water disinfection. •OH was found to be the main oxidation species under exposure to 38 and 80 kHz in the presence of ZnO [30], whereas $•O_2^-$ and •OH were primarily generated by the 1 MHz + ZnO process [31]. ZnO-based composites, such as Bi_2O_3-ZnO-B_2O_3 [32,33], are also fabricated and used for piezo-catalysis.

To our knowledge, DFUS coupled with a piezo-catalyst, such as ZnO, has not been applied for microbial inactivation neither for water disinfection nor for SDT. In our previous work, a high synergistic effect of DFUS (120 + 1700 kHz) coupled with persulfate was found for bacterial inactivation in water [34]. The choice of these frequencies is due to the wide use and low price of commercial converters and producers that make them attractive for full-scale application. Specifically, the first frequency (120 kHz) is applied for the precision cleaning of complicated industrial parts, and the second frequency (1700 kHz) is commonly used for making water mist in nebulizers and humidifying the indoor air.

The present study is the first to explore a potential synergy of high-frequency DFUS in the presence of persulfate and ZnO nano- and microparticles towards the piezo-catalytic inactivation of *Escherichia coli* and *Enterococcus faecalis* in synthetic water. The comparison of the piezo-catalytic performance of nano- and microparticles represents considerable interest due to their different toxicological properties. Furthermore, a novel hybrid system, which simultaneously exploits DFUS, ZnO and persulfate, was also examined for enhanced inactivation efficiency.

2. Materials and Methods

2.1. Test Bacteria

Bacterial strains *Enterococcus faecalis* B 4053 and *Escherichia coli* K-12 were supplied by the State Research Institute of Genetics and Selection of Industrial Microorganisms (Genetika, Moscow, Russia). Cultures of *E. faecalis* and *E. coli* were grown aerobically with shaking at 37 °C and 180 rpm (Biosan ES-20, Riga, Latvia) in tryptic soy broth (Merck, Darmstadt, Germany) and nutrient broth (SRCAMB, Obolensk, Russia), respectively. After overnight culturing, the cells were harvested by centrifugation at 1690× *g* for 5 min (C2006, Centurion Scientific, Chichester, West Sussex, UK) and washed twice with phosphate-buffered saline (PBS, pH 7.4, Rosmedbio Ltd., St. Petersburg, Russia). Washed cells were then resuspended in PBS to obtain a stock suspension containing approximately 10^8 CFU/mL. For disinfection experiments, 1.5 L deionized water was contaminated with a 150 µL aliquot of cell suspension to obtain synthetic water with an initial cell count of 10^5 CFU/mL.

2.2. *Single- and Dual-Frequency Ultrasonication*

Inactivation experiments were performed in an orthogonal rectangle-shaped ultrasonic setup (Figure 1). Briefly, it represents a 4 L stainless steel water reservoir with four 120 kHz converters (50 W each, Fan Ying Sonic, Granbosonic Ultrasonic Producer 300 W/120 kHz, Shenzhen, China) on its long opposite sides and six 1700 kHz converters (total power 150 W, MSX, model ZCX-RM6D48F, Jiaxing, China) as a single unit, which was placed on the bottom of the reservoir. The water to be disinfected was thermostated at 20 ± 2 °C by a water jacket.

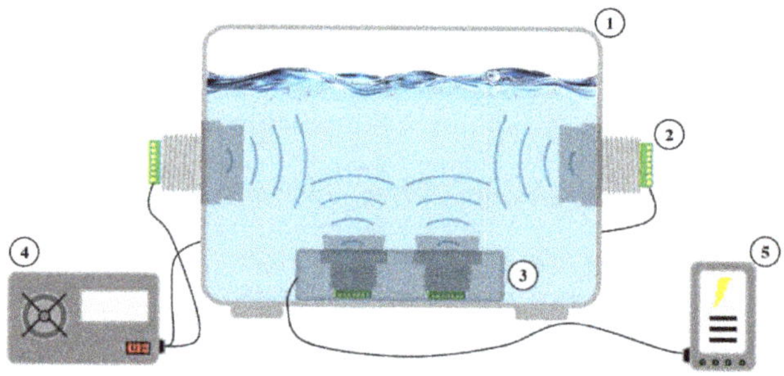

Figure 1. A diagram of dual-frequency ultrasonic setup. 1—water reservoir, 2—ceramic converter (120 kHz), 3—ceramic converter unit (1700 kHz), 4—ultrasonic producer (120 kHz), 5—power source (48 V).

Synthetic water containing 10^5 CFU/mL of *E. coli* or *E. faecalis* was irradiated in dual-frequency mode with added 1 g/L ZnO nano- or microparticles (99.7%, Hebei Shengyin Packaging Material Co., Ltd., Shijiazhuang, China) and 100 mg/L potassium persulfate (Vekton, St. Petersburg, Russia). Micro- and nanoparticles were 1 µm and 50 nm in size, respectively. Generally, the following hybrid systems were designed in this study:

- 120/1700 kHz/ZnO (50 nm);
- 120/1700 kHz/ZnO (1 µm);
- 120/1700 kHz/ZnO (50 nm)/$S_2O_8{}^{2-}$;
- 120/1700 kHz/ZnO (1 µm)/$S_2O_8{}^{2-}$.

To evaluate the synergistic effect of DFUS in terms of inactivation, target bacteria were also sequentially treated via single-frequency at 120 and 1700 kHz in the above piezo-catalytic systems. The CFUs were enumerated via the serial dilution technique in triplicate after incubation on tryptic soy agar (*E. faecalis*) and nutrient agar (*E. coli*) plates at 37 °C for 24 h. The obtained data were presented as plots of the log reduction (Lg(N/N_0)) versus irradiation time (min). Each data point is the mean value (±SD) from 3 to 5 replicates. The statistical treatment of the data was carried out with the Statistica 10.0 software program.

Radical scavenging tests were conducted under DFUS conditions using *p*-chlorobenzoic acid (*p*CBA) as a probe compound (Supplementary Material, Text S1).

3. Results and Discussion

3.1. Single-Frequency Piezo-Catalytic Inactivation

The efficiency of single-frequency inactivation was compared in terms of log reduction between two frequencies (120 and 1700 kHz), ZnO- and $S_2O_8^{2-}$-mediated processes and ZnO particle sizes (50 nm and 1 μm). The target bacteria were resistant to ultrasonic exposure at 120 or 1700 kHz alone, while adding an oxidant or catalyst caused a measurable inactivation in most cases. Comparing two frequencies, 1700 kHz performed better than 120 kHz in the presence of ZnO or $S_2O_8^{2-}$ (1700 kHz/ZnO and 1700 kHz/$S_2O_8^{2-}$ systems), and 5-log (total) reduction was attained with this frequency and persulfate (Figure 2).

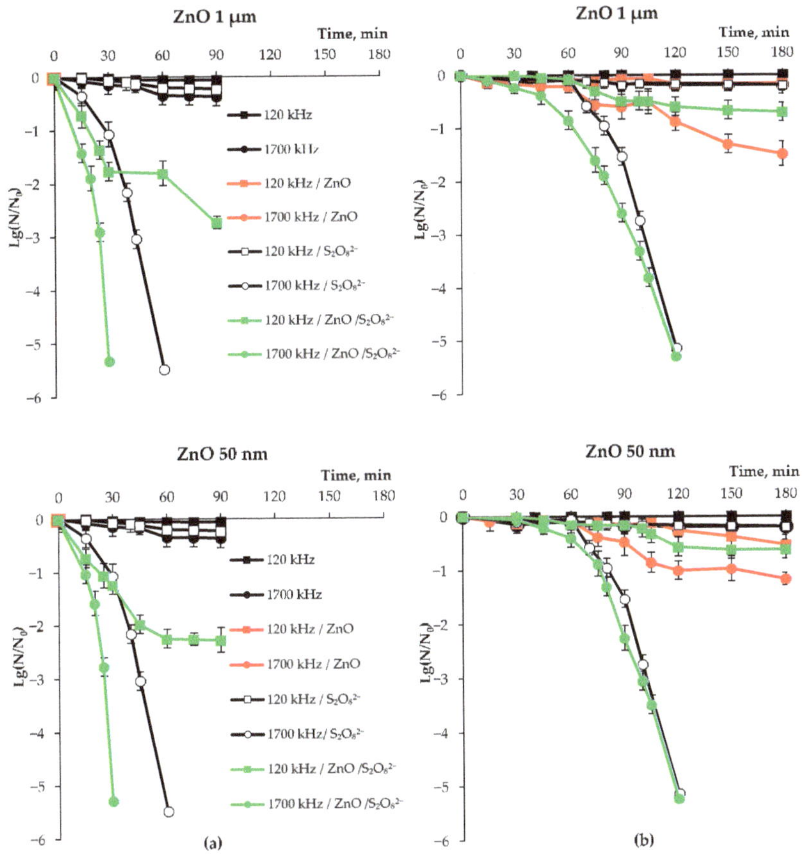

Figure 2. Single-frequency piezo-catalytic inactivation of *E. coli* (**a**) and *E. faecalis* (**b**) in the presence of ZnO micro- and nanoparticles and persulfate. $N_0 = 10^5$ CFU/mL, $[ZnO]_0 = 1$ g/L, $[S_2O_8^{2-}]_0 = 0.1$ g/L. Error bars represent ±SD.

The obtained result suggests that more ROS are produced under ZnO or $S_2O_8^{2-}$ activation with 1700 kHz. This is consistent with the data of Vighetto et al. (2019) [35], who found that ROS exposure at 1 MHz in the presence of ZnO-NH_2 nanocrystals (20 nm) was one order of magnitude higher than that obtained at 150 kHz. The authors proposed that the applied ultrasonic conditions were even sufficient to initiate the acoustic cavitation of gas nanobubbles trapped at the catalyst surface. Earlier, Mason et al. (2011) [36] proved that the efficiency of •OH production increases as the frequency is increased and more radical reactions occur at higher frequencies. This is explained by decreasing the radius of microbubbles with increasing frequency, which accelerates their collapse and yields more ROS. Particularly, the microbubble radius at 500 kHz was ~25-fold lower than that measured at 20 kHz, and higher production of H_2O_2 and •OH was observed [37].

Under dark conditions, i.e., upon contact of cells with ZnO micro- and nanoparticles without ultrasonic exposure, *E. coli* was reduced by ~1-log and *E. faecalis* was not inactivated (Supplementary Material, Figure S1). In 120 kHz/ZnO mode, *E. faecalis* remained resistant, whereas *E. coli* was inactivated by 1.8–1.9-log at both particle sizes. Considering the 1700 kHz/ZnO process, the log reduction of *E. coli* with 1 μm ZnO was significantly higher than that observed with 50 nm (4.3-log vs. 2.6-log after 90 min exposure). Meanwhile, this finding was not significant for *E. faecalis* (1.5-log vs. 1.2-log after 180 min exposure). It is known that ZnO toxicity increases with decreasing the particle size and ZnO nanoparticles are more toxic to bacteria than microparticles [38]. Furthermore, ZnO nanoparticles were reported to be the most toxic to different bacteria among the nano-sized TiO_2, Al_2O_3 and SiO_2 [39]. However, piezo-catalytic inactivation in the absence of persulfate appears to have an inverse relationship, and the microparticles make a greater contribution to ROS generation. This is supported by the fact that the piezoelectric properties of ZnO microparticles (~10 μm) are significantly higher than those of nanoparticles (80–100 nm) [40]. It is known that piezoelectric properties are quantitatively described by the piezoelectric coefficient (usually written "d_{33}"). Li et al. (2015) [41] found that d_{33} depends on the ZnO particle size, increasing from 8.36 pm/V at 800 nm to 46.97 pm/V at 1.5 μm. Accordingly, the piezoelectric properties of ZnO microparticles are several times superior to nanoparticles, and therefore the former generates more radicals.

The comparison of the 1700 kHz/$S_2O_8^{2-}$ and 1700 kHz/ZnO modes showed the higher efficiency of the persulfate-based process by which *E. coli* and *E. faecalis* were completely inactivated after 60 and 120 min exposure, respectively. In all cases, the inactivation of *E. faecalis*, which is the more resistant Gram-positive bacterium, required prolonged irradiation time as compared to *E. coli*. The hybrid system {1700 kHz/ZnO/$S_2O_8^{2-}$} was the most efficient for eradicating *E. coli* by 5-log for 30 min treatment (Figure 2). On the contrary, the performance of this system for the total inactivation of *E. faecalis* was similar to that of the 1700 kHz/$S_2O_8^{2-}$ process. Notably, when using 1 μm ZnO, the lag period was shorter and the log reduction was higher up to 90 min exposure.

The hybrid system {120 kHz/ZnO/$S_2O_8^{2-}$} was much less efficient and showed on average a 2.5- and about a 1-log reduction in *E. coli* (90 min) and *E. faecalis* (180 min), respectively. No effect of particle size was observed when comparing the performance of hybrid US/ZnO/$S_2O_8^{2-}$ processes, presumably due to the dominating effect of activated persulfate. This can be explained by the greater yield of ROS (SO_4•$^-$ and •OH) from activated persulfate as compared to ZnO (•OH). Wen et al. (2023) [42] also reported that SO_4•$^-$ and •OH (as a product of SO_4•$^-$ hydrolysis) were produced under the piezo-activation of Co-ZnO nanorods and persulfate.

In summary, single-frequency experiments revealed that piezo-catalytic inactivation with 1700 kHz was more efficient than that with 120 kHz. The performance of 1700 kHz-based processes can be ranked for *E. coli* as follows: 1700 kHz < 1700 kHz/ZnO < 1700 kHz/$S_2O_8^{2-}$ < 1700 kHz/ZnO/$S_2O_8^{2-}$. For *E. faecalis*, the 1700 kHz/$S_2O_8^{2-}$ process performed comparably in the absence and presence of ZnO that assumes the hidden contribution of the catalyst.

3.2. Dual-Frequency Piezo-Catalytic Inactivation and Synergistic Effect

Figure 3 shows that DFUS alone can slightly inactivate selected bacteria. Specifically, *E. coli* was inactivated by 1.7-log, whereas *E. faecalis* did not exhibit measurable inactivation after 90 min exposure. To enhance the generation of ROS and explore the potential synergistic effect between two frequencies, dual-frequency piezo-catalytic inactivation was investigated using ZnO particles and persulfate separately and simultaneously. As can be seen from Figure 3, the hybrid system {120/1700 kHz/ZnO/$S_2O_8^{2-}$} provided the fastest total inactivation of *E. coli* (25 min).

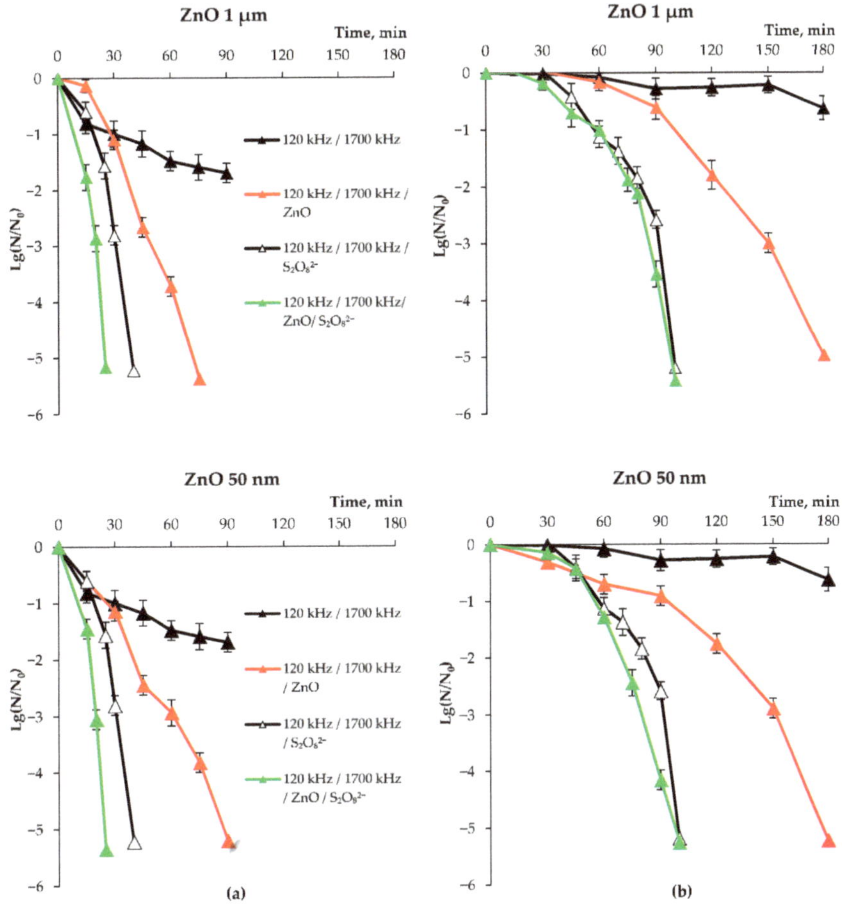

Figure 3. Dual-frequency piezo-catalytic inactivation of *E. coli* (**a**) and *E. faecalis* (**b**) in the presence of ZnO micro- and nanoparticles and persulfate. N_0 = 10^5 CFU/mL, $[ZnO]_0$ = 1 g/L, $[S_2O_8{}^{2-}]_0$ = 0.1 g/L. Error bars represent ±SD.

Similar to single-frequency disinfection, the inactivation kinetics of *E. faecalis* via the DFUS/$S_2O_8{}^{2-}$/ZnO and DFUS/$S_2O_8{}^{2-}$ processes were similar, and a 5-log reduction was achieved after the same exposure time (100 min). In turn, the DFUS/$S_2O_8{}^{2-}$ process was also more efficient than the DFUS/ZnO process and the difference was more pronounced for *E. faecalis*. As observed under single-frequency ZnO/$S_2O_8{}^{2-}$ conditions, no difference between the 50 nm and 1 μm particle sizes was found in the {DFUS/ZnO/$S_2O_8{}^{2-}$} system. In the absence of persulfate (DFUS/ZnO), the performance of 1 μm for inactivating *E. coli* was also higher than that observed with 50 nm. Specifically, the complete (5-log) inactivation of *E. coli* was attained faster using ZnO microparticles. No effect of particle size was observed for *E. faecalis*, and its 5-log reduction required the same time (180 min) (Figure 3).

The synergistic effect of DFUS in efficient systems was evaluated through the synergistic coefficient (SC). An SC value larger than 1 indicates synergy, whereas an additive and an antagonistic effect are observed when the SC is equal to and less than 1, respectively [6]. A synergy occurs if log reduction after DFUS treatment is higher than the sum of log reductions obtained after single-frequency treatment. The log reductions in single-frequency mode were obtained for the same exposure periods, which were needed for achieving a 5-log reduction in DFUS mode (Table 1). A fixed time for evaluating a synergy of DFUS was also previously used by Adelnia et al. (2020) [43]. Given that approach, the synergistic coefficient was determined relative to the 5-log reduction times for the DFUS process (1):

$$SC = \frac{\text{Log reduction}(120 + 1700\text{ kHz})}{\text{Log reduction}(120\text{ kHz}) + \text{Log reduction}(1700\text{ kHz})} \quad (1)$$

Table 1. The synergistic coefficients of piezo-catalytic inactivation using DFUS (120 + 1700 kHz), ZnO particles and persulfate.

Strain	System	ZnO Particle Size	5-Log Reduction Time, min	Sum of Log Reductions after Single-Frequency Treatment	SC
E. coli K-12	DFUS/ZnO/$S_2O_8{}^{2-}$	50 nm	25	3.8	1.3
		1 μm	25	4.3	1.2
	DFUS/ZnO	50 nm	90	4.4	1.2
		1 μm	75	4.4	1.2
	DFUS/$S_2O_8{}^{2-}$	no ZnO	40	2.3	2.3
E. faecalis B 4053	DFUS/ZnO/$S_2O_8{}^{2-}$	50 nm	100	3.3	1.5
		1 μm	100	3.8	1.4
	DFUS/ZnO	50 nm	180	1.7	3.0
		1 μm	180	1.7	2.9
	DFUS/$S_2O_8{}^{2-}$	no ZnO	100	2.9	1.8

Generally, the synergistic effect for inactivating *E. faecalis* by ZnO-based processes was higher than that for *E. coli*, which is a more susceptible organism and inactivated faster (Table 1). Interestingly, SC values were similar for micro- and nanoparticles within the same system. In the case of the DFUS/$S_2O_8{}^{2-}$ process, *E. coli* exhibited higher synergy due to a greater difference between dual- and single-frequency (1700 kHz/$S_2O_8{}^{2-}$) inactivation curves (Supplementary Material, Figure S2). The highest synergistic effect (SC = 3.0) was found for *E. faecalis* in the {DFUS/ZnO} system (Supplementary Material, Figure S3), whereas the SC value for *E. coli* was lower by a factor of 2.5. This is explained by the contribution of the 120 kHz/ZnO system (1.8–1.9 log reduction), which reduced the synergistic effect of the DFUS mode for *E. coli* (Supplementary Material, Figure S2).

It is known that the dual-frequency ultrasonication of water enhances acoustic cavitation due to the production of new (combination) frequencies, as discussed in detail previously [6,44]. Briefly, under the nonlinear interaction of two acoustic waves at different frequencies in water, the combination frequencies are produced alongside the main and additional frequencies (harmonics, subharmonics, and ultraharmonics). The combination frequencies represent a sum or a difference of two main frequencies, main frequencies, harmonics and so on. These new frequencies make the oscillating microbubbles unstable and increase the probability of their collapse [45], which enhances cavitation and generates more ROS. Specifically, Lei et al. (2020) [46] explored the system {20 + 43 kHz/persulfate} towards the degradation of petroleum hydrocarbons and confirmed that DFUS generates more $SO_4{\bullet}^-$ and •OH than single-frequency US/persulfate.

A synergistic effect can be attributed to the boosted generation of ROS, thus intensifying the inactivation processes. The increased generation of •OH and $SO_4{\bullet}^-$ was also reported under the activation of persulfate with low-frequency DFUS for the synergistic degradation of per- and polyfluoroalkyl substances [47]. Other related research dealt with the single-frequency system {US/ZnO/$S_2O_8{}^{2-}$}, which has attracted attention for its cost efficiency. This system was majorly induced by •OH and $SO_4{\bullet}^-$ [48,49]; other ROS, such as $^1O_2$ and $\bullet O_2{}^-$, were also identified [42,50,51]. These ROS destroy the cellular membrane by damaging proteins and lipids. The examination of the cell wall of *E. coli* by scanning electron microscopy (SEM) showed its disruption after exposure to •OH from US/ZnO [52]. Our •OH scavenging tests showed that *p*CBA was degraded faster by the {120 + 1700 kHz/ZnO/$S_2O_8{}^{2-}$} process as compared to the {120 + 1700 kHz} process

(Supplementary Material, Figure S4). This indicates that more ROS were produced in the presence of ZnO and persulfate.

We propose that cell permeabilization via acoustic cavitation (sonoporation) also plays an important role in inactivation. Zhang et al. (2012) [53] investigated different types of sonoporation in detail using SEM and found that the morphological changes of the cell wall and its poration were lethal to cells. Recently, Ali et al. (2023) [54] reviewed the data on exposure to ROS and ultrasound and reported that the radicals were also produced inside the cell (intracellular ROS). Sonoporation facilitates the penetration of ZnO particles [15] and extracellular ROS (generated outside the cell) into sonoporated cells [3] that ultimately accelerate their apoptosis. In turn, the penetrated ZnO particles also generate intracellular ROS which damage the organelles [15].

In summary, a synergistic effect of DFUS in the presence of ZnO and $S_2O_8^{2-}$ is supposed to be driven by the simultaneous action of two different mechanisms of inactivation via (1) generated ROS and (2) sonoporation with associated effects. A comparison of DFUS-based processes showed that the contribution of ZnO was also hidden in the DFUS/ZnO/$S_2O_8^{2-}$ process for *E. faecalis*. For *E. coli*, the efficiency of DFUS-based processes increased in the order: DFUS < DFUS/ZnO < DFUS/$S_2O_8^{2-}$ < DFUS/ZnO/$S_2O_8^{2-}$.

4. Conclusions

This study revealed the synergistic inactivation of *E. coli* and *E. faecalis* in synthetic water via dual-frequency ultrasound at 120 and 1700 kHz with added persulfate and ZnO particles. Apparently, the intensification of inactivation processes under DFUS exposure is driven by at least two key processes, attack by increased ROS yield and sonoporation, which ultimately create a synergistic effect. Microparticles, which are more favorable to the aquatic environment, were found to be more efficient than nanoparticles for inactivating *E. coli* via ZnO-mediated processes. The present study supplies the integration of high-frequency DFUS, persulfate and piezo-catalyst ZnO as a novel approach for further development in water disinfection.

Supplementary Materials: The following supporting information can be downloaded at: https://www.mdpi.com/article/10.3390/w15162937/s1, Figure S1: Inactivation of bacteria in the presence of ZnO micro- and nanoparticles under dark conditions (control); Figure S2: Single- and dual-frequency inactivation of *E. coli* in the presence of ZnO micro- and nanoparticles and persulfate; Figure S3: Single- and dual-frequency inactivation of *E. faecalis* in the presence of ZnO micro- and nanoparticles and persulfate; Figure S4: Dual-frequency degradation of *p*CBA in the absence and presence of ZnO microparticles and persulfate.

Author Contributions: Investigation, validation, formal analysis, I.T. and E.K.; writing—original draft preparation, data curation, G.M.; writing—review and editing, project administration, conceptualization, methodology, supervision, V.B. All authors have read and agreed to the published version of the manuscript.

Funding: This research was funded by Russian Science Foundation, grant number 22-24-00482.

Data Availability Statement: Data is contained within the article and Supplementary Material.

Conflicts of Interest: The authors declare no conflict of interest.

References

1. Sathishkumar, P.; Mangalaraja, R.V.; Anandan, S. Review on the recent improvements in sonochemical and combined sonochemical oxidation processes—A powerful tool for destruction of environmental contaminants. *Ren. Sustain. Energy Rev.* **2016**, *55*, 426–454. [CrossRef]
2. Ince, N.H. Ultrasound-assisted advanced oxidation processes for water decontamination. *Ultrason. Sonochem.* **2018**, *40*, 97–103. [CrossRef]
3. Matafonova, G.; Batoev, V. Review on low- and high-frequency sonolytic, sonophotolytic and sonophotochemical processes for inactivating pathogenic microorganisms in aqueous media. *Water Res.* **2019**, *166*, 115085. [CrossRef]
4. Yap, H.C.; Pang, Y.L.; Lim, S.; Abdullah, A.Z.; Ong, H.C.; Wu, C.-H. A comprehensive review on state-of-the-art photo-, sono-, and sonophotocatalytic treatments to degrade emerging contaminants. *Int. J. Environ. Sci. Technol.* **2019**, *16*, 601–628. [CrossRef]

5. Dehghani, M.H.; Karri, R.R.; Koduru, J.R.; Manickam, S.; Tyagi, I.; Mubarak, N.M. Recent trends in the applications of sonochemical reactors as an advanced oxidation process for the remediation of microbial hazards associated with water and wastewater: A critical review. *Ultrason. Sonochem.* **2023**, *94*, 106302. [CrossRef] [PubMed]
6. Matafonova, G.; Batoev, V. Dual-frequency ultrasound: Strengths and shortcomings to water treatment and disinfection. *Water Res.* **2020**, *182*, 116016. [CrossRef] [PubMed]
7. Wu, X.; Mason, T.J. Evaluation of power ultrasonic effects on algae cells at a small pilot scale. *Water* **2017**, *9*, 470. [CrossRef]
8. Zou, H.; Tang, H. Comparison of different bacteria inactivation by a novel continuous-flow ultrasound/chlorination water treatment system in a pilot scale. *Water* **2019**, *11*, 258. [CrossRef]
9. Zou, H.; Wang, L. The disinfection effect of a novel continuous-flow water sterilizing system coupling dual-frequency ultrasound with sodium hypochlorite in pilot scale. *Ultrason. Sonochem.* **2017**, *36*, 246–252. [CrossRef]
10. Alamolhoda, M.; Mokhtari-Dizaji, M.; Barati, A.H.; Hasanzadeh, H. Comparing the in vivo sonodynamic effects of dual- and single-frequency ultrasound in breast adenocarcinoma. *J. Med. Ultrason.* **2012**, *39*, 115–125. [CrossRef] [PubMed]
11. Serpe, L.; Giuntini, F. Sonodynamic antimicrobial chemotherapy: First steps towards a sound approach for microbe inactivation. *J. Photochem. Photobiol. B* **2015**, *150*, 44–49. [CrossRef] [PubMed]
12. Rengeng, L.; Qianyu, Z.; Yuehong, L.; Zhongzhong, P.; Libo, L. Sonodynamic therapy, a treatment developing from photodynamic therapy. *Photodiagnosis Photodyn. Ther.* **2017**, *19*, 159–166. [CrossRef] [PubMed]
13. Tabatabaei, Z.S.; Rajabi, O.; Nassirli, H.; Noghreiyan, A.V.; Sazgarnia, A. A comparative study on generating hydroxyl radicals by single and two-frequency ultrasound with gold nanoparticles and protoporphyrin IX. *Australas. Phys. Eng. Sci. Med.* **2019**, *42*, 1039–1047. [CrossRef] [PubMed]
14. Fan, L.; Muhammad, A.I.; Ismail, B.B.; Liu, D. Sonodynamic antimicrobial chemotherapy: An emerging alternative strategy for microbial inactivation. *Ultrason. Sonochem.* **2021**, *75*, 105591. [CrossRef] [PubMed]
15. Zhang, L.; Qi, H.; Yan, Z.; Gu, Y.; Sun, W.; Zewde, A.A. Sonophotocatalytic inactivation of *E. coli* using ZnO nanofluids and its mechanism. *Ultrason. Sonochem.* **2017**, *34*, 232–238. [CrossRef] [PubMed]
16. Bayrami, A.; Alioghli, S.; Rahim Pouran, S.; Habibi-Yangjeh, A.; Khataee, A.; Ramesh, S. A facile ultrasonic-aided biosynthesis of ZnO nanoparticles using *Vaccinium arctostaphylos* L. leaf extract and its antidiabetic, antibacterial, and oxidative activity evaluation. *Ultrason. Sonochem.* **2019**, *55*, 57–66. [CrossRef]
17. Wang, G.; Wu, W.; Zhu, J.-J.; Peng, D. The promise of low-intensity ultrasound: A review on sonosensitizers and sonocatalysts by ultrasonic activation for bacterial killing. *Ultrason. Sonochem.* **2021**, *79*, 105781. [CrossRef]
18. Su, K.; Tan, L.; Liu, X.; Cui, Z.; Zheng, Y.; Li, B.; Han, Y.; Li, Z.; Zhu, S.; Liang, Y.; et al. Rapid photo-sonotherapy for clinical treatment of bacterial infected bone implants by creating oxygen deficiency using sulfur doping. *ACS Nano* **2020**, *14*, 2077–2089. [CrossRef]
19. Wang, Y.; Sun, Y.; Liu, S.; Zhi, L.; Wang, X. Preparation of sonoactivated TiO_2-DVDMS nanocomposite for enhanced antibacterial activity. *Ultrason. Sonochem.* **2020**, *63*, 104968. [CrossRef]
20. Pourhajibagher, M.; Bahador, A. Synergistic biocidal effects of metal oxide nanoparticles-assisted ultrasound irradiation: Antimicrobial sonodynamic therapy against *Streptococcus mutans* biofilms. *Photodiagnosis Photodyn. Ther.* **2021**, *35*, 102432. [CrossRef]
21. Wu, M.; Zhang, Z.; Liu, Z.; Zhang, J.; Zhang, Y.; Ding, Y.; Huang, T.; Xiang, D.; Wang, Z.; Dai, Y.; et al. Piezoelectric nanocomposites for sonodynamic bacterial elimination and wound healing. *Nano Today* **2021**, *37*, 101104. [CrossRef]
22. Xu, Q.; Xiu, W.; Li, Q.; Zhang, Y.; Li, X.; Ding, M.; Yang, D.; Mou, Y.; Dong, H. Emerging nanosonosensitizers augment sonodynamic-mediated antimicrobial therapies. *Mater. Today Bio* **2023**, *19*, 100559. [CrossRef] [PubMed]
23. Ninomiya, K.; Noda, K.; Ogino, C.; Kuroda, S.; Shimizu, N. Enhanced OH radical generation by dual-frequency ultrasound with TiO_2 nanoparticles: Its application to targeted sonodynamic therapy. *Ultrason. Sonochem.* **2014**, *21*, 289–294. [CrossRef] [PubMed]
24. Wang, Z.L.; Song, J. Piezoelectric nanogenerators based on zinc oxide nanowire arrays. *Science* **2006**, *312*, 242–246. [CrossRef]
25. Jiang, Z.; Tan, X.; Huang, Y. Piezoelectric effect enhanced photocatalysis in environmental remediation: State-of-the-art techniques and future scenarios. *Sci. Tot. Environ.* **2022**, *806*, 150924. [CrossRef]
26. Ma, W.; Lv, M.; Cao, F.; Fang, Z.; Feng, Y.; Zhang, G.; Yang, Y.; Liu, H. Synthesis and characterization of ZnO-GO composites with their piezoelectric catalytic and antibacterial properties. *J. Environ. Chem. Eng.* **2022**, *10*, 107840. [CrossRef]
27. Kumar, S.; Sharma, M.; Frömling, T.; Vaish, R. Antibacterial ferroelectric materials: Advancements and future directions. *J. Ind. Eng. Chem.* **2021**, *97*, 95–110. [CrossRef]
28. Zhao, Y.; Low, Z.-X.; Pan, Y.; Zhong, Z.; Gao, G. Universal water disinfection by piezoelectret aluminium oxide-based electroporation and generation of reactive oxygen species. *Nano Energy* **2021**, *92*, 106749. [CrossRef]
29. Li, J.; Liu, X.; Zhao, G.; Liu, Z.; Cai, Y.; Wang, S.; Shen, C.; Hu, B.; Wang, X. Piezoelectric materials and techniques for environmental pollution remediation. *Sci. Total Environ.* **2023**, *869*, 161767. [CrossRef]
30. Daraei, H.; Maleki, A.; Mahvi, A.H.; Zandsalimi, Y.; Alaei, L.; Gharibi, F. Synthesis of ZnO nano-sono-catalyst for degradation of reactive dye focusing on energy consumption: Operational parameters influence, modeling, and optimization. *Desalin. Water Treat.* **2014**, *52*, 6745–6755. [CrossRef]
31. Tamboia, G.; Campanini, M.; Vighetto, V.; Racca, L.; Spigarelli, L.; Canavese, G.; Cauda, V. A comparative analysis of low intensity ultrasound effects on living cells: From simulation to experiments. *Biomed. Microdev.* **2022**, *24*, 35. [CrossRef] [PubMed]
32. Porwal, C.; Verma, S.; Chauhan, S.V.; Vaish, R. Bismuth zinc borate- Polyacrylonitrile nanofibers for photo-piezocatalysis. *J. Ind. Eng. Chem.* **2023**, *124*, 358–367. [CrossRef]

33. Porwal, C.; Sharma, M.; Vaish, R.; Chauhan, V.S.; ben Ahmed, S.; Hwang, W.; Park, H.K.B.; Sung, T.H.; Kumar, A. Piezocatalytic dye degradation using Bi_2O_3-ZnO-B_2O_3 glass-nanocomposites. *J. Mater. Res. Technol.* **2022**, *21*, 2028–2037. [CrossRef]
34. Garkusheva, N.; Tsenter, I.; Kobunova, E.; Matafonova, G.; Batoev, V. Dual-frequency ultrasonic inactivation of *Escherichia coli* and *Enterococcus faecalis* using persulfate: A synergistic effect. *Water* **2022**, *14*, 2604. [CrossRef]
35. Vighetto, V.; Ancona, A.; Racca, L.; Limongi, T.; Troia, A.; Canavese, G.; Cauda, V. The synergistic effect of nanocrystals combined with ultrasound in the generation of reactive oxygen species for biomedical applications. *Front. Bioeng. Biotechnol.* **2019**, *7*, 374. [CrossRef]
36. Mason, T.J.; Cobley, A.J.; Graves, J.E.; Morgan, D. New evidence for the inverse dependence of mechanical and chemical effects on the frequency of ultrasound. *Ultrason. Sonochem.* **2011**, *18*, 226–230. [CrossRef]
37. Hua, I.; Hoffmann, M.R. Optimization of ultrasonic irradiation as an advanced oxidation technology. *Environ. Sci. Technol.* **1997**, *31*, 2237–2243. [CrossRef]
38. Ann, L.C.; Mahmud, S.; Seeni, A.; Bakhori, S.K.M.; Sirelkhatim, A.; Mohamad, D.; Hasan, H. Structural morphology and in vitro toxicity studies of nano- and micro-sized zinc oxide structures. *J. Environ. Chem. Eng.* **2015**, *3*, 436–444. [CrossRef]
39. Jiang, W.; Mashayekhi, H.; Xing, B. Bacterial toxicity comparison between nano- and micro-scaled oxide particles. *Environ. Pollut.* **2009**, *157*, 1619–1625. [CrossRef]
40. Zhang, X.; Le, M.-Q.; Zahhaf, O.; Capsal, J.-F.; Cottinet, P.-J.; Petit, L. Enhancing dielectric and piezoelectric properties of micro-ZnO/PDMS composite-based dielectrophoresis. *Mater. Des.* **2020**, *192*, 108783. [CrossRef]
41. Li, T.; Li, Y.T.; Qin, W.W.; Zhang, P.P.; Chen, X.Q.; Hu, X.F.; Zhang, W. Piezoelectric size effects in a zinc oxide micropillar. *Nanoscale Res. Lett.* **2015**, *10*, 394. [CrossRef] [PubMed]
42. Wen, Y.; Chen, J.; Gao, X.; Liu, W.; Che, H.; Liu, B.; Ao, Y. Two birds with one stone: Cobalt-doping induces to enhanced piezoelectric property and persulfate activation ability of ZnO nanorods for efficient water purification. *Nano Energy* **2023**, *107*, 108173. [CrossRef]
43. Adelnia, A.; Mokhtari-Dizaji, M.; Hoseinkhani, S.; Bakhshandeh, M. The effect of dual-frequency ultrasound waves on B16F10 melanoma cells: Sonodynamic therapy using nanoliposomes containing methylene blue. *Skin Res. Technol.* **2021**, *27*, 376–384. [CrossRef]
44. Zhang, Y.; Zhang, Y.; Li, S. Combination and simultaneous resonances of gas bubbles oscillating in liquids under dual-frequency acoustic excitation. *Ultrason. Sonochem.* **2017**, *35*, 431–439. [CrossRef]
45. Ye, L.; Zhu, X.; Liu, Y. Numerical study on dual-frequency ultrasonic enhancing cavitation effect based on bubble dynamic evolution. *Ultrason. Sonochem.* **2019**, *59*, 104744. [CrossRef]
46. Lei, Y.-J.; Zhang, J.; Tian, Y.; Yao, J.; Duan, O.-S.; Zuo, F. Enhanced degradation of total petroleum hydrocarbons in real soil by dual-frequency ultrasound-activated persulfate. *Sci. Tot. Environ.* **2020**, *748*, 141414. [CrossRef]
47. Lei, Y.-J.; Tian, Y.; Sobhani, Z.; Naidu, R.; Fang, C. Synergistic degradation of PFAS in water and soil by dual-frequency ultrasonic activated persulfate. *Chem. Eng. J.* **2020**, *388*, 124215. [CrossRef]
48. Hu, S.-B.; Li, L.; Luo, M.-Y.; Yun, Y.-F.; Chang, C.-T. Aqueous norfloxacin sonocatalytic degradation with multilayer flower-like ZnO in the presence of peroxydisulfate. *Ultrason. Sonochem.* **2017**, *38*, 446–454. [CrossRef] [PubMed]
49. Liu, F.; Yi, P.; Wang, X.; Gao, H.; Zhang, H. Degradation of Acid Orange 7 by an ultrasound/ZnO-GAC/persulfate process. *Sep. Pur. Technol.* **2018**, *194*, 181–187. [CrossRef]
50. Subramani, A.K.; Rani, P.; Wang, P.-H.; Chen, B.-Y.; Mohan, S.; Chang, C.-T. Performance assessment of the combined treatment for oxytetracycline antibiotics removal by sonocatalysis and degradation using *Pseudomonas aeruginosa*. *J. Environ. Chem. Eng.* **2019**, *7*, 103215. [CrossRef]
51. Zhang, M.; Tao, H.; Zhai, C.; Yang, J.; Zhou, Y.; Xia, D.; Comodi, G.; Zhu, M. Twin-brush ZnO mesocrystal for the piezo-activation of peroxymonosulfate to remove ibuprofen in water: Performance and mechanism. *Appl. Catal. B* **2023**, *326*, 122399. [CrossRef]
52. Anju, S.G.; Bright Singh, I.S.; Yesodharan, E.P.; Yesodharan, S. Investigations on semiconductor sonocatalysis for the removal of pathological micro-organisms in water. *Desalin. Water Treat.* **2015**, *54*, 3161–3168. [CrossRef]
53. Zhang, J.-Z.; Saggar, J.K.; Zhou, Z.-L.; Hu, B. Different effects of sonoporation on cell morphology and viability. *Bosn. J. Basic Med. Sci.* **2012**, *12*, 64–68. [CrossRef] [PubMed]
54. Ali, A.; Chen, L.; Nasir, M.S.; Wu, C.; Guo, B.; Yang, Y. Piezocatalytic removal of water bacteria and organic compounds: A review. *Environ. Chem. Lett.* **2023**, *21*, 1075–1092. [CrossRef]

water

Article

Thermal Polymerisation Synthesis of g-C_3N_4 for Photocatalytic Degradation of Rhodamine B Dye under Natural Sunlight

Muxi Zhang [1,2,3], Meiyan Xing [4], Bin Dong [1,2,3,4], Hongxia Zhang [1,2,3,*], Xiaojie Sun [1,2,3,*], Qiuhong Li [1,2,3], Xueshuang Lu [1,2,3], Jingjing Mo [1,2,3] and Hongxiang Zhu [5]

1 College of Environmental Science and Engineering, Guilin University of Technology, Guilin 541006, China; zhangmuxi2022@163.com (M.Z.); dongbin@tongji.edu.cn (B.D.); m18868006306@163.com (Q.L.); 17344519765@163.com (X.L.); 18579411886@163.com (J.M.)
2 Guangxi Key Laboratory of Environmental Pollution Control Theory and Technology, Guilin University of Technology, Guilin 541006, China
3 Guangxi Collaborative Innovation Center for Water Pollution Control and Water Safety in Karst Area, Guilin University of Technology, Guilin 541006, China
4 School of Environmental Science and Engineering, Tongji University, Shanghai 200092, China; xingmeiyan@tongji.edu.cn
5 Modern Industry College of Ecology and Environmental Protection, Guilin University of Technology, Guilin 541006, China; zhx@gxu.edu.cn
* Correspondence: zhx75@glut.edu.cn (H.Z.); sunxiaojie@glut.edu.cn (X.S.); Tel.: +86-1507-8329-789 (X.S.)

Abstract: The photocatalytic performance of g-C_3N_4 materials prepared by different precursors for Rhodamine B (RhB) dye degradation was studied. Their crystal structure, morphologies, chemical compositions, functional groups, and optical and photoelectrochemical performances of prepared g-C_3N_4 were analysed and characterised using X-ray diffraction, scanning electron microscopy, X-ray photoelectron spectroscopy, Fourier transform infrared spectroscopy, UV-Vis diffuse reflectance spectra, photoluminescence, and electrochemical workstations. The degradation of RhB dye in the presence of visible light and sunlight was utilised to assess the photocatalytic efficiency of the g-C_3N_4 photocatalyst. The results of the photocatalytic comparison experiment showed that the g-C_3N_4 photocatalyst prepared with urea as a precursor (UCN) has the best photocatalytic performance, achieving 99.61% removal in 40 min. In addition, the photocatalyst UCN can completely degrade 10 mg/L RhB dye within 20 min under sunlight, demonstrating its potential for practical applications under natural sunlight conditions. After four cycles, the degradation rate remains above 99%, demonstrating excellent stability and reusability. Due to its lower average pore number, larger BET-specific surface area and volume of pores, UCN provides more activity spaces and facilitates the adsorption of pollutant molecules, thereby enhancing photocatalytic activity. It was established through the active substance trapping studies that the main reactive species involved in the photocatalytic degradation process of RhB dye is $\bullet O_2^-$. This study showed that g-C_3N_4 synthesised with urea as the precursor has better photocatalytic performance in the degradation of RhB dye.

Keywords: g-C_3N_4; urea; melamine; photocatalysts; Rhodamine B

Citation: Zhang, M.; Xing, M.; Dong, B.; Zhang, H.; Sun, X.; Li, Q.; Lu, X.; Mo, J.; Zhu, H. Thermal Polymerisation Synthesis of g-C_3N_4 for Photocatalytic Degradation of Rhodamine B Dye under Natural Sunlight. *Water* **2023**, *15*, 2903. https://doi.org/10.3390/w15162903

Academic Editor: Mehrab Mehrvar

Received: 8 July 2023
Revised: 31 July 2023
Accepted: 9 August 2023
Published: 11 August 2023

1. Introduction

With the continuous advancement of China's industrialisation and urbanisation, environmental pollution, such as water pollution, has become an increasingly severe problem, a crucial factor restricting economic development and social progress [1]. Many efforts have been made to eliminate contaminants from the water. Recently, photocatalysis technology was investigated for removing organic dyes from water because of its mild reaction conditions, low price, cleanliness, and high efficiency [2,3]. Budnyak et al. [4] synthesised the photoactive lignin/$Bi_4O_5Br_2$/BiOBr bioinorganic complex site and found that 18.9% of

$Bi_4O_5Br_2$/BiOBr exhibited photocatalytic degradation capabilities towards cationic methylene blue and Rhodamine B (RhB) dyes under 20 W blue light irradiation. Li et al. [5] fabricated a C-BiOCl composite photocatalyst and found that it exhibited superior photocatalytic activity with RhB dye degradation irradiated for only 6 min under a 300 W Xe arc lamp. Zhang et al. [3] synthesised ZSM-5/$Bi_4O_5Br_2$ photocatalysts and found that 1ZSM-5/$Bi_4O_5Br_2$ had an excellent photocatalytic performance. The RhB dye removal rate reached 99.86% within 25 min under 300 W Xe arc lamp irradiation. Yang et al. [6] synthesised a nitrogen-deficient porous g-C_3N_4 photocatalyst that can completely degrade RhB dye in just 20 min under simulated visible light irradiation.

The photocatalytic process depends on highly active substances, such as superoxide radicals ($\bullet O_2^-$), hydroxyl radicals ($\bullet$OH), and holes (h^+) produced by the photocatalyst after absorbing light energy to decompose target pollutants [7]. Various photocatalysts, including TiO_2-based [8], Bi-based [9,10], Mo-based [11], Zn-based [12], and Ag-based [13] semiconductors, have been developed and effectively removed RhB dye. Nevertheless, many of the aforementioned photocatalysts might give off precious elemental metals, introducing the risk of additional contamination caused by the release of metals from the photocatalysts. Therefore, metal-free photocatalysts are environmentally friendly alternatives that have prospective widespread usage [14].

Graphite carbon nitride (g-C_3N_4) has generated considerable interest as a metal-free photocatalyst because of its straightforward and affordable synthesis; energy bandwidth for solar power collection; thermal stability; trustworthy chemical inertness; and electrochemical, multipurpose, and nicely organised photocatalytic characteristics [15–17]. g-C_3N_4 has the in-plane tri-s-triazine units, and the sp^2-hybridised carbon and nitrogen form a π-conjugated system, advantageous for modifying materials based on nanosheets in electrical and optical electronics [18,19]. g-C_3N_4 has good photocatalytic performance because of the narrow band gap and capacity to take in light that is visible, which can effectively absorb and use the light of wavelengths below 460 nm [20]. The g-C_3N_4 band gap is approximately 2.70 eV, and the conduction band (CB) potential is negative (-1.3 eV), providing a strong reduction ability for photogenerated electrons [21]. The synthesis methods of g-C_3N_4 primarily include deposition [22,23], template [24], solvothermal [25], and thermal polymerisation [26,27]. Thermal polymerisation is frequently employed because of its affordable, easy availability of raw materials, and simple operation. A few molecules with a high nitrogen content, such as cyanamide [28], dicyandiamide [29], melamine [30], and urea [31], served as a single source of molecular antecedents to fabricate g-C_3N_4. Ma et al. [32] prepared g-C_3N_4 from various raw materials using thermal polycondensation, and the shapes of the products obtained using mono-cyanamide, dicyandiamide, and melamine were spherical, irregular, and layered, respectively. Liu et al. [33] prepared 3D interconnected g-C_3N_4 via one-step thermal polymerisation using melamine and cyanuric acid.

The occurrence of catalytic reactions necessitates adsorption, and numerous studies have highlighted that optimal adsorption can yield enhanced catalytic efficiency [34,35]. Zhang et al. [36] loaded Cu on an oxide semiconductor surface to improve wastewater remediation, and the results showed that Cu particle deposition can improve the adsorption activity of a single oxide and thus improve the photocatalytic activity. Luo et al. [37] demonstrated that the adsorption process promotes the photodegradation of pollutants by g-C_3N_4, while the photodegradation of dyes supported on g-C_3N_4 regenerates its adsorption capacity many times, indicating that the photocatalytic process also enhances pollutant adsorption on g-C_3N_4. Yu et al. [38] confirmed that a modified g-C_3N_4/MgZnAl-calcined layered double hydroxide composite (MCN/cLDH) possesses a large S_{BET}-specific surface area, which enhances its adsorption capacity and improves the separation efficiency of photogenerated carriers. Furthermore, MCN/cLDH nanostructures exhibit enhanced synergistic degradation through adsorption-photocatalysis to remove oxytetracycline (95.73%) from artificial seawater. Ni et al. [39] demonstrated that the addition of g-C_3N_4 to CCN (Co_3O_4 QDs/3D g-C_3N_4) catalyst led to an increase in tetracycline adsorption, indicating that an optimal amount of g-C_3N_4 as a Co_3O_4 QDs carrier can enhance adsorption and expose more active sites, thereby achieving the best synergistic effect between adsorption and oxidative degradation. Liu et al. [40] prepared an efficient three-dimensional (3D) hollow porous C-doped polymeric carbon nitride (CPCN) catalyst, and found that the formed 3D hollow porous structure gave the CPCN catalyst a large specific surface area and numerous exposed active sites, promoted the substrate adsorption reaction and maintained its structural stability, and thus improved its photolytic water hydrogen production and pollutant elimination activity.

In this study, melamine and urea were utilised as precursors for the preparation of g-C_3N_4 (MCN and UCN, respectively) through thermal polymerisation. The photocatalytic degradation ability and practical application potential of MCN and UCN were assessed by comparing the change in RhB dye concentration under visible light and sunlight irradiation. The crystal structure, morphology, chemical composition, functional groups, optics, photoelectric chemical properties, and stability of g-C_3N_4 were investigated. To understand how RhB molecules behaved in the reaction procedure, three-dimensional excitation-emission matrix fluorescence spectra (3D EEMs) were used. In comparison to MCN, UCN has better photocatalytic capabilities for degrading RhB dye when exposed to visible light and sunlight. This study serves as a reference for research on the synthesis and properties of g-C_3N_4.

2. Experimental

2.1. Materials

Urea (H_2NCONH_2), melamine ($C_3H_6N_6$), isopropanol (C_3H_8O, IPA), benzoquinone ($C_6H_4O_2$, BQ), Rhodamine B (RhB, $C_{28}H_{31}ClN_2O_3$), and sodium oxalate ($Na_2C_2O_4$, SO) were purchased from Xilong Scientific Co., Ltd. (Shantou, China). All reagents are analytical grade and can be used without further purification.

2.2. Synthesis of g-C_3N_4

g-C_3N_4 was synthesised using the thermal polymerisation process with melamine and urea as the reactive precursors, and 15 g melamine and urea were placed, respectively, into a 50 mL covered crucible in a muff furnace. The heating rate was set to 5°/min; the target temperature was 550 °C, maintained for 3 h; and the substance was ground for 5 min to obtain g-C_3N_4 photocatalytic materials, named MCN and UCN.

2.3. Material Characterisation

The crystal structure information of MCN and UCN was recorded using powder X-ray diffraction (XRD) on a PANalytical X'Pert^3 Powder diffraction instrument. The specific functional groups of MCN and UCN were measured using Fourier transform infrared spectroscopy (FTIR) (IS10, Thermo Fisher Scientific, Madison, WI, USA), and the morphology and structure were evaluated using scanning electron microscopy (SEM) (JSM-7900F Plus, Nidec Corporation, Kyoto, Japan). The surface chemical composition and electronic state of MCN and UCN were determined using X-ray photoelectron spectroscopy (XPS) (Thermo Scientific K-Alpha Nexsa, Waltham, MA, USA), and the N_2 adsorption–desorption isotherms and Brunauer–Emmett–Teller (BET) surface area were evaluated using a specific surface area analyser (ASAP 2460). The optical properties were obtained using diffuse reflectance spectra (DRS) via a 3600 UV-Vis spectrophotometer. The photoluminescence (PL) spectra of MCN and UCN were studied using an FLS 980 fluorescence spectrophotometer (Edinburgh Instruments, Livingston, UK) employing an excitation wavelength of 325 nm. Electrochemical impedance (EIS) and photocurrent response tests were conducted on MCN and UCN using an electrochemical workstation (CHI660E, Shanghai Chenhua Instrument Co., Ltd., Shanghai, China). A standard three-electrode system is adopted (sample-coated conductive glass is the working electrode, platinum electrode is the opposite electrode, and Ag/AgCl is the reference electrode). The working electrode is prepared as follows: 10 mg of powder sample is weighed and dispersed in 1 mL of ethanol solution, followed by the addition of 50 μL of Nafion solution. The mixture is then subjected to ultrasonic treatment for 30 min to achieve a homogeneous suspension. Subsequently, 150 μL of the suspension is dropped onto an ITO glass substrate and dried at room temperature for testing. The measurement process was conducted in a 0.1 M Na_2SO_4 aqueous solution. Raman analysis was performed using a confocal Raman microscope (CRM) (Alpha300R, WITec GmbH, Ulm, Germany) equipped with a TEM single-frequency laser (λ = 532 nm, laser power = 40 mW, WITec GmbH, Germany). Elemental analysis of MCN and UCN was performed using the PerkinElmer 24,000 Series II CHNS/O analyser. The EEM fluorescence spectrum was acquired using a fluorescence spectrophotometer (Aqualog-UV-800).

2.4. Photocatalytic Performance Measurement

The photocatalytic performance was assessed by comparing the photocatalytic degradation of RhB dye (dissolved in deionised water) using MCN and UCN. The procedure was as follows: 50 mg of MCN (or UCN) was weighed and placed into a solution in a 250 mL beaker (100 mL, 10 mg/L, 20 mg/L, and 30 mg/L RhB dye). To ensure a balance between adsorption and desorption, each suspension was agitated in the darkness for 60 min and then exposed to a 300 W Xenon lamp ($\lambda \geq 420$ nm) or direct sunlight. Experiments using sunlight illumination were carried out outside between 13:00 pm and 14:00 pm (2023/5; weather: fine; 26–35 °C). During the photocatalytic procedure, 5 mL of suspension was taken at predetermined time intervals, and the photocatalyst powder was separated using a 0.45 μm filter membrane. The RhB dye concentration (λ_{max} = 554 nm) in the suspension at different irradiation time intervals was measured using a 6100A UV-Vis spectrophotometer (Shanghai Yuanxi Instruments Co., Ltd., Shanghai, China), and the Lambert–Beer law was used to calculate the RhB removal efficiency.

3. Results and Discussions

3.1. Crystal Structure and Functional Group Analysis

The crystal properties and composition of the synthesised photocatalysts were characterised using the XRD technique. The diffraction pattern in Figure 1a displays two distinct peaks at 2θ values of 13.18° and 27.86°, which can be attributed to the (100) and (002) planes of $g\text{-}C_3N_4$, respectively (JCPDS NO. 01-087-1526) [41]. The (100) crystal plane indicates the periodic arrangement of the structural units in the layer, and the (002) crystal plane corresponds to the interlayer stacking of aromatic segments. The positions of these two diffraction peaks correlate with those previously reported for $g\text{-}C_3N_4$ [42]. Compared with MCN, UCN's peak intensity and width of the (002) crystal plane is weaker and broader, indicating that urea as a precursor can inhibit the crystal growth of $g\text{-}C_3N_4$. The average crystal diameters were calculated using Scherrer's equation: $D = K\lambda/\beta\cos\theta$, where D is the average crystal diameter (nm); K is the Scherrer constant (0.89); λ is the wavelength of the X-ray (1.5406°A); β is the full width at half-maximum intensity and θ corresponds to the angle of maximum peak [3,43], and the calculated average crystallite sizes of MCN and UCN are equal to 3.67 and 1.77 nm, respectively.

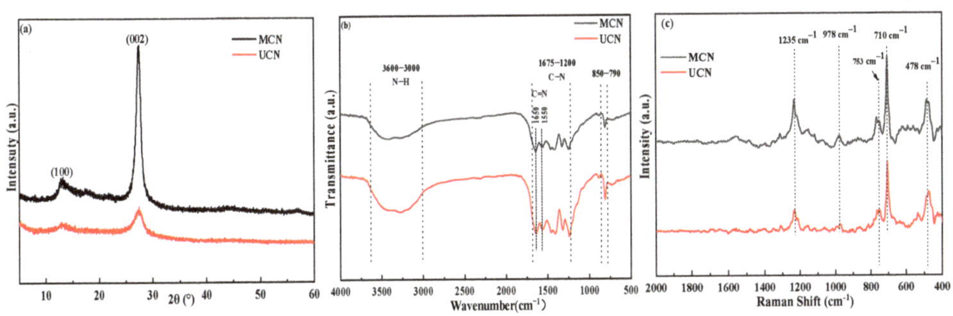

Figure 1. (**a**) XRD patterns, (**b**) FTIR spectra, and (**c**) Raman spectra of MCN and UCN.

The intrinsic chemical structure of the samples can be detected using FTIR spectra. Figure 1b shows that the characteristic peak value at 810 cm^{-1} corresponds to the out-of-plane bending vibration of the triazine ring [44], and the broader band of approximately 3000 to 3600 cm^{-1} corresponds to the N–H stretching vibration [45]. The characteristic band at 1675–1200 cm^{-1} ascribes to the stretching vibration mode of C–N and C=N [46,47]. The tensile vibrations of the bonds in the prepared photocatalyst can be revealed via Raman spectroscopy, as depicted in Figure 1c. The peaks observed at 478 cm^{-1}, 710 cm^{-1}, 753 cm^{-1}, 978 cm^{-1}, and 1235 cm^{-1} correspond to the tensile vibrations of C–C and C–N heterocycles [48–50].

3.2. Microstructure and Compositions Analysis

SEM was performed on MCN and UCN to study the typical morphology and detailed structure of the as-prepared samples. Figure 2a shows that MCN presents a hierarchical structure with layers stacked tightly into blocks. UCN becomes thinner and presents a more dispersed and uniform porous structure (Figure 2c). The degree of the UCN sample's interlayer stacking structure is low, and the interlayer structure is loose, which can increase the surface area and reactive sites [51]. The results are consistent with those of BET. SEM-EDS was used to analyse the distribution and content percentage of elements on the surface of the photocatalyst. C and N are uniformly distributed on the surface of the photocatalyst (Figure 2b,d). The results of the elemental analysis of MCN and UCN are shown in Table 1. The C/N of MCN and UCN are 0.56 and 0.61, respectively.

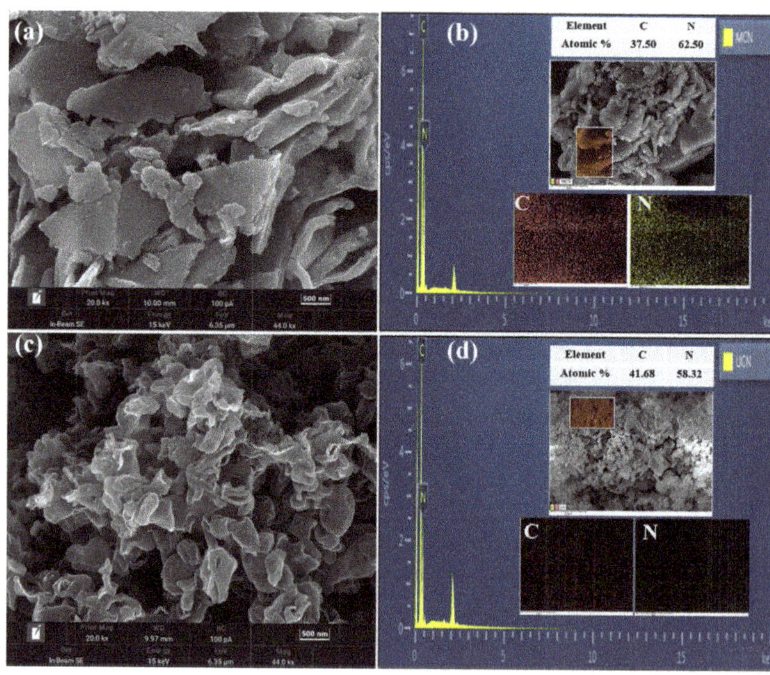

Figure 2. SEM patterns of (**a**) MCN and (**c**) UCN; EDS mapping images of (**b**) MCN and (**d**) UCN.

Table 1. Elemental analysis of MCN and UCN.

Samples	C (%)	N (%)	C/N
MCN	33.93	60.97	0.56
UCN	34.73	57.32	0.61

XPS was used to analyse and compare the surface elemental composition of MCN and UCN. The XPS survey spectrum shows that C and N existed in MCN and UCN (Figure 3a). The C 1s spectra are fitted into two peaks at 288.1 and 284.8 eV (Figure 3b). The primary C 1s peak at 288.1 eV is associated with the sp^2-bonded carbon (N–C=N), and the peak at 284.8 eV corresponds to C–C [31]. The N1s XPS spectrum in Figure 3c displays four peaks at 398.5, 400.4, 401.3, and 404.2 eV [52]. These peaks correspond to C=N–C, ternary N ($N(-C)_3$), N-H bond [7,46], and π-excitations [31], respectively.

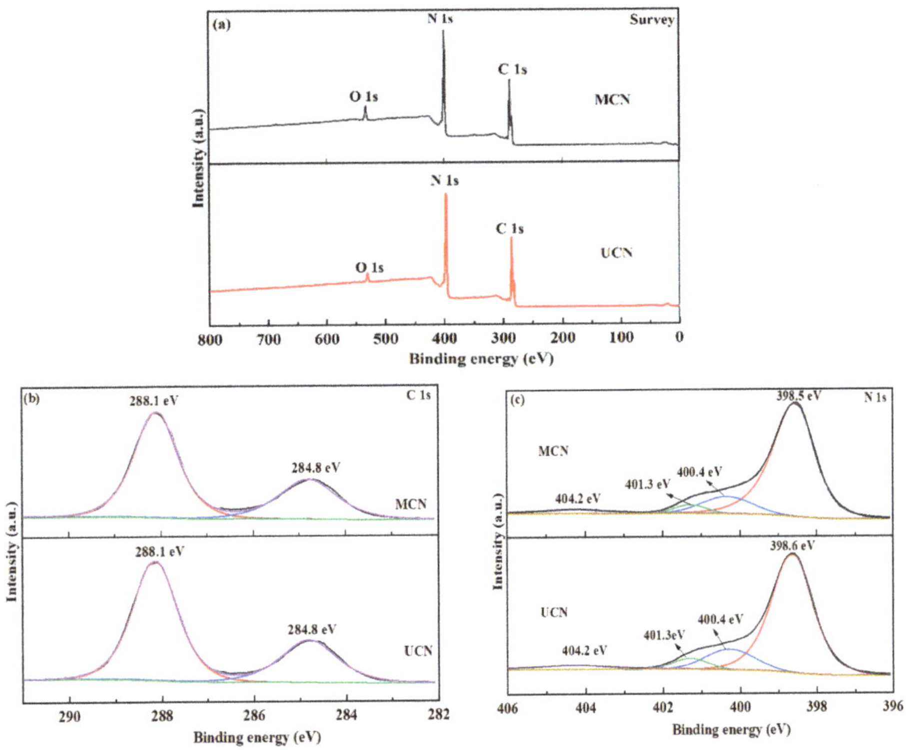

Figure 3. XPS spectra: (**a**) survey scan, (**b**) C 1s, and (**c**) N 1s of MCN and UCN.

3.3. BET Surface Area and Pore Size Distribution Analysis

The BET surface area and the distribution of pore sizes for the MCN and UCN were measured using the N_2 adsorption–desorption isotherms (Figure 4). The adsorption–desorption isotherms of the MCN and UCN are classical type IV with a hysteresis loop of type H_3, indicating that the catalyst has a mesoporous structure [53]. The BET surface area of UCN is 60.03 $m^2 \cdot g^{-1}$, more than eight times that of MCN. Table 2 shows that UCN exhibits a larger specific surface area, greater pore volume, and smaller average pore size, thereby providing an increased number of active sites that contribute to the adsorption of pollutant molecules [54]. Therefore, the mesoporous structure and large BET surface area of UCN are expected to enhance the adsorption capacity and photocatalytic degradation efficiency of pollutants.

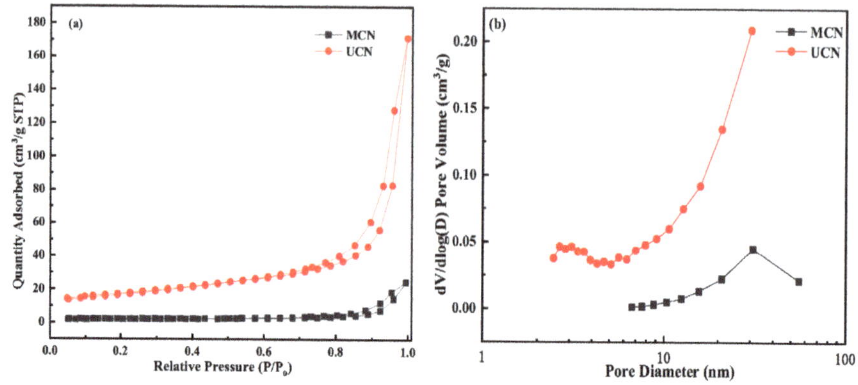

Figure 4. (**a**) N_2 adsorption–desorption isotherms and (**b**) pore size distribution curves of MCN and UCN.

Table 2. The results of the N_2 adsorption–desorption analysis for MCN and UCN.

Samples	BET Surface Area ($m^2 \cdot g^{-1}$)	Pore Volume ($cm^3 \cdot g^{-1}$)	Average Pore Size (nm)
MCN	7.4408	0.002278	20.4142
UCN	60.0322	0.005151	17.6493

3.4. UV–Vis Spectroscopy and Photochemical Characteristics

To understand the light absorption capacity of MCN and UCN, UV–vis absorption spectra were analysed, as shown in Figure 5. The sharp innate absorption edge for MCN and UCN are at around 480 and 460 nm, respectively. The absorption edge's shift is primarily due to the porous structure's strong quantum constraint causing the nanosheets to become thinner and smaller in the UCN [55]. The Kubelka–Munk formula was used for estimation to understand the MCN and UCN energy bands better. This is based on $\alpha h\upsilon = A\,(h\upsilon - Eg)^{n/2}$, where a, h, n, A and *Eg* are the absorption coefficient, the Planck constant, light frequency, a constant and band gap energy [56], respectively. The value of n depends on the type of optical transition exhibited by the semiconductor. For $g\text{-}C_3N_4$ with a direct band gap, the value of n is equal to 1 [21,57]. The *Eg* of MCN and UCN are 2.70 and 2.79 eV, respectively, similar to the study by Hong et al. [58].

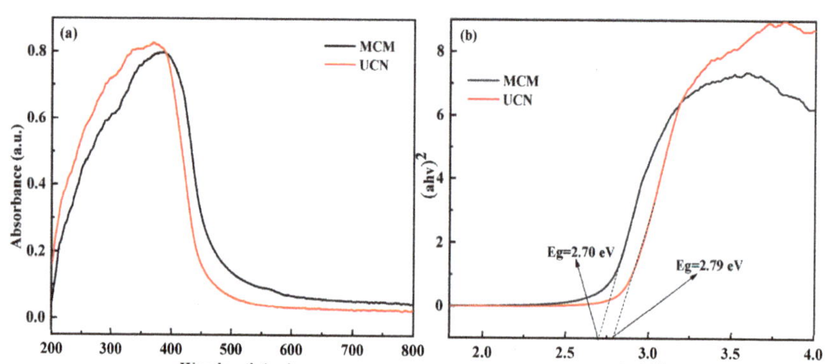

Figure 5. (**a**) UV-vis diffuse reflectance spectra and (**b**) plots of $(\alpha h\upsilon)^2$–$h\upsilon$ of MCN and UCN.

Recombining photogenerated carriers within the photocatalyst releases fluorescence energy, and a high fluorescence signal indicates a high photogenerated carrier recombination rate [9,54]. Therefore, PL analysis was used to evaluate the efficiency of photogenerated electron–hole pair separation. Figure 6a shows that MCN and UCN have distinct signal peaks near 480 nm. The UCN fluorescence intensity is significantly weaker than that of MCN, indicating that UCN aids in the separation of photogenerated electron-hole pairs and thus improves photocatalytic activity. Furthermore, the photocarrier interface separation and migration of g-C_3N_4 were measured using the Nyquist plots of EIS. Generally, a smaller radius of the arc on the EIS Nyquist plot indicates more efficient interface charge transfer and lower photogenerated electron–hole pair recombination [59]. In Figure 6b, the EIS Nyquist plot of UCN shows a smaller arc radius than that of MCN, indicating that UCN has a lower interfacial charge transfer resistance and can separate photoinduced carriers more efficiently. The photocurrent responses of MCN and UCN are shown in Figure 6c. The higher the photocurrent density, the better the electron–hole pair separation performance [60]. The photocurrent response of UCN is higher, which further indicates that UCN has a stronger electron–hole pair separation ability.

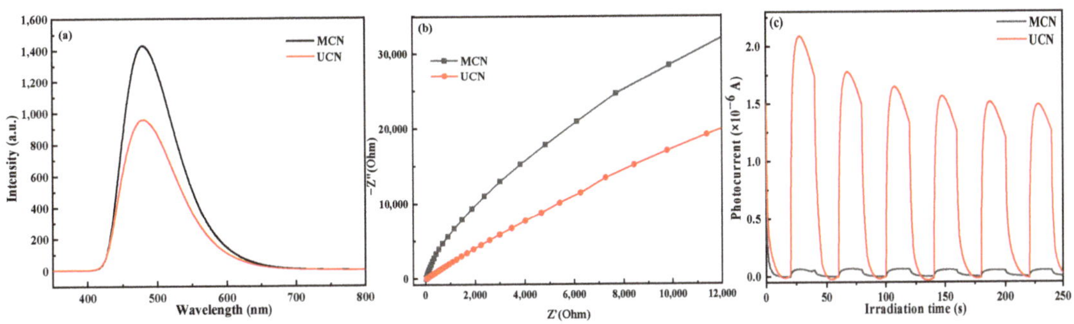

Figure 6. (**a**) PL spectra, (**b**) EIS plots, and (**c**) photocurrent responses of MCN and UCN.

3.5. Photocatalytic Degradation Performance

The degradation of RhB dye was used to gauge both the adsorption and photocatalytic abilities of MCN and UCN. Figure 7 demonstrates that UCN exhibits superior removal efficiency in treating RhB dye at different concentrations under visible light or sunlight, with equilibrium being achieved within 20 min. After 40 min of visible light exposure (Figure 7a), the removal efficiency of UCN towards 30 mg/L RhB dye reaches up to 99.61%, which is higher by 20.31% compared to MCN. MCN has the best effect on the treatment of RhB dye with a concentration of 10 mg/L under visible light, and the removal rate is 98.58% after 40 min of illumination, but the removal efficiency is still lower than that of UCN (99.36%). Additionally, the applicability of the prepared photocatalyst under natural sunlight was also investigated. As shown in Figure 7c, the photocatalyst UCN outperforms the laboratory-simulated visible light catalytic system by completely degrading 10 mg/L of RhB dye in 20 min. This is mostly because that natural sunlight emits not only visible light but also ultraviolet radiation, which can potentially enhance the generation of charge carriers by the catalyst and facilitate the degradation of RhB dye [61]. In order to visualise the photocatalytic degradation of RhB dyes via UCN and MCN, a degradation diagram excluding the adsorption effect was made. As shown in Figure 7b,d, under simulated visible light and real sunlight irradiation, the effect of UCN is still better after deducting the adsorption effect, which indicates the adsorption–photocatalytic synergy. In conclusion, the utilisation of urea yields superior results in the preparation of g-C_3N_4.

Figure 8 shows the UV-Vis absorption spectra for the photodegradation of RhB dye in the UCN photocatalyst. The maximum absorption peak of RhB dye at 554 nm (λ_{max}) gradually decreased with the photodegradation process. After 5 min of visible light irradiation, the UCN photocatalyst decomposed approximately 96.15% of the RhB dye. The primary absorption peak's position was significantly reduced, indicating that UCN has excellent photocatalytic performance.

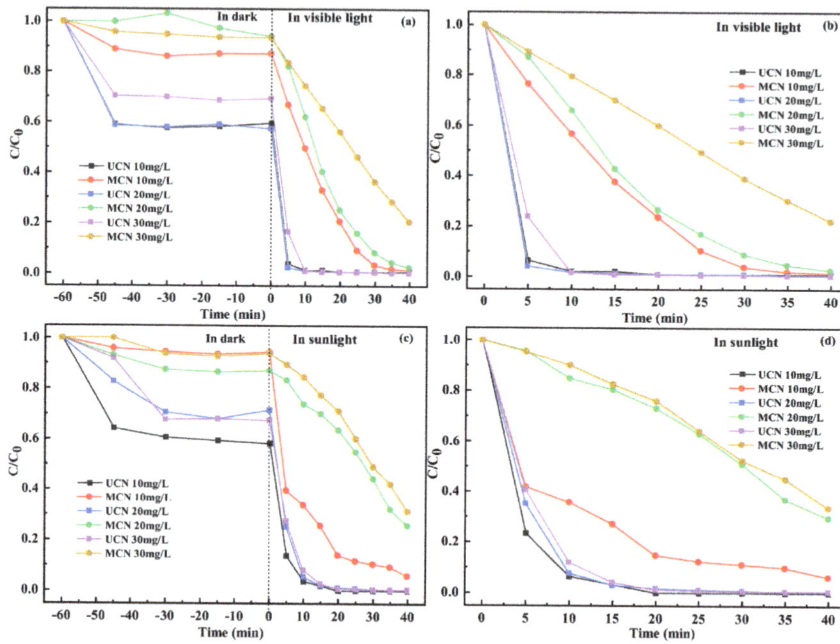

Figure 7. (**a**) Adsorption and photocatalytic degradation, (**b**) photocatalytic degradation of RhB dye under visible light; (**c**) adsorption and photocatalytic degradation, (**d**) photocatalytic degradation of RhB dye under sunlight.

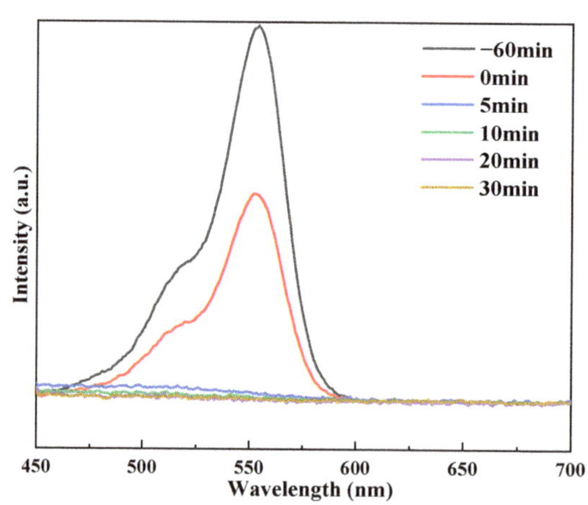

Figure 8. UV-Vis spectral changes of RhB dye in UCN.

In addition, 3D EEMs technology was employed to elucidate the alterations of RhB dye after photocatalysis. Before visible light illumination, the characteristic fluorescence signal appeared at Ex/Em = 460–590/540–700 nm and 200–425/550–670 nm. After being stirred in darkness for 30 min (Figure 9a), the position and shape of the fluorescence peak remained relatively stable, but its intensity slightly decreased, indicating that the RhB molecule was only adsorbed on the catalyst without decomposition. After 40 min, the fluorescence intensity began to decrease. This indicates that the RhB has broken down into pieces and molecular fragments [62]. This is consistent with the results shown in Figure 8.

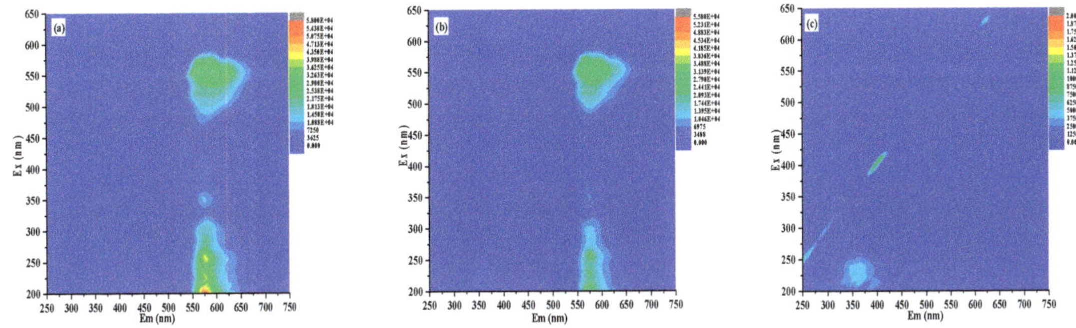

Figure 9. The 3D EEMs results of the RhB solution: (**a**) the original solution, (**b**) after dark adsorption for 60 min, and (**c**) after illumination for 40 min of UCN.

The stability and recycling properties of the UCN were verified by carrying out four cycle tests for the photocatalytic degradation of RhB dye. After each photocatalytic degradation, the powder material was extracted and recycled, cleaned three times with deionised water, and dried for the next cycle experiment. Compared with the first time (99.61%), the removal rate shows a slight change (Figure 10a). In four cycles, the treatment rates were 99.58%, 99.49%, 99.36%, and 99.35%. The results indicate that UCN exhibits exceptional stability and reusability. The photostability of UCN was further confirmed by measuring the XRD patterns of UCN after four recycling tests on RhB (Figure 10b). The XRD results show that there was no change in the crystal structure of UCN and that all the characteristic peaks are present.

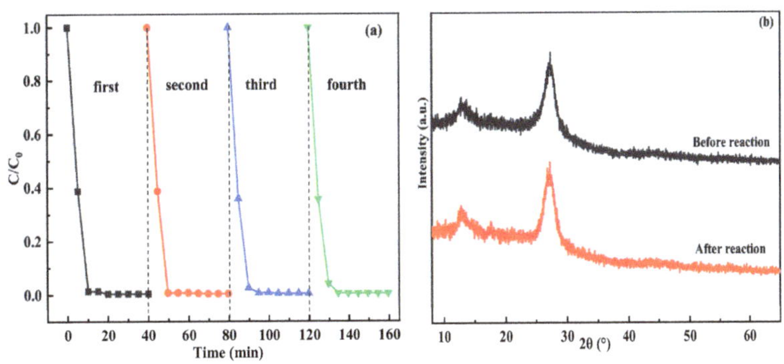

Figure 10. (**a**) Cycling experiments of the photocatalytic RhB dye degradation under visible light and (**b**) XRD patterns before and after photocatalytic RhB dye degradation for UCN.

Furthermore, the photocatalytic property of g-C_3N_4 was compared with other materials explored by other researchers in removing RhB dye, and the results are demonstrated in Table 3. Although the g-C_3N_4 studied in this work is relatively basic, its degradation rate of RhB dye exceeds that of other composite photocatalysts. The results demonstrate that g-C_3N_4 exhibits superior photocatalytic performance compared with most materials over a specific time, thereby establishing a robust foundation for further research.

Table 3. Comparison of RhB dye treatment efficiency of different photocatalysts.

Photocatalyst (Dosages)	RhB (mg/L)	Time (min)	Rate (%)	Light Source	Refs.
$g\text{-}C_3N_4$ (50 mg)	10 20 30	20	99.36 99.39 99.35	300 W Xe lamp	This paper
$g\text{-}C_3N_4$ (50 mg)	10 20 30	20	100 98.78 99.15	Direct sunlight	This paper
ZnO-$g\text{-}C_3N_4$ (40%)-Go (15%) (15 mg)	10	100	98	350 W Xe lamp	[63]
$g\text{-}C_3N_4/Bi_4O_5Br_2$-75 (50 mg)	10	10	92	72 W LED lamp	[64]
$TlSnI_3/g\text{-}C_3N_4$ (100 mg)	5	120	84.60	150 W Osram bulb	[65]
$TiO_2/g\text{-}C_3N_4$ (50 mg)	10	150	99.30	350 W Xe lamp	[66]
Co@ZnSQDs/$g\text{-}C_3N_4$/MWCNT (220 mg)	10	75	96	500 W halogen lamp	[67]
$g\text{-}C_3N_4/ZnO/Cu_2O$ (50 mg)	30	100	91.4	500 W halogen lamp	[68]

3.6. Photocatalytic Mechanism

Trapping tests were conducted to study the various species implicated in photodegradation and the primarily responsible species. The primary active species of UCN photocatalytic degradation were detected by adding IPA, BQ, and SO to RhB dye as scavengers of •OH, $•O_2^-$, and h^+ [43]. Figure 11 shows that the degree of removal of RhB dye (41.56%) declined significantly with the addition of BQ, while the scavenging effect of IPA (99.58%) and SO (98.53%) was weaker than that of BQ. Therefore, in the reaction process of UCN degradation of RhB dye, $•O_2^-$ serves as the primary reactive species.

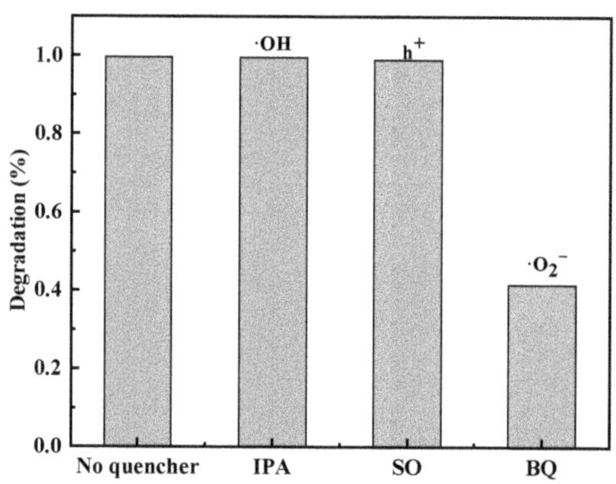

Figure 11. Trapping experiment of active species for UCN photocatalytic degradation of RhB under visible light.

The valence band (VB) potentials of a semiconductor can be determined using the empirical equation $E_{VB} = X - E^e + 0.5E_g$, where X, E^e, and E_g represent the semiconductor's electronegativity (the X value for $g\text{-}C_3N_4$ is 4.72 eV [69]), the energy of free electrons on the hydrogen scale (4.5 eV), and the semiconductor's bandgap, respectively [70]. Moreover, the CB potentials can be calculated using $E_{CB} = E_{VB} + E_g$ [71]. In the DRS analysis results (Figure 5b), the E_g of UCN and MCN are 2.79 and 2.70 eV, respectively. Therefore, the CB and VB potentials of UCN are calculated to be −1.17 eV (E_{CB}) and 1.62 eV (E_{VB}), while the CB and VB potentials of MCN are −1.13 eV (E_{CB}) and 1.57 eV (E_{VB}), respectively. According to these results, Figure 12 illustrates the plausible photocatalytic mechanism involved in the process of RhB photocatalytic degradation. UCN and MCN tend to produce e^- and h^+ under visible light radiation. Since the CB potential of UCN and MCN is lower than $E^{\theta}(O_2/•O_2^-) = -0.33$ eV, the e^- in the CB of UCN and MCN can capture the adsorbed O_2

on the catalyst surface and reduce it to give $\bullet O_2^-$. Since the E_{VB} of UCN and MCN are more negative than $E^{\theta}(\bullet OH/H_2O) = 2.27$ eV [71], •OH cannot be generated on the VB of UCN and MCN. The results correlate with the trapping experiments.

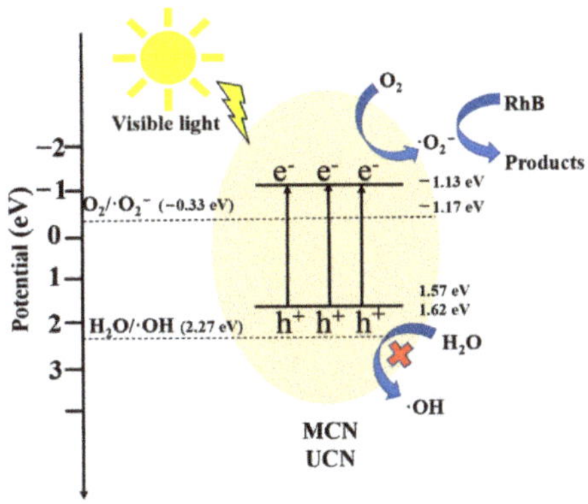

Figure 12. Photocatalytic mechanism diagram of MCN and UCN under visible light radiation.

4. Conclusions

MCN and UCN were prepared via thermal polymerisation using melamine and urea as precursors. Compared to MCN, UCN exhibits superior photocatalytic degradation performance towards RhB dye. Under the irradiation of visible light, the removal rate of RhB dye at 10 mg/L within 20 min reaches 99.36%, while complete degradation can be achieved under sunlight. The photocatalytic efficiency of UCN is enhanced by the layered structures and a large specific surface area, which enrich the reaction sites, improve the transfer of photoinduced charges, and strengthen the absorbance of visible light. The prepared UCN exhibits exceptional stability, even after undergoing four cycles of photocatalytic reaction. Moreover, the trapping experiments established that $\bullet O_2^-$ was the primary reactive species in the photocatalytic degradation of RhB dye under visible light radiation. Therefore, the as-synthesised UCN are promising photocatalysts for environmental applications such as the treatment of industrial wastewater under natural sunlight. This study highlights the potential of UCN as a photocatalyst for RhB dye removal in wastewater treatment and lays a foundation for further exploration of $g\text{-}C_3N_4$.

Author Contributions: Conceptualisation, software, data curation, formal analysis, writing—original draft, M.Z.; writing—review and editing, M.X., B.D., H.Z. (Hongxiang Zhu), Q.L., X.L. and J.M.; supervision, writing—review and editing, X.S. and H.Z. (Hongxia Zhang); funding acquisition, X.S. All authors have read and agreed to the published version of the manuscript.

Funding: This work was financially supported by the National Natural Science Foundation of China (No. 52070049) and the Innovation Project of Guangxi Graduate Education (No. YCSW2023354).

Data Availability Statement: Data are available upon request.

Conflicts of Interest: The authors declare no conflict of interest.

References

1. Tian, Y.; Zhang, J.; Wang, W.; Liu, J.; Zheng, X.; Li, J.; Guan, X. Facile assembly and excellent elimination behavior of porous BiOBr-g-C_3N_4 heterojunctions for organic pollutants. *Environ. Res.* **2022**, *209*, 112889. [CrossRef] [PubMed]
2. Mahdieh, V.; Iman, K.; Farzaneh, S.; Parastoo, J. Green synthesis of a novel magnetic Fe_3O_4@SiO_2/TiO_2@WO_3 nanocomposite for methylene blue removal under UV and visible light irradiations. *Res. Chem. Intermed.* **2023**, *49*, 1909–1924.
3. Zhang, M.; Sun, X.; Wang, C.; Wang, Y.; Tan, Z.; Li, J.; Xi, B. Photocatalytic degradation of rhodamine B using $Bi_4O_5Br_2$-doped ZSM-5. *Mater. Chem. Phys.* **2022**, *278*, 125697. [CrossRef]
4. Budnyak, T.M.; Onwumere, J.; Pylypchuk, I.V.; Jaworski, A.; Chen, J.; Rokicińska, A.; Lindström, M.E.; Kuśtrowski, P.; Sevastyanova, O.; Slabon, A. LignoPhot: Conversion of hydrolysis lignin into the photoactive hybrid lignin/$Bi_4O_5Br_2$/BiOBr composite for simultaneous dyes oxidation and Co^{2+} and Ni^{2+} recycling. *Chemosphere* **2021**, *279*, 130538. [CrossRef] [PubMed]

5. Li, Z.; Ma, B.; Zhang, X.; Sang, Y.; Liu, H. One-pot synthesis of BiOCl nanosheets with dual functional carbon for ultra-highly efficient photocatalytic degradation of RhB. *Environ. Res.* **2020**, *182*, 109077. [CrossRef]
6. Yang, X.; Zhang, L.; Wang, D.; Zhang, Q.; Zeng, J.; Zhang, R. Facile synthesis of nitrogen-defective g-C_3N_4 for superior photocatalytic degradation of rhodamine B. *RSC Adv.* **2021**, *11*, 30503–30509. [CrossRef] [PubMed]
7. Zhou, J.; Liu, W.; Cai, W. The synergistic effect of Ag/AgCl@ZIF-8 modified g-C_3N_4 composite and peroxymonosulfate for the enhanced visible-light photocatalytic degradation of levofloxacin. *Sci. Total Environ.* **2019**, *696*, 133962. [CrossRef] [PubMed]
8. Cui, S.; Li, X.; Li, Y.; Zhao, H.; Wang, Y.; Li, N.; Li, X.; Li, G. Synthesis of CdS/m-TiO_2 mesoporous spheres and their application in photocatalytic degradation of rhodamine B under visible light. *Chem. Res. Chin. Univ.* **2017**, *33*, 436–441. [CrossRef]
9. Wang, C.; Sun, X.; Zhang, M.; Wang, Y.; Tan, Z.; Li, J.; Xi, B. Ultrasound-assisted room-temperature in situ precipitation synthesis of BC doped $Bi_4O_5Br_2$ for enhanced photocatalytic activity in pollutants degradation under visible light. *J. Alloys Compd.* **2021**, *889*, 161609. [CrossRef]
10. Zhang, G.-Q.; Chang, N.; Han, D.-Q.; Zhou, A.-Q.; Xu, X.-H. The enhanced visible light photocatalytic activity of nanosheet-like Bi_2WO_6 obtained by acid treatment for the degradation of rhodamine B. *Mater. Lett.* **2010**, *64*, 2135–2137. [CrossRef]
11. Hu, C.; Xu, M.; Zhang, J.; Zhou, Y.; Hu, B.; Yu, G. Recyclable MoO_3 nanobelts for photocatalytic degradation of Rhodamine B by near infrared irradiation. *Int. J. Chem. Kinet.* **2018**, *51*, 3–13. [CrossRef]
12. Nguyen, L.T.T.; Vo, D.-V.N.; Nguyen, L.T.H.; Duong, A.T.T.; Nguyen, H.Q.; Chu, N.M.; Nguyen, D.T.C.; Van Tran, T. Synthesis, characterization, and application of $ZnFe_2O_4$@ZnO nanoparticles for photocatalytic degradation of Rhodamine B under visible-light illumination. *Environ. Technol. Innov.* **2021**, *25*, 102130. [CrossRef]
13. Lin, X.; Li, Y. Preparation of TiO_2/Ag[BMIM]Cl Composites and Their Visible Light Photocatalytic Properties for the Degradation of Rhodamine B. *Catalysts* **2021**, *11*, 661. [CrossRef]
14. Liu, W.; Li, Y.; Liu, F.; Jiang, W.; Zhang, D.; Liang, J. Visible-light-driven photocatalytic degradation of diclofenac by carbon quantum dots modified porous g-C_3N_4: Mechanisms, degradation pathway and DFT calculation. *Water Res.* **2018**, *151*, 8–19. [CrossRef] [PubMed]
15. Prabavathi, S.L.; Saravanakumar, K.; Mamba, G.; Muthuraj, V. 1D/2D $MnWO_4$ nanorods anchored on g-C_3N_4 nanosheets for enhanced photocatalytic degradation ofloxacin under visible light irradiation. *Colloids Surf. A. Physicochem. Eng. Asp.* **2019**, *581*, 123845.
16. Song, Y.; Gu, J.; Xia, K.; Yi, J.; Che, H.; She, X.; Chen, Z.; Ding, C.; Li, H.; Xu, H. Construction of 2D SnS_2/g-C_3N_4 Z-scheme composite with superior visible-light photocatalytic performance. *Appl. Surf. Sci.* **2019**, *467/468*, 56–64. [CrossRef]
17. Orooji, Y.; Ghanbari, M.; Amiri, O.; Salavati-Niasari, M. Facile fabrication of silver iodide/graphitic carbon nitride nanocomposites by notable photo-catalytic performance through sunlight and antimicrobial activity. *J. Hazard. Mater.* **2020**, *389*, 122079. [CrossRef] [PubMed]
18. Li, H.; Wang, Z.; Lu, Y.; Liu, S.; Chen, X.; Wei, G.; Ye, G.; Chen, J. Microplasma electrochemistry (MIPEC) methods for improving the photocatalytic performance of g-C_3N_4 in degradation of RhB. *Appl. Surf. Sci.* **2020**, *531*, 147307. [CrossRef]
19. Yang, Z.; Zhang, Y.; Schnepp, Z. Soft and hard templating of graphitic carbon nitride. *J. Mater. Chem. A* **2015**, *3*, 14081–14092. [CrossRef]
20. Wang, W.; Li, G.; An, T.; Chan, D.K.L.; Yu, J.C.; Wong, P.K. Photocatalytic hydrogen evolution and bacterial inactivation utilizing sonochemical-synthesized g-C_3N_4/red phosphorus hybrid nanosheets as a wide-spectral-responsive photocatalyst: The role of type I band alignment. *Appl. Catal. B Environ.* **2018**, *238*, 126–135. [CrossRef]
21. Zhong, S.; Zhou, H.; Shen, M.; Yao, Y.; Gao, Q. Rationally designed a g-C_3N_4/BiOI/$Bi_2O_2CO_3$ composite with promoted photocatalytic activity. *J. Alloys Compd.* **2021**, *853*, 157307. [CrossRef]
22. Cao, C.; Wang, H. Carbon nitride films deposited from organic solutions by electrodeposition. *Diam. Relat. Mater.* **1999**, *8*, 1786–1789. [CrossRef]
23. Viehland, J.; Lin, S.; Feldman, B.J.; Kilgore, K.; Jones, M.T. Search for the nitrogen dangling bond in amorphous hydrogenated carbon nitride. *Solid State Commun.* **1991**, *80*, 597–599. [CrossRef]
24. Wei, H.; McMaster, W.A.; Tan, J.Z.Y.; Cao, L.; Chen, D.; Caruso, R.A. Mesoporous TiO_2/g-C_3N_4 Microspheres with Enhanced Visible-Light Photocatalytic Activity. *J. Phys. Chem. C. Nanomater. Interfaces* **2017**, *121*, 22114–22122. [CrossRef]
25. Bai, Y.-J.; Lü, B.; Liu, Z.-G.; Li, L.; Cui, D.-L.; Xu, X.-G.; Wang, Q.-L. Solvothermal preparation of graphite-like C_3N_4 nanocrystals. *J. Cryst. Growth* **2003**, *247*, 505–508. [CrossRef]
26. Niu, P.; Zhang, L.; Liu, G.; Cheng, H.-M. Graphene-Like Carbon Nitride Nanosheets for Improved Photocatalytic Activities. *Adv. Funct. Mater.* **2012**, *22*, 4763–4770. [CrossRef]
27. Andreyev, A.; Akaishi, M.; Golberg, D. Sodium flux-assisted low-temperature high-pressure synthesis of carbon nitride with high nitrogen content. *Chem. Phys. Lett.* **2003**, *372*, 635–639. [CrossRef]
28. Goettmann, F.; Fischer, A.; Antonietti, M.; Thomas, A. Chemical Synthesis of Mesoporous Carbon Nitrides Using Hard Templates and Their Use as a Metal-Free Catalyst for Friedel-Crafts Reaction of Benzene. *Angew. Chem.* **2006**, *45*, 4467–4471. [CrossRef] [PubMed]
29. Zhang, J.; Guo, F.; Wang, X. An Optimized and General Synthetic Strategy for Fabrication of Polymeric Carbon Nitride Nanoarchitectures. *Adv. Funct. Mater.* **2013**, *23*, 3008–3014. [CrossRef]
30. Yan, S.C.; Li, Z.S.; Zou, Z.G. Photodegradation Performance of g-C_3N_4 Fabricated by Directly Heating Melamine. *Langmuir* **2009**, *25*, 10397–10401. [CrossRef] [PubMed]

31. Liu, J.; Zhang, T.; Wang, Z. Simple pyrolysis of urea into graphitic carbon nitride with recyclable adsorption and photocatalytic activity. *J. Mater. Chem. A* **2011**, *21*, 14398–14401. [CrossRef]
32. Ma, Y.; Liu, E.; Hu, X.; Tang, C.; Wan, J.; Li, J.; Fan, J. simple process to prepare few-layer g-C_3N_4 nanosheets with enhanced photocatalytic activities. *Appl. Surf. Sci.* **2015**, *358*, 246–251. [CrossRef]
33. Dong, L.; Chunling, L.; Jiayu, G.; Congyue, Z.; Qian, Z.; Fengquan, Z.; Tianjun, N.; Weidong, W. 3D interconnected g-C_3N_4 hybridized with 2D Ti_3C_2 MXene nanosheets for enhancing visible light photocatalytic hydrogen evolution and dye contaminant elimination. *Appl. Surf. Sci.* **2021**, *579*, 152180.
34. Xu, F.; Cheng, G.; Song, S.; Wei, Y.; Chen, R. Insights into Promoted Adsorption Capability of Layered BiOCl Nanostructures Decorated with TiO_2 Nanoparticles. *ACS Sustain. Chem. Eng.* **2016**, *4*, 12. [CrossRef]
35. Xiong, J.; Gan, Y.; Zhu, J.; Li, W.; Gao, C.; Wei, Y.; Cheng, G.; Li, Z.; Dou, S. Insights into the structure-induced catalysis dependence of simply engineered one-dimensional zinc oxide nanocrystals towards photocatalytic water purification. *Inorg. Chem. Front.* **2017**, *4*, 2075–2087. [CrossRef]
36. Zhang, M.; Xiong, J.; Yang, H.; Wen, Z.; Chen, R.; Cheng, G. Surface Potential/Wettability and Interface Charge Transfer Engineering of Copper-Oxide (Cu–MOx, M = W, Ti, and Ce) Hybrids for Efficient Wastewater Treatment through Adsorption–Photocatalysis Synergy. *Ind. Eng. Chem. Res.* **2020**, *59*, 35. [CrossRef]
37. Luo, Y.; Wei, X.; Gao, B.; Zou, W.; Zheng, Y.; Yang, Y.; Zhang, Y.; Tong, Q.; Dong, L. Synergistic adsorption-photocatalysis processes of graphitic carbon nitrate (g-C_3N_4) for contaminant removal: Kinetics, models, and mechanisms. *Chem. Eng. J.* **2019**, *375*, 122019. [CrossRef]
38. Yu, Y.; Chen, D.; Xu, W.; Fang, J.; Sun, J.; Liu, Z.; Chen, Y.; Liang, Y.; Fang, Z. Synergistic adsorption-photocatalytic degradation of different antibiotics in seawater by a porous g-C_3N_4/calcined-LDH and its application in synthetic mariculture wastewater. *J. Hazard. Mater.* **2021**, *416*, 126183. [CrossRef]
39. Ni, T.; Zhang, H.; Yang, Z.; Zhou, L.; Pan, L.; Li, C.; Yang, Z.; Liu, D. Enhanced adsorption and catalytic degradation of antibiotics by porous 0D/3D Co_3O_4/g-C_3N_4 activated peroxymonosulfate: An experimental and mechanistic study. *J. Colloid Interface Sci.* **2022**, *625*, 466–478. [CrossRef] [PubMed]
40. Liu, D.; Li, C.; Zhao, C.; Zhao, Q.; Niu, T.; Pan, L.; Xu, P.; Zhang, F.; Wu, W.; Ni, T. Facile synthesis of three-dimensional hollow porous carbon doped polymeric carbon nitride with highly efficient photocatalytic performance. *Chem. Eng. J.* **2022**, *438*, 135623. [CrossRef]
41. Wang, Y.; Cao, Y.; Liu, Y.; Yang, P. Effect of nonmetal element dopants on photo- and electro-chemistry performance of ultrathin g-C_3N_4 nanosheets. *Int. J. Hydrogen Energy* **2020**, *45*, 16519–16527. [CrossRef]
42. Di, X.; Shuo, Z.; Tianyu, W.; Mei, H.; Yiyan, C.; Panpan, Z.; Fangchao, C. Efficient RhB degradation using $MnFe_2O_4$/g-C_3N_4 composites under visible light irradiation. *Opt. Mater.* **2022**, *124*, 111965.
43. Zhang, M.; Xing, M.; Dong, B.; Sun, X.; Zhang, H.; Wang, C.; Zhu, H. Preparation of $BiVO_4/CO_3^{2-}$-$Bi_2O_2CO_3$ heterojunctions for enhanced photocatalytic activity in the degradation of levofloxacin under visible light. *J. Alloys Compd.* **2023**, *965*, 171471. [CrossRef]
44. Minjie, Z.; Aiyue, J.; Yujuan, C.; Mengyu, T.; Yulian, W.; Zhigang, T. Synergistic effects of Au/g-C_3N_4/montmorillonite composite nanocatalysts for enhanced photocatalytic performance with visible light illumination. *J. Nanoparticle Res.* **2023**, *25*, 87.
45. Li, D.; Huang, J.; Li, R.; Chen, P.; Chen, D.; Cai, M.; Liu, H.; Feng, Y.; Lv, W.; Liu, G. Synthesis of a carbon dots modified g-C_3N_4/SnO_2 Z-scheme photocatalyst with superior photocatalytic activity for PPCPs degradation under visible light irradiation. *J. Hazard. Mater.* **2021**, *401*, 123257.1–123257.13. [CrossRef] [PubMed]
46. Van, K.N.; Huu, H.T.; Nguyen Thi, V.N.; Le Thi, T.L.; Truong, D.H.; Truong, T.T.; Dao, N.N.; Vo, V.; Tran, D.L.; Vasseghian, Y. Facile construction of S-scheme SnO_2/ g-C_3N_4 photocatalyst for improved photoactivity. *Chemosphere* **2021**, *289*, 133120. [CrossRef]
47. Ghafuri, H.; Rashidizadeh, A.; Zaron, T.M. Facile preparation of CuS-g-C_3N_4/Ag nanocomposite with improved photocatalytic activity for the degradation of rhodamine B. *Polyhedron* **2020**, *179*, 114368. [CrossRef]
48. Ma, W.; Wang, N.; Guo, Y.; Yang, L.; Lv, M.; Tang, X.; Li, S. Enhanced photoreduction CO_2 activity on g-C_3N_4: By synergistic effect of nitrogen defective-enriched and porous structure, and mechanism insights. *Chem. Eng. J.* **2020**, *388*, 124288. [CrossRef]
49. Zhu, Z.; Fan, W.; Liu, Z.; Yu, Y.; Dong, H.; Huo, P.; Yan, Y. Fabrication of the metal-free biochar-based graphitic carbon nitride for improved 2-Mercaptobenzothiazole degradation activity. *J. Photochem. Photobiol. A Chem.* **2018**, *358*, 284–293. [CrossRef]
50. Xu, Z.; Zhang, J.; Zhong, J.; Wu, J.; Li, M. Reductively-induced carbon vacancies facilitate visible light-driven hydrogen evolution enhancement of g-C_3N_4. *Mater. Sci. Semicond. Process.* **2023**, *161*, 107459. [CrossRef]
51. Shi, J.; Chen, T.; Guo, C.; Liu, Z.; Feng, S.; Li, Y.; Hu, J. The bifunctional composites of AC restrain the stack of g-C_3N_4 with the excellent adsorption-photocatalytic performance for the removal of RhB. *Colloids Surf. A. Physicochem. Eng. Asp.* **2019**, *580*, 123701. [CrossRef]
52. Bai, J.; Yin, C.; Xu, H.; Chen, G.; Ni, Z.; Wang, Z.; Li, Y.; Kang, S.; Zheng, Z.; Li, X. Facile urea-assisted precursor pre-treatment to fabricate porous g-C_3N_4 nanosheets for remarkably enhanced visible-light-driven hydrogen evolution. *J. Colloid Interface Sci.* **2018**, *532*, 280–286. [CrossRef] [PubMed]
53. Zhu, J.; Xiong, Y.; Mu, X.; Wan, J.; Li, T.; Jin, Y.; Li, R. Cobalt-Doped g-C_3N_4 Nanosheets for One-Pot Synthesis of Imines under Mild Conditions. *ACS Appl. Nano Mater.* **2023**, *6*, 7. [CrossRef]

54. Guo, Y.; Li, C.; Guo, Y.; Wang, X.; Li, X. Ultrasonic-assisted synthesis of mesoporous g-C_3N_4/Na-bentonite composites and its application for efficient photocatalytic simultaneous removal of Cr(VI) and RhB. *Colloids Surf. A. Physicochem. Eng. Asp.* **2019**, *578*, 123624. [CrossRef]
55. Liu, X.; Pang, F.; He, M.; Ge, J. Confined reaction inside nanotubes: New approach to mesoporous g-C_3N_4 photocatalysts. *Nano Res.* **2017**, *10*, 3638–3647. [CrossRef]
56. Liqun, Y.; Xiaoli, J.; Chuan, L.; Chenghua, D.; Haiquan, X.; Ka Him, C.; Po Keung, W. Thickness-ultrathin and bismuth-rich strategies for BiOBr to enhance photoreduction of CO_2 into solar fuels. *Appl. Catal. B Environ.* **2016**, *187*, 281–290.
57. Feng, W.; Fang, J.; Zhou, G.; Zhang, L.; Lu, S.; Wu, S.; Chen, Y.; Ling, Y.; Fang, Z. Rationally designed Bi@BiOCl/g-C_3N_4 heterostructure with exceptional solar-driven photocatalytic activity. *Mol. Catal.* **2017**, *434*, 69–79. [CrossRef]
58. Hong, Y.; Li, C.; Zhang, G.; Meng, Y.; Yin, B.; Zhao, Y.; Shi, W. Efficient and stable Nb_2O_5 modified g-C_3N_4 photocatalyst for removal of antibiotic pollutant. *Chem. Eng. J.* **2016**, *299*, 74–84. [CrossRef]
59. Wang, J.; Yang, Z.; Gao, X.; Yao, W.; Wei, W.; Chen, X.; Zong, R.; Zhu, Y. Core-shell g-C_3N_4@ZnO composites as photoanodes with double synergistic effects for enhanced visible-light photoelectrocatalytic activities. *Appl. Catal. B* **2017**, *217*, 169–180. [CrossRef]
60. Pei, Q.; Mengjie, L.; Gang, C.; Wei, L.; Lijun, L.; Jinyan, X. Co-implantation of oxygen vacancy and well-dispersed Cu cocatalyst into TiO_2 nanoparticles for promoting solar-to-hydrogen evolution. *Int. J. Hydrogen Energy* **2022**, *48*, 933–942.
61. Chen, X.; Yao, L.; He, J.; Li, J.; Xu, S.; Li, N.; Zhu, Y.; Chen, X.; Zhu, R. Enhanced degradation of tetracycline under natural sunlight through the synergistic effect of Ag_3PO_4/MIL-101(Fe) photocatalysis and Fenton catalysis: Mechanism, pathway, and toxicity assessment. *J. Hazard. Mater.* **2023**, *449*, 131024. [CrossRef]
62. Guo, H.; Niu, C.-G.; Wen, X.-J.; Zhang, L.; Liang, C.; Zhang, X.-G.; Guan, D.-L.; Tang, N.; Zeng, G.-M. Construction of highly efficient and stable ternary AgBr/Ag/$PbBiO_2Br$ Z-scheme photocatalyst under visible light irradiation: Performance and mechanism insight. *J. Colloid Interface Sci.* **2017**, *513*, 852–865. [CrossRef]
63. Zhang, J.; Li, J.; Liu, X. Ternary nanocomposite ZnO- g-C_3N_4–Go for enhanced photocatalytic degradation of RhB. *Opt. Mater.* **2021**, *119*, 111351. [CrossRef]
64. Yi, F.; Ma, J.; Lin, C.; Wang, L.; Zhang, H.; Qian, Y.; Zhang, K. Insights into the enhanced adsorption/photocatalysis mechanism of a $Bi_4O_5Br_2$/g-C_3N_4 nanosheet. *J. Alloys Compd.* **2020**, *821*, 153557. [CrossRef]
65. Fatemeh, Y.; Qahtan, A.Y.; Mojgan, G.; Masoud, S.-N. Fabrication of $TlSnI_3$/C_3N_4 nanocomposites for enhanced photodegradation of toxic contaminants below visible light and investigation of kinetic and mechanism of photocatalytic reaction. *J. Mol. Liq.* **2022**, *349*, 118443.
66. Jiang, Z.; Zhu, C.; Wan, W.; Qian, K.; Xie, J. Constructing graphite-like carbon nitride modified hierarchical yolk–shell TiO_2 spheres for water pollution treatment and hydrogen production. *J. Mater. Chem. A* **2015**, *4*, 1806–1818. [CrossRef]
67. Danish, M.; Muneer, M. Facile synthesis of highly efficient Co@ZnSQDs/g-C_3N_4/MWCNT nanocomposites and their photocatalytic potential for degradation of RhB dye: Efficiency, degradation kinetics, and mechanism pathway. *Ceram. Int.* **2021**, *47*, 13043–13056. [CrossRef]
68. Renji, R.; Shanmugam, V.; Asokan, S.; Palanisamy, P.; Sanjeevamuthu, S.; Vairamuthu, R.; Jeyaperumal Kalyana, S.; Manickam, S.; Mohd, S.; AlFaify, S. Investigation on novel Cu_2O modified g-C_3N_4/ZnO heterostructures for efficient photocatalytic dye degradation performance under visible-light exposure. *Colloid Interface Sci. Commun.* **2021**, *44*, 100480.
69. Luo, J.; Zhou, X.; Ma, L.; Xu, X. Enhancing visible-light photocatalytic activity of g-C_3N_4 by doping phosphorus and coupling with CeO_2 for the degradation of methyl orange under visible light irradiation. *RSC Adv.* **2015**, *5*, 68728–68735. [CrossRef]
70. Zhang, L.; Wang, G.; Xiong, Z.; Tang, H.; Jiang, C. Fabrication of flower-like direct Z-scheme beta-Bi_2O_3/g-C_3N_4 photocatalyst with enhanced visible light photoactivity for Rhodamine B degradation. *Appl. Surf. Sci.* **2018**, *436*, 162–171. [CrossRef]
71. Huang, L.; Liu, J.; Li, P.; Li, Y.; Wang, C.; Shu, S.; Song, Y. CQDs modulating Z-scheme g-C_3N_4/BiOBr heterostructure for photocatalytic removing RhB, BPA and TC and E. coli by LED light. *J. Alloys Compd.* **2022**, *895*, 162637. [CrossRef]

water

Perspective

Tracking and Utilizing Sargassum, an Abundant Resource from the Caribbean Sea

Meng-Chou Lee [1,2], Han-Yang Yeh [1], Chun-Ming Chang [3], Yu-Fu Liou [3], Fan-Hua Nan [1] and Jane Wungen-Sani [1,*]

[1] Department of Aquaculture, National Taiwan Ocean University, Keelung City 20224, Taiwan; mengchoulee@email.ntou.edu.tw (M.-C.L.); 20833001@mail.ntou.edu.tw (H.-Y.Y.)
[2] Center of Excellence for the Ocean, National Taiwan Ocean University, Keelung City 20224, Taiwan
[3] Advanced Research Center for Earth Sciences, National Central University, Taoyuan City 328003, Taiwan; glazejimmy@gmail.com (C.-M.C.); liouyufu@g.ncu.edu.tw (Y.-F.L.)
* Correspondence: 21133009@mail.ntou.edu.tw; Tel.: +886-2-246222192 (ext. 5239)

Abstract: Due to climate change and its associated factors, there has been an increased influx of pelagic brown algae biomass drifting freely in the Caribbean Sea in recent years. Its use as an industrial recyclable material is feasible, although pelagic Sargassum species have Arsenic (As) heavy metal content; among 531 tested pesticide residues applied to vegetables, fruit, infant food, herbs, and spices, zero were found in Sargassum. Sargassum also contains sulfated polysaccharides and oligosaccharides, which are known to be beneficial immunomodulators. Our results thus suggest Sargassum to be a functionally useful organic material in small quantities as an additive in animal feed. With improved chemical extraction methods, it could also be highly effective in adjusted proportions in cosmetics and for other industrial uses. A viable solution for predicting and tracking the large-scale movements of algal masses is provided. Furthermore, a method for addressing increasing Sargassum influx is proposed via its use as an industrial recyclable material following composition analysis, evaluation and safety assessments for cosmetic use, and research and design of new beauty products and other functional cosmetics.

Keywords: *Sargassum*; Sargasso Sea; algal bloom; arsenic; satellite; numerical simulation

Citation: Lee, M.-C.; Yeh, H.-Y.; Chang, C.-M.; Liou, Y.-F.; Nan, F.-H.; Wungen-Sani, J. Tracking and Utilizing Sargassum, an Abundant Resource from the Caribbean Sea. *Water* **2023**, *15*, 2694. https://doi.org/10.3390/w15152694

Academic Editor: Christos S. Akratos

Received: 27 June 2023
Revised: 14 July 2023
Accepted: 18 July 2023
Published: 26 July 2023

1. Introduction

Climate change and its associated factors has resulted in the abundant growth and dispersal of brown algae drifting in great masses across the Atlantic Ocean. These occurrences have been progressively increasing since 2011 [1]. The influx of *Sargassum sp.* in the Caribbean Sea is influenced by climate-induced fluctuations in sea surface currents and increasing sea surface temperatures, which are associated with anthropogenic factors, such as deforestation in the Amazon, as well as leaching and runoff from agricultural sources [2]. The ocean thus acts as a reservoir for human-induced carbon, which in turn impacts ecosystems and resources that have important societal functions [3–5]. Climate variations can have direct and indirect influences on water environments through different pathways, such as by affecting pollutant quantities in flowing water or changing their toxicity in different circumstances [6–8].

Heightened amounts of algae can cause numerous environmental and ecological problems, and interdisciplinary collaboration may help provide solutions to effectively manage this issue and avoid more detrimental effects on coastal environments in the Caribbean. Primarily made up of water, organic components, and trace toxic material, the composition of *Sargassum* sp. can vary depending on lifespan, environmental factors, and geographic distribution [1,9–11]. Of the three groups of macroalgae that are distinguished by their pigments (i.e., Phaeophyta, brown; Rhodophyta, red; and Chlorophyta, green), brown algae, in particular, has been widely researched as a plausible heavy metal bioindicator species [12]. Phaeophyta is also known to accumulate the most arsenic compared to Rhodophyta and

Chlorophyta [6]. In addition, only when toxic component concentrations are excessively increased are *Sargassum* considered hazardous [13]. Some known disadvantages created by the influx and mass accumulation of *Sargassum* along coastal areas, aside from its offensive odor, is that it is conducive to the proliferation of sandflies, which in turn negatively impact recreational beach activities and the navigation of fishing vessels. Harmful pesticide residues are known to be quite toxic for some algal species and other aquatic fauna and flora, so frequent monitoring of these environments is fundamental for controlling pesticide concentrations within all biota [14].

Algae are low-energy foods containing a wide variety of compounds with multifunctional potential for use in nutrition and technology. *Sargassum* species are known to contain sulfated polysaccharides and oligosaccharides, which are highly recognized immunomodulators that have positive effects on human and animal health and well-being [15–17]. With widespread availability, *Sargassum* can also offer abundant biomass for use in biogas plants [18,19]. The bioactive compounds of the genus *Sargassum*, including fucoxanthin, fucoidan, and alginates, are all highly valued commercially and are rich in essential minerals; with potassium and low nitrogen-phosphorous content, it can furthermore be used as a complimentary ingredient in fertilizer [13,20–22].

We aim to provide an effective solution for predicting and tracking the large-scale movements of algal biomass, and to also propose a method of addressing *Sargassum* influx through its use as an industrial recyclable material following composition analysis, evaluation and safety assessments for cosmetic use, and research and design of new beauty products and functional cosmetics.

2. Materials and Methods

The authors collected *Sargassum* samples in late October 2022 from the eastern coasts of four countries in the Caribbean Sea: Saint Lucia, Saint Vincent and the Grenadines, Saint Kitts and Nevis, and Belize. The samples were air-dried and ground into a powder, then sent to the Algal Cultivation and Biotech Laboratory at The National Taiwan Ocean University. The Traceability Certification and Inspection Center at The National Taiwan Ocean University analyzed the samples for four heavy metals: lead (Pb), cadmium (Cd), mercury (Hg), and arsenic (As). Additionally, the samples were analyzed for 531 pesticides by Eurofins Scientific Taiwan, a division of the Eurofins Scientific Group, in their food testing laboratory. All measured data in this study is outlined and shown in Table 1b. By detecting the heavy metal content present, we sought to determine the functional usability of pelagic *Sargassum*, establish limits of detection (LOD) for values presently not determined by renowned regulations such as the European Union (EU) and the United States (US), and to confirm suitability for use according to the standard limitations presently regulated. Determining *Sargassum* sp. suitability for industrial applications and investigating processing techniques helps allow for the recycling and use of this existing highly abundant resource.

Table 1. (**a**) Algal sample analysis (LOD) of heavy metal content. (**b**) Standard LOD heavy metal content (mg kg^{-1}, dry weight).

(**a**)

Items		**Heavy Metal (mg kg^{-1}, Dry Weight)**				
Country	**Species**	**Pb (Lead)**	**Cd (Cadmium)**	**Hg (Mercury)**	**Total Arsenic**	**Inorganic Arsenic**
St. Lucia	*Sargassum* sp.	0.03	0.38	N.D.	109.18	6.26
St. Vincent	*Sargassum* sp.	0.08	0.35	N.D.	111.29	8.79
St. Kitts	*Sargassum* sp.	N.D.	0.29	N.D.	106.11	6.15
Belize	*Sargassum* sp.	0.05	0.39	N.D.	84.09	5.13
St. Kitts	*Eucheuma* sp.	0.07	0.18	N.D.	3.67	N.D.
Method detection limit		0.01	0.01	0.001	0.02	0.05

Table 1. *Cont.*

(b)					
Categories	Heavy Metal (mg kg^{-1}, Dry Weight)				
	Pb (Lead)	Cd (Cadmium)	Hg (Mercury)	Total Arsenic	Inorganic Arsenic
EU food rule	3	3	0.1	N.D.	N.D.
USA food rule	0.01	3	1	-	3
Taiwan food rule	1	1	0.5	-	1.0
Taiwan cosmetic Products rule	10	5	1	3	-
Taiwan animal Feed rule	5	1 (cows, goats, excluding young animals); 0.5 others	0.1	2	-
Taiwan aquafeed rule	5	3 (marine fish and shrimps); 1 other	0	10	-
Taiwan soil rule	500	5	5	60	-

Notes: N.D.: Not Determined. Source: EU and USA Standards for heavy metal data: report of the expert meeting on food safety for seaweed: current status and future perspectives, Rome, 28–29 October 2021, MDL: the traceability certification and inspection center at the National Taiwan Ocean University Taiwan, October 2022 [23,24].

3. Results

3.1. Heavy Metal Accumulation in Sargassum and Its Ecological Implications

Hazardous heavy metal pollutants are widespread throughout the marine environment, with macroalgae known to be efficient bio-indicator species for heavy metal accumulation [12,25]. Ecologically, certain metals are essential for macroalgal growth, including iron, manganese, zinc, and copper. However, other naturally occurring metals like mercury and lead are non-essential elements, and are not required for growth or other physiological functions; when found in excess, these may become highly toxic [26]. Although certain heavy metals such as cadmium, lead, mercury, and arsenic do occur naturally in the environment, their concentrations vary in different locations of the ocean; macroalgae are known to uptake these elements through the process of biosorption, which reduces their toxic effects and lowers the numbers of free ions present [18,25,26]. These heavy metals are also known food safety hazards, and present safety concerns for seafood products. Concentrations of these metals in seaweeds have been used to test metal pollution along ocean coastlines [23].

Various human-induced activities play a major role in the availability of heavy metals in the open ocean. According to Clark (2001), $HAsO_4^{-2}$ concentrations in the Atlantic Ocean range from 1.27 to 2.10 µg L^{-1} in aerated water at a salinity level of 35% [26,27]. With a confirmed absorption concentration almost 180 times more than seawater, it is assumed that macroalgae absorb arsenic as a phosphorus substitute for growth and other physiological requirements [18,25–29]. As a biosorbent species, *Sargassum* can eliminate heavy metals like lead, mercury, zinc, and copper ions in the ocean through physisorption, microprecipitation exchange, and chemisorption [13,27]. Internally, concentrations of heavy metals have a significant impact on most algal species' ability to fluoresce depending on metal speciation [26,30].

Sargassum samples obtained from four locations in the Caribbean were tested for heavy metal content. Our analysis indicates that arsenic absorption is dependent on seawater pollution sources, which influence the bioavailability and speciation of other heavy metals (Table 1). For instance, bivalves are also found to accumulate chemical contaminants, including As, and can serve as an indicator species of climate variability [6,31]. A significant rise in temperature influences enzyme activity, and may promote cell wall enlargement through accelerated metabolic processes during adsorption [6,32].

Bioaccumulation of harmful pesticides has been widely researched, and climate change-influenced scenarios are proving to be of greater and greater relevance [2]. Our results show that no pesticide residue was detected in *Sargassum* among the 531 tested pesticides that are known to be applied to vegetables, fruit, infant food, spices, and herbs. Although pelagic algae may be susceptible to anthropogenic activities such as agricultural runoff, deforestation, and other associated events, the bioaccumulation of pesticide residue was negligible [6,33]. This suggests that various uses of *Sargassum* could be highly feasible with efficient chemical extraction of the heavy metal content, including as an animal feed additive or for use within the cosmetic industry [2].

The European Union (EU) legislation Commission Regulation (EC) No. 1881/2006 outlines the acceptable standards for inorganic and total arsenic, cadmium, lead, and mercury for seaweed (macroalgae) use in food products for animals, but no legislation has been developed for human consumption [23,24]. Similarly, the Food and Drug Association (FDA) of the United States (US) establishes the following acceptable limits: "Heavy metals as lead (Pb) not more than 10 parts per million, Arsenic (As) not more than 3 parts per million, Mercury (Hg) not more than 1 part per million." However, the list does not include cadmium, the limit for which has yet to be determined [20,23,34]. Similarly, standard limits for products other than food and supplements remain to be determined and documented. Accordingly, EU legislation Commission Regulation (EC) No. 396/2005 also states an applicable default maximum residue limit (MRL) for pesticide residue of 0.01 mg/kg for most seaweeds [23,24]. Our LOD testing and analysis show that other than arsenic, the heavy metal and pesticide content in *Sargassum* fall within these established requirements. Our findings, therefore, support the functional utilization of pelagic *Sargassum*. With effective extraction methods and processing techniques, it is highly suitable for a wide variety of applications.

Although human-induced, climate change-related activities can influence algal growth, abundance, and distribution, residual chemicals from pesticides have not been found to accumulate in the cell structure of *Sargassum* [26]. Moreover, bioactive compounds of the *Sargassum* genus, including fucoxanthin, fucoidan, and alginates, are considered to be of high value commercially [20,22]. Advanced extraction processes can effectively remove heavy metal substances from the algal biomass, thereby enhancing the commercial value of *Sargassum* as an organic material for industrial applications [21,22]. *Sargassum* inundation is known to inflict damage on nearshore ecosystems and beaches, and, when left to decompose, produces a highly unpleasant odor that poses safety risks to human health through the release of toxic gases, such as sulfuric acid and ammonia [29,35].

The heavy metal content of *Sargassum* sp. was analyzed (Table 1a) at four specific locations: St. Lucia, St. Vincent, St. Kitts and Nevis, and Belize, in comparison to *Eucheuma* sp. Analysis showed higher levels of arsenic content than the recognized standards. Algal cells can bind to heavy metals and are subject to increased pH levels in the environment [6,36]. Effective measures to extract the arsenic content are thus needed to adhere to EU and USA stipulated heavy metal content standard requirements (Table 1b) for full usage of the algae in industrial applications such as consumable food, cosmetic products, or animal and aquaculture feed. Pretreatment methods should be determined by algal species type and the intended product use.

3.2. Sargassum Distribution and Predictive Measures via Ocean Climate Circulation

Sargassum sp. biomass' presence and timing are dependent on sea surface currents and wind direction. These large, widely distributed biomasses are further subject to Stokes drift, a confined variation of Langmuir circulation, that may influence the responses of *Sargassum* to temperature and nutrient availability [37,38]. Essentially all living components within an ecosystem are influenced by climatic conditions and the combined action of sea surface currents, wind, and wave action; all of these also determine the spatial coverage of pelagic *Sargassum* sp. [18,38]. Due to wider distribution and more frequent occurrences of algal biomass, a part of the Atlantic Ocean between North America and the west of Africa has been named the Sargasso Sea. The *Sargassum* that occur in the Sargasso Sea are holopelagic and can effectively reproduce on the high seas [39]. Periodically, these brown algal mats drift along the coastlines of most Caribbean Island nations—it has become a frequent phenomenon each winter due to stronger trade winds, which drive great amounts of drift toward coastlines [40,41]. These floating mats provide an ecologically important niche that many different marine species depend on, although they can also transport non-native species.

Ideally, offshore collection of *Sargassum* is preferable, because once it has arrived at the beach, even if it is collected within one day, damage has already been inflicted [2,13]. Studies have shown a great variation in the species of *Sargassum* within locations, so predictions can be made by measuring the drift of the *Sargassum* biomass with respect to ocean flux through satellite observation [39,40,42]. Predictive measurements and forecasting can be derived from current ocean circulation modeling systems, including SMA Tools (*Sargassum* Monitoring Tool), which can provide a five-day predictive forecast of drifting *Sargassum* [39,40]. In general, there are discrepancies even in the best solutions for predicting influx or tracking movement by satellite.

The dominant pelagic species of *Sargassum*, *S. natans* and *S. fluitans*, comprise much of the drifting seaweed biomass and have a widespread distribution within the North Atlantic Ocean, starting from the Gulf of Mexico to the Sargasso Sea [43,44]. These prolific algal blooms are highly detrimental to local economies and surrounding environments [2]. Solutions for the management of *Sargassum* require spatial and temporal predictive management measures; remote sensing techniques provide a cost-effective solution by delivering important information on the locality of algal accumulation for collection and removal [2,40]. Similar techniques can also be useful on a wider scale for information on the spatial distribution of frequent *Sargassum* landings [31,40]. The affected communities are required to anticipate surges in beach landings and implement strategies for their collection to effectively manage the *Sargassum* influx [44]. The uncertainty and limited availability of precise forecasting models pose challenges to the adequate management of *Sargassum* incursions each year.

In a related scenario, floating populations of *S. horneri*, a large marine brown alga of the Korean strait, have drifted in larger quantities towards the north of Taiwan via the cold current flowing south along the coast of China [45]. Drifting *S. horneri* biomass is speculated to be caused by stronger Kuroshio currents that create an Ekman drift, thus transporting massive amounts of algae toward Taiwan [46,47]. Similarly, the canary current flows from the North to the West in the Atlantic Ocean, where it meets the south equatorial current. In doing so, it pushes the algal mass towards the east coast of Mexico and the Caribbean Sea. Larger aggregations of *Sargassum* have been found frequently in the central part of the coast during times of increased influx from March to October [13,48].

Anthropogenic activities, such as the uncharacteristic nutrient enrichment related to deforestation in the Amazon and changes in African dust deposition patterns in the Sahara Desert, are inferred to encourage an increase in the nutrient flow from terrestrial sources specific to the Caribbean Sea [1,47]. These contribute to the overgrowth of *Sargassum* blooms, and their prolific nature suggests that changes in ocean temperature alongside stronger monsoon periods can influence the growth, distribution, and sightings of floating algae, similar to *S. horneri* biomass movements in northern Taiwan [45]. It is, therefore, essential to be able to predict spatial distribution patterns induced by temporal variations due to climate change. Studies have also determined that surface currents, altered by wind patterns, cause a redistribution of pelagic *Sargassum*, with the major currents found in the Atlantic Ocean influencing distribution in the Great Atlantic *Sargassum* Belt [13]. Predictive six-month forecasts are highly viable through the use of various drift observations and data obtained from ocean models, which are dependent on satellite imagery detecting *Sargassum* across a wider area of the Atlantic Ocean and the Caribbean Sea [40].

In situ applications use remote sensing to develop predictions based on maps obtained by high spatial resolution satellite imagery [40,49]. Satellite images with manual recognition (Figure 1) are comparable to assessments of forest biomass, which has proven effective for forest management and monitoring in the context of carbon budgeting and reduced emissions [50]. Similarly, using map-making or ground truth masks validates maps of land and sea and provides a method for classification and prediction. Segmentation masks, which are generated automatically in real-time, can also enhance the input layers with extra information to predict future frames, thereby presenting opportunities to create knowledge under uncertain conditions [51,52].

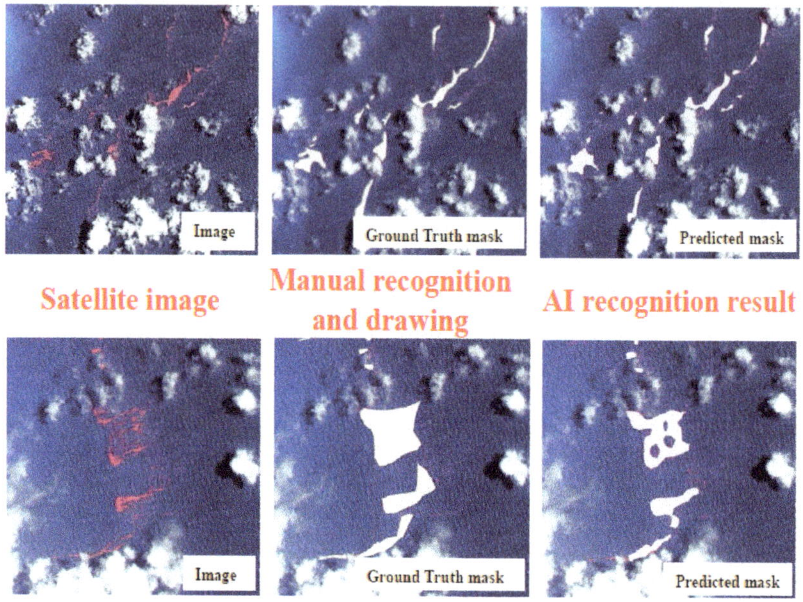

Figure 1. AI satellite images of floating algae on the eastern coast of St. Lucia, captured by Optical satellite Sentinel-2 on 22 August 2022.

4. Discussion

Cost-Effective Management and Opportunities for Functional Utilization of Sargassum

Marine macroalgae occur in highly complex and diverse ecosystems, which can offer ideal conditions for new active molecules to be developed and allows for a variety of applications [26]. Macroalgae are eukaryotic organisms comprised of chlorophyll and carbohydrates; 80% of seaweed harvested is consumed by humans, and these algae contribute quality proteins with high amounts of vitamins and minerals, dietary fiber, and essential unsaturated fatty acids, which altogether provide nutrition that is beneficial to human health. [18,53–55]. The *Sargassum* genus has several important components in the thallus, and with advances in extraction processes to effectively remove heavy metal substances from the algal biomass, the commercial value of *Sargassum* can be further enhanced. The specific content and characteristics of the constituents in *Sargassum* may vary depending on species and growth conditions.

Important constituents of the genus *Sargassum*:

1. Fucoxanthin is a natural pigment belonging to the class of carotenoids called fucoxanthins. It imparts a deep brown color to the genus *Sargassum* and possesses antioxidant and anti-inflammatory properties. Fucoxanthin has been studied as a potential anticancer and antiobesity compound;
2. Fucoidan is one of the most significant bioactive compounds found in *Sargassum*. It is a sulfated polysaccharide with various biological activities, including anti-inflammatory, anticoagulant, antioxidant, antitumor, and immune-modulating properties. Fucoidan has been extensively researched and is considered to have potential applications in pharmaceuticals and functional foods;
3. Alginates are polysaccharides composed primarily of α-L-guluronic acid and β-D-mannuronic acid. They are widely present in the cell walls of *Sargassum*. Alginates

have gel-forming abilities, and find extensive applications in food, pharmaceutical, and cosmetic industries as thickeners, stabilizers, and emulsifiers [20–22].

Sourcing *Sargassum* from the ocean surface through sustainable collection methods facilitated by floating nets lessens the negative ecological impacts of beach landings [31]. Initially, processing includes washing with fresh water, drying, grinding, and vacuum sealing and packaging for storage or transport. The development and production of biodegradable materials and functional textile materials and feed for both aquaculture and agricultural uses remain important, and recently seaweed has been found to be a highly useful raw material. In addition to playing an important role in diets, it can also be functionally useful for creating value-added goods, in energy production methods, and in several fields, such as cosmetics, agriculture, and construction [18,56–58]. Developing new materials from seaweed, such as textiles, can also help create environmentally friendly alternatives to reduce the negative impacts of non-degradable materials, such as single-use plastics that pollute bodies of water. Using *Sargassum* could therefore help reduce fossil fuel usage, the production of plastic waste, logging, and pulp production, which can, in turn, reduce the accidental ingestion of non-degradable materials by marine organisms. *Sargassum* biomass landings onshore present larger problems, with higher costs associated with recurrent blooms [19,47]. U-shaped barriers can provide a suitable recovery method with the least detriment to beaches, and allow for an efficient offshore capture method [31]. *Sargassum* collection is dependent on the oceanographic and geomorphological attributes of the location, and the accumulated *Sargassum* in these barriers are likely to decompose or sink within 24–48 h [13,31].

As discussed, alternative uses for this abundant biomass can provide countless benefits both for communities and the environment [44,58]. With a high nutritional content, *Sargassum* can be used as fertilizer for soil and plants as well as food for consumption, while extracting their phytochemicals has the potential for medicinal use [16,55]. The main regulation categories listed in Table 1b outline utilization rules for human consumption, and there has been an increased interest in marine algae recently as a bioactive source useful for many applications in processing and developing healthier functional food groups [59]. Algae can enrich the nutritional properties of food; cereal-based food products, such as pasta, are low in protein and essential amino acids, so incorporating algae could greatly enhance its nutritional properties [18]. Studies have shown that eating algae-based pasta resulted in improved amino acid and fatty acid profiles, with better total phenolic content, antioxidant activity, and a higher fucoxanthin and fucosterol content [18,59]. Similar benefits of algae-enriched bread and noodle products can also be expected [18]. Furthermore, there is a higher feasibility of *Sargassum* sp. producing compounds that can enhance food flavors, such as carbonyls, alcohols esters, aldehydes, and esters, because these compounds occur naturally with less toxic waste under minimal production conditions [60].

Composition analysis has confirmed seaweed to be suitable for energy production, such as a biofuel, due to its high sugar content; even so, pretreatment is recommended for exploring *Sargassum* use for anaerobic biogas production [16,18,36,57,58]. Although bioenergy production from *Sargassum* has great potential and utilizes an abundant resource, economic viability also depends on the production of higher-value products, similar to bioethanol production [59,61]. Other commercially viable solutions aside from algae-based functional foods, nutraceuticals, and cosmeceuticals can include environmentally friendly solutions such as biodegradable materials, plant fiber panels for buildings, textiles, and toiletries, although these possibilities depend on creating sustainable and innovative technologies and products with specific product markets [61,62]. With an eco-friendly decomposition rate of over 150 years, macroalgae can be used as a building material: it has the potential to provide quality insulation, and offers good acoustics, humidity control, and visual comfort while being non-toxic and fireproof with less associated CO_2 emissions [62].

Pretreatments are generally required to release the usable organic matter through the breakdown of the cell walls, and *S. natans* and *S. fluitans* have been found to produce small quantities of lignin compounds in response to stressors related to climate change, such as heat and light [1,13,61]. *Sargassum*, with its low to zero lignin content and good nutrient properties, can be beneficial as feed for anaerobic digestion; for instance, the polyphenols found in brown algae are found to have an influence on methanogen activity in ruminant animals, mitigating the effects of methane (CH_4) production and thus reducing problematic greenhouse gas emissions in livestock supply chains [13,63,64]. Interestingly, a previous study determined that the usage of brown algae as livestock feed can supply greater energy content compared to other seaweed varieties due to its low ash content [63]. Extracts from *Sargassum* have also been used in crop production as seed treatments, in foliar applications, and for the purpose of soil drenching [13,65]. Similarly, there are viable possibilities for *Sargassum* filtration systems due to their high biosorption potential to effectively remove organic dyes and heavy metal ions [13]. The alginates and fucoidans derived from algae can benefit the pharmaceutical, cosmetic, and food industries [35]. Moreover, other practical uses include sodium alginate production for eco-friendly bioplastics; use as an emulsifier and stabilizer; and use for its antioxidant, anti-inflammatory, and anticoagulant properties, all of which can help treat or protect against viruses, cancer, and many other diseases [17,31,59].

The implications of large-scale utilization of *Sargassum* have not been fully determined; however, ecologically, the prolific nature of increasing pelagic *Sargassum* influx to coastal shores requires solutions and alternative approaches to alleviate the negative impacts on coastal environments, essential ecosystems, and people's livelihoods and tourism, which are important concerns for the Caribbean Island nations [2,31]. Furthermore, large-scale cultivation is promising for increasing the supply of usable material in the event that it is not available or accessible in nature; this cultivation can be achieved with long lines near coastal areas of the open ocean, and would have the twofold benefit of not only keeping up with an abundant supply of resources for food and non-food applications, but also mitigating issues such as eutrophication and nutrient enrichment from anthropogenic activities, which are highly problematic in coastal ecosystems [66–68].

5. Conclusions

There have been numerous studies undertaken on the functional uses of *Sargassum*, and management methods using satellite data and numerical simulations can provide solutions for forecasting and tracking large amounts of algae, allowing for effective collection and recovery of large quantities of algal biomass, which is a valuable organic resource and offers great potential for industrial applications. Developing more advanced processing techniques can open further avenues for industries in various uses of *Sargassum*.

Author Contributions: Conceptualization, M.-C.L.; Resources, C.-M.C., Y.-F.L., F.-H.N.; writing—original draft preparation, J.W.-S.; writing—review and editing, M.-C.L. and H.-Y.Y.; Supervision, M.-C.L. All authors have read and agreed to the published version of the manuscript.

Funding: This research received no external funding.

Data Availability Statement: The data that support the findings of this study are available from the first author, [MC. Lee], upon reasonable request.

Acknowledgments: The authors would like to express their sincere gratitude to the editor and the anonymous reviewers for their valuable insights, comments, and suggestions, which greatly contributed to the improvement of this manuscript. Furthermore, the authors would like to extend their special thanks to Fan-Hua Nan's team at the Traceability Certification and Inspection Center in NTOU for their invaluable assistance in conducting the heavy metal analysis. The authors would like to acknowledge Eurofins Co., LTD for their collaboration and assistance in conducting the pesticide analysis. Lastly, the authors are grateful to the International Cooperation and Development Fund (Taiwan ICDF) for their assistance in sample collection and shipping in this study.

Conflicts of Interest: The authors declare no conflict of interest.

References

1. Orozco-González, J.G.; Amador-Castro, F.; Gordillo-Sierra, A.R.; García-Cayuela, T.; Alper, H.S.; Carrillo-Nieves, D. Opportunities Surrounding the Use of *Sargassum* Biomass as Precursor of Biogas, Bioethanol, and Biodiesel Production. *Front. Mar. Sci.* **2022**, *8*, 791054. [CrossRef]
2. Lee, M.-C. Situations and Solutions of *Sargassum* Influxes—Saint Lucia, Saint Vincent and the Grenadines, Saint Kitts and Nevis, and Belize. Unpublished Work. 2022.
3. Fox-Kemper, B.; Adcroft, A.; Böning, C.W.; Chassignet, E.P.; Curchitser, E.; Danabasoglu, G.; Eden, C.; England, M.H.; Gerdes, R.; Greatbatch, R.J.; et al. Challenges and Prospects in Ocean Circulation Models. *Front. Mar. Sci.* **2019**, *6*, 65. [CrossRef]
4. Khatiwala, S.; Primeau, F.; Hall, T. Reconstruction of the history of anthropogenic CO_2 concentrations in the ocean. *Nature* **2009**, *462*, 346–349. [CrossRef]
5. Khatiwala, S.; Tanhua, T.; Mikaloff Fletcher, S.; Gerber, M.; Doney, S.C.; Graven, H.D.; Gruber, N.; McKinley, G.A.; Murata, A.; Ríos, A.F.; et al. Global ocean storage of anthropogenic carbon. *Biogeosciences* **2013**, *10*, 2169–2191. [CrossRef]
6. Lee, M.-C.; Libatique, M.J.H.; Yeh, H.-Y.; Chloe Lung, W.Q. Increasing arsenic accumulation as an implication of climate change: A case study using red algae. *Bull. Environ. Contam. Toxicol.* **2022**, *108*, 839–847. [CrossRef]
7. Wu, Q.; Qi, J.; Xia, X. Long-term variations in sediment heavy metals of a reservoir with changing trophic states: Implications for the impact of climate change. *Sci. Total Environ.* **2017**, *609*, 242–250. [CrossRef]
8. Yuan, Y.; Zhang, C.; Zeng, G.; Liang, J.; Guo, S.; Huang, L.; Wu, H.; Hua, S. Quantitative assessment of the contribution of climate variability and human activity to streamflow alteration in Dongting Lake, China. *Hydrol. Process.* **2016**, *30*, 1929–1939. [CrossRef]
9. Davis, D.; Simister, R.; Campbell, S.; Marston, M.; Bose, S.; McQueen-Mason, S.J.; Gomez, L.D.; Gallimore, W.A.; Tonon, T. Biomass composition of the golden tide pelagic seaweeds *Sargassum fluitans* and *S. natans* (morphotypes I and VIII) to inform valorisation pathways. *Sci. Total Environ.* **2021**, *762*, 143134. [CrossRef]
10. Kordjazi, M.; Etemadian, Y.; Shabanpour, B.; Pourashouri, P. Chemical composition antioxidant and antimicrobial activities of fucoidan extracted from two species of brown seaweeds(*Sargassum ilicifolium* and *S. angustifolium*) around Qeshm Island. *Iran. J. Fish. Sci.* **2018**, *18*, 457–475. [CrossRef]
11. Yaich, H.; Garna, H.; Besbes, S.; Paquot, M.; Blecker, C.; Attia, H. Chemical composition and functional properties of Ulva lactuca seaweed collected in Tunisia. *Food Chem.* **2011**, *128*, 895–901. [CrossRef]
12. Al-Homaidan, A.A.; Al-Ghanayem, A.A.; Al-Qahtani, H.S.; Al-Abbad, A.F.; Alabdullatif, J.A.; Alwakeel, S.S.; Ameen, F. Effect of sampling time on the heavy metal concentrations of brown algae: A bioindicator study on the Arabian Gulf coast. *Chemosphere* **2021**, *263*, 127998. [CrossRef]
13. López Miranda, J.L.; Celis, L.B.; Estévez, M.; Chávez, V.; van Tussenbroek, B.I.; Uribe-Martínez, A.; Cuevas, E.; Rosillo Pantoja, I.; Masia, L.; Cauich-Kantun, C.; et al. Commercial Potential of Pelagic *Sargassum* spp. in Mexico. *Front. Mar. Sci.* **2021**, *8*, 768470. [CrossRef]
14. Polat, A.; Polat, S.; Simsek, A.; Kurt, T.T.; Ozyurt, G. Pesticide residues in muscles of some marine fish species and seaweeds of Iskenderun Bay (Northeastern Mediterranean), Turkey. *Environ. Sci. Pollut. Res.* **2018**, *25*, 3756–3764. [CrossRef]
15. Sanjeewa, K.K.A.; Kang, N.; Ahn, G.; Jee, Y.; Kim, Y.-T.; Jeon, Y.-J. Bioactive potentials of sulfated polysaccharides isolated from brown seaweed *Sargassum* spp in related to human health applications: A review. *Food Hydrocoll.* **2018**, *81*, 200–208. [CrossRef]
16. Leandro, A.; Pacheco, D.; Cotas, J.; Marques, J.C.; Pereira, L.; Gonçalves, A.M.M. Seaweed's Bioactive Candidate Compounds to Food Industry and Global Food Security. *Life* **2020**, *10*, 140. [CrossRef]
17. Hwang, J.; Yadav, D.; Lee, P.C.; Jin, J.-O. Immunomodulatory effects of polysaccharides from marine algae for treating cancer, infectious disease, and inflammation. *Phytother. Res.* **2022**, *36*, 761–777. [CrossRef]
18. Zhao, Y.; Bourgougnon, N.; Lanoisellé, J.-L.; Lendormi, T. Biofuel Production from Seaweeds: A Comprehensive Review. *Energies* **2022**, *15*, 9395. [CrossRef]
19. Milledge, J.J.; Maneein, S.; Arribas López, E.; Bartlett, D. *Sargassum* Inundations in Turks and Caicos: Methane Potential and Proximate, Ultimate, Lipid, Amino Acid, Metal and Metalloid Analyses. *Energies* **2020**, *13*, 1523. [CrossRef]
20. Holdt, S.L.; Kraan, S. Bioactive compounds in seaweed: Functional food applications and legislation. *J. Appl. Phycol.* **2011**, *23*, 543–597. [CrossRef]
21. Plaza, M.; Cifuentes, A.; Ibáñez, E. In the search of new functional food ingredients from algae. *Trends Food Sci. Technol.* **2008**, *19*, 31–39. [CrossRef]
22. Zhang, R.; Zhang, X.; Tang, Y.; Mao, J. Composition, isolation, purification and biological activities of *Sargassum fusiforme* polysaccharides: A review. *Carbohydr. Polym.* **2020**, *228*, 115381. [CrossRef]
23. Publication, F. *Report of the Expert Meeting on Food Safety for Seaweed—Current Status and Future Perspectives*; FAO: Rome, Italy, 2022.
24. Publication, E. Commission Regulation (EC) No 1881/2006 of 19 December 2006 setting maximum levels for certain contaminants in foodstuffs. *Off. J. Eur. Union* **2006**, *364*, 5–24.
25. Libatique, M.J.H.; Lee, M.-C.; Yeh, H.-Y.; Jhang, F.-J. Total and inorganic arsenic biosorption by *Sarcodia suiae* (Rhodophyta), as affected by controlled environmental conditions. *Chemosphere* **2020**, *248*, 126084. [CrossRef]
26. Hurd, C.L.; Harrison, P.J.; Bischof, K.; Lobban, C.S. *Seaweed Ecology and Physiology*, 2nd ed.; Cambridge University Press: Cambridge, UK, 2014.
27. Clark, R.B. Marine Pollution and Sea Life: (ed. M. Ruivo). Fishing News (Books) Ltd., London, by arrangement with the Food and Agriculture Organization of the UN (1972). xxiv + 624 pp. £17.75. *Mar. Pollut. Bull.* **1973**, *4*, 112. [CrossRef]

28. Nischwitz, V.; Pergantis, S.A. First report on the detection and quantification of arsenobetaine in extracts of marine algae using HPLC-ES-MS/MS. *Analyst* **2005**, *130*, 1348–1350. [CrossRef]
29. Robalds, A.; Naja, G.M.; Klavins, M. Highlighting inconsistencies regarding metal biosorption. *J. Hazard. Mater.* **2016**, *304*, 553–556. [CrossRef]
30. Baumann, H.A.; Morrison, L.; Stengel, D.B. Metal accumulation and toxicity measured by PAM—Chlorophyll fluorescence in seven species of marine macroalgae. *Ecotoxicol. Environ. Saf.* **2009**, *72*, 1063–1075. [CrossRef]
31. Chávez, V.; Uribe-Martínez, A.; Cuevas, E.; Rodríguez-Martínez, R.E.; van Tussenbroek, B.I.; Francisco, V.; Estévez, M.; Celis, L.B.; Monroy-Velázquez, L.V.; Leal-Bautista, R.; et al. Massive Influx of Pelagic *Sargassum* spp. on the Coasts of the Mexican Caribbean 2014–2020: Challenges and Opportunities. *Water* **2020**, *12*, 2908. [CrossRef]
32. Ramelow, G.; Fralick, D.; Zhao, Y.-f. Factors affecting the uptake of aqueous metal ions by dried seaweed biomass. *Microbios* **1992**, *72*, 81–93.
33. Maulvault, A.L.; Camacho, C.; Barbosa, V.; Alves, R.; Anacleto, P.; Fogaça, F.; Kwadijk, C.; Kotterman, M.; Cunha, S.C.; Fernandes, J.O.; et al. Assessing the effects of seawater temperature and pH on the bioaccumulation of emerging chemical contaminants in marine bivalves. *Environ. Res.* **2018**, *161*, 236–247. [CrossRef]
34. U.S. Food and Drug Administration. *FDA Total Diet Study TDS*; U.S. Food and Drug Administration: Silver Spring, MD, USA, 2022.
35. Libatique, M.J.H.; Yeh, H.-Y.; Nan, F.-H.; Lee, M.-C. Phosphate affects susceptivity of *Sarcodia suae* to arsenate. *Eur. J. Phycol.* **2023**, *58*, 45–57. [CrossRef]
36. Resiere, D.; Mehdaoui, H.; Névière, R.; Mégarbane, B. *Sargassum invasion* in the Caribbean: The role of medical and scientific cooperation. *Rev. Panam. Salud Pública* **2019**, *43*, e52. [CrossRef]
37. Sulaymon, A.H.; Mohammed, A.A.; Al-Musawi, T.J. Competitive biosorption of lead, cadmium, copper, and arsenic ions using algae. *Environ. Sci. Pollut. Res.* **2013**, *20*, 3011–3023. [CrossRef]
38. Freitas, R.; Salamanca, L.; Velez, C.; Wrona, F.J.; Soares, A.M.V.M.; Figueira, E. Multiple stressors in estuarine waters: Effects of arsenic and salinity on *Ruditapes philippinarum*. *Sci. Total Environ.* **2016**, *541*, 1106–1114. [CrossRef]
39. Jouanno, J.; Benshila, R.; Berline, L.; Soulié, A.; Radenac, M.H.; Morvan, G.; Diaz, F.; Sheinbaum, J.; Chevalier, C.; Thibaut, T.; et al. A NEMO-based model of *Sargassum* distribution in the tropical Atlantic: Description of the model and sensitivity analysis (NEMO-Sarg1.0). *Geosci. Model Dev.* **2021**, *14*, 4069–4086. [CrossRef]
40. Marsh, R.; Oxenford, H.A.; Cox, S.-A.L.; Johnson, D.R.; Bellamy, J. Forecasting seasonal *Sargassum* events across the tropical Atlantic: Overview and challenges. *Front. Mar. Sci.* **2022**, *9*, 914501. [CrossRef]
41. NOAA. *What is Eutrophication*; NOAA: Washington, DC, USA, 2023.
42. Berline, L.; Ody, A.; Jouanno, J.; Chevalier, C.; André, J.-M.; Thibaut, T.; Ménard, F. Hindcasting the 2017 dispersal of *Sargassum* algae in the Tropical North Atlantic. *Mar. Pollut. Bull.* **2020**, *158*, 111431. [CrossRef]
43. Gavio, B.; Rincón-Díaz, M.N.; Santos-Martínez, A. Massive quantities of pelagic *Sargassum* on the shores of san andres island, southwestern caribbean. *Acta Biológica Colomb.* **2015**, *20*, 239–241. [CrossRef]
44. Louime, C.; Fortune, J.; Gervais, G. *Sargassum* Invasion of Coastal Environments: A Growing Concern. *Am. J. Environ. Sci.* **2017**, *13*, 58–64. [CrossRef]
45. Lin, S.-M.; Huang, R.; Ogawa, H.; Liu, L.-C.; Wang, Y.-C.; Chiou, Y. Assessment of germling ability of the introduced marine brown alga, *Sargassum horneri*, in Northern Taiwan. *J. Appl. Phycol.* **2017**, *29*, 2641–2649. [CrossRef]
46. Fine, M.L. Faunal variation on pelagic *Sargassum*. *Mar. Biol.* **1970**, *7*, 112–122. [CrossRef]
47. Amador-Castro, F.; García-Cayuela, T.; Alper, H.S.; Rodriguez-Martinez, V.; Carrillo-Nieves, D. Valorization of pelagic *Sargassum* biomass into sustainable applications: Current trends and challenges. *J. Environ. Manag.* **2021**, *283*, 112013. [CrossRef]
48. Fidai, Y.A.; Dash, J.; Tompkins, E.L.; Tonon, T. A systematic review of floating and beach landing records of *Sargassum* beyond the Sargasso Sea. *Environ. Res. Commun.* **2020**, *2*, 122001. [CrossRef]
49. García-Sánchez, M.; Graham, C.; Vera, E.; Escalante-Mancera, E.; Álvarez-Filip, L.; van Tussenbroek, B.I. Temporal changes in the composition and biomass of beached pelagic *Sargassum* species in the Mexican Caribbean. *Aquat. Bot.* **2020**, *167*, 103275. [CrossRef]
50. Hojas-Gascon, L.; Eva, H. *Field Guide for Forest Mapping with High Resolution Satellite Data*; Monitoring Deforestation and Forest Degradation in the Context of the UN-REDD Programme. The Tanzania REDD+ Initiative; Joint Research Center, Publications Office of the European Union: Luxembourg, 2014.
51. Yin, Q.; Hu, Q.; Liu, H.; Zhang, F.; Wang, Y.; Lin, Z.; An, W.; Guo, Y. Detecting and Tracking Small and Dense Moving Objects in Satellite Videos: A Benchmark. *IEEE Trans. Geosci. Remote Sens.* **2022**, *60*, 5612518. [CrossRef]
52. Hsu, M.C.; Shyur, J.C.; Watanabe, H. Pseudo Ground Truth Segmentation Mask to Improve Video Prediction Quality. In Proceedings of the 2020 IEEE 9th Global Conference on Consumer Electronics (GCCE), Kobe, Japan, 13–16 October 2020; pp. 831–832.
53. Fleurence, J. Chapter 5—Seaweeds as Food. In *Seaweed in Health and Disease Prevention*; Fleurence, J., Levine, I., Eds.; Academic Press: San Diego, CA, USA, 2016; pp. 149–167.
54. Levine, I. Chapter 1—Algae: A Way of Life and Health. In *Seaweed in Health and Disease Prevention*; Fleurence, J., Levine, I., Eds.; Academic Press: San Diego, CA, USA, 2016; pp. 1–5.

55. Bocanegra, A.; Bastida, S.; Benedi, J.; Rodenas, S.; Sanchez-Muniz, F.J. Characteristics and Nutritional and Cardiovascular-Health Properties of Seaweeds. *J. Med. Food* **2009**, *12*, 236–258. [CrossRef] [PubMed]
56. Barbot, Y.N.; Al-Ghaili, H.; Benz, R. A Review on the Valorization of Macroalgal Wastes for Biomethane Production. *Mar. Drugs* **2016**, *14*, 120. [CrossRef] [PubMed]
57. Balina, K.; Romagnoli, F.; Blumberga, D. Seaweed biorefinery concept for sustainable use of marine resources. *Energy Procedia* **2017**, *128*, 504–511. [CrossRef]
58. Baghel, R.S.; Suthar, P.; Gajaria, T.K.; Bhattacharya, S.; Anil, A.; Reddy, C.R.K. Seaweed biorefinery: A sustainable process for valorising the biomass of brown seaweed. *J. Clean. Prod.* **2020**, *263*, 121359. [CrossRef]
59. Domínguez, H. 1—Algae as a source of biologically active ingredients for the formulation of functional foods and nutraceuticals. In *Functional Ingredients from Algae for Foods and Nutraceuticals*; Domínguez, H., Ed.; Woodhead Publishing: Sawston, UK, 2013; pp. 1–19.
60. Amanjyoti; Nehra, M.; Khan, M. Algae as a Nutritional and Functional Food Source. *Madridge J. Food Technol.* **2022**, *7*, 189–199. [CrossRef]
61. Konda, N.V.S.N.M.; Singh, S.; Simmons, B.A.; Klein-Marcuschamer, D. An Investigation on the Economic Feasibility of Macroalgae as a Potential Feedstock for Biorefineries. *BioEnergy Res.* **2015**, *8*, 1046–1056. [CrossRef]
62. Widera, B. Possible Application of Seaweed as Building Material in the Modern Seaweed House on Laesø. In Proceedings of the 30th International Plea Conference, Ahmedabad, India, 16–18 December 2014.
63. Park, K.Y.; Jo, Y.H.; Ghassemi Nejad, J.; Lee, J.C.; Lee, H.G. Evaluation of nutritional value of *Ulva* sp. and *Sargassum horneri* as potential eco-friendly ruminants feed. *Algal Res.* **2022**, *65*, 102706. [CrossRef]
64. Alzate-Gaviria, L.; Domínguez-Maldonado, J.; Chablé-Villacís, R.; Olguin-Maciel, E.; Leal-Bautista, R.M.; Canché-Escamilla, G.; Caballero-Vázquez, A.; Hernández-Zepeda, C.; Barredo-Pool, F.A.; Tapia-Tussell, R. Presence of Polyphenols Complex Aromatic "Lignin" in *Sargassum* spp. from Mexican Caribbean. *J. Mar. Sci. Eng.* **2021**, *9*, 6. [CrossRef]
65. Chinnadurai, K.; Venkatesalu, V. Utilization of seaweed *Sargassum myriocystum* extracts as a stimulant of seedlings of *Vigna mungo* (L.) Hepper. *Span. J. Agric. Res.* **2012**, *10*, 466. [CrossRef]
66. Fernandes, D.R.P.; de Oliveira, V.P.; Yoneshigue Valentin, Y. Seaweed biotechnology in Brazil: Six decades of studies on natural products and their antibiotic and other biological activities. *J. Appl. Phycol.* **2014**, *26*, 1923–1937. [CrossRef]
67. Li, J.; Liu, Y.; Liu, Y.; Wang, Q.; Gao, X.; Gong, Q. Effects of temperature and salinity on the growth and biochemical composition of the brown alga *Sargassum fusiforme* (Fucales, Phaeophyceae). *J. Appl. Phycol.* **2019**, *31*, 3061–3068. [CrossRef]
68. Hwang, E.K.; Park, C.S.; Baek, J.M. Artificial seed production and cultivation of the edible brown alga, *Sargassum fulvellum* (Turner) C. In Agardh: Developing a new species for seaweed cultivation in Korea. In Proceedings of the Eighteenth International Seaweed Symposium, Bergen, Norway, 20–25 June 2004; pp. 25–31.

water

MDPI

Article

Alkaline Prehydrolysis Prolongs Resin Life and Enhances the Adsorption of Phenolic Compounds

Chao Gu [1], Kailing Yan [1], Lei Bo [1], Xin Zhou [1], Yiwen He [2], Jiacheng Feng [2] and Jinyi Qin [2,*]

[1] CHN Energy Shendong Coal Group Co., Ltd., Ordos 017209, China; sgbtljl@163.com (C.G.); sgbt123456@163.com (K.Y.); 2017900357@chd.edu.cn (L.B.); 2020128081@chd.edu.cn (X.Z.)

[2] School of Civil Engineering, Chang'an University, Xi'an 710054, China; 2020228103@chd.edu.cn (Y.H.); 022228101@chd.edu.cn (J.F.)

* Correspondence: jinyi.qin@chd.edu.cn

Abstract: Phenolic compounds in oil wastewater are highly toxic and refractory. Hydrolysis at pH 12 for 12 h makes these compounds more vulnerable to attack and destruction. Under the binding and precipitation of polyaluminum chloride (PAC) at pH 8, the chemical oxygen demand (COD) was significantly reduced by 38%. The simulation found that hydrolysis + flocculation was a complex multistep process. The COD removal rate was mainly controlled by the prehydrolysis process. The metabolic pathway suggested that the m-cresol produced in the factory was oxidized to low water-soluble aldehyde. Alkaline hydrolysis converted the aldehyde into m-toluene-methanol, which was conducive to being captured by the PAC. Fourier transform infrared spectroscopy (FTIR) showed that the alkaline dehydrogenation of two m-methylphenyl carbinols produced a molecule that was removed by the resin. Generally, the particle size of the residue after alkaline hydrolysis was 6.4–8.3 nm, which was included in the pore size range of the resin. Therefore, the resin adsorption capacity for the hydrolyzed substances increased to 47,000 mg L^{-1}, with 93% renewability. In short, hybrid technology reduces the concentration burden of the resin inflow and controls the molecular size of adsorbed substances for repurification. It strengthens the treatment effect of high COD wastewater and provides innovative ideas for extending the service life of resin.

Keywords: prehydrolysis; phenolic compounds; resin sorption; particle size; renewable

Citation: Gu, C.; Yan, K.; Bo, L.; Zhou, X.; He, Y.; Feng, J.; Qin, J. Alkaline Prehydrolysis Prolongs Resin Life and Enhances the Adsorption of Phenolic Compounds. *Water* **2023**, *15*, 2566. https://doi.org/10.3390/w15142566

Academic Editors: José Alberto Herrera-Melián, Weiying Feng, Fang Yang and Jing Liu

Received: 13 June 2023
Revised: 30 June 2023
Accepted: 10 July 2023
Published: 13 July 2023

1. Introduction

Phenolic compounds are prototype toxic substances. Butylated hydroxytoluene (BHT) and its derivatives are common highly toxic and refractory organic compounds in petrochemical wastewater. Phenolic substances in wastewater mainly include phenol, cresol, and 6-tert-Butyl-m-cresol, which originate from petroleum, petrochemical, coal chemical, phenol, and phenolic resin production [1]. For each ton of phenolic resin production, 750 L of wastewater containing 600–42,000 mgL^{-1} of phenol needs to be discharged [2]. Phenolic compounds are toxic to biological individuals and can enter the human body through respiration and skin contact, causing teratogenesis and carcinogenesis. Various countries have successively listed phenolic compounds as the priority pollutants in water control [3].

The commonly used methods for treating phenol-containing wastewater include biochemical methods [1,4], advanced oxidation, electrochemical methods [5], and adsorption [1]. However, phenolic compounds exceeding a concentration of 1000 mgL^{-1} will inhibit the biotreatment capacity of the bioreactor [6]. The intermediates of BHT degradation, such as BHT-Q and BHT-tert-butanol, are more toxic than BHT; therefore, their concentration will increase compared with the influent [7]. Fenton, photoelectric Fenton, electro-Fenton, and other oxidation methods have obvious treatment effects; however, they are difficult to control due to their high energy consumption and harsh reaction conditions [8]. Resin adsorption has the advantages of treating high concentrations with no secondary pollution and regenerating valuable substances. However, desorption treatment is required after adsorption saturation; therefore, it is necessary to reduce the feed concentration of the resin, and the molecular size of the adsorbed resin is required to facilitate its desorption. In the case of alkaline prehydrolysis, it loosens the phenolic structures, breaking the C-O bonds [9], and promotes the generation of hydroxyl radicals in the advanced oxidation process [10], which effectively removes the phenolics [11]. In practice, it is difficult for a single technology to economically and effectively treat wastewater to meet the standard; therefore, it is imperative to develop a combined process.

To prolong the service life of the resin, it is necessary to identify methods to strengthen its adsorption and desorption; therefore, alkaline hydrolysis was used to pretreat the pollutants. Referring to the effect of the advanced oxidation of phenolic compounds, the present study focused on the process of reducing the content of the phenolic compounds and breaking their molecular structure by alkaline prehydrolysis. The pathways of substance decomposition by advanced oxidation and alkaline hydrolysis were investigated separately using liquid chromatography–mass spectrometry (LC–MS). Fourier transform infrared spectroscopy (FTIR) was used to study the effect of the alkaline hydrolysis on the surface functional groups of the flocs and their ability to capture the phenol. Finally, based on the molecular weight obtained from the alkaline hydrolysis of the substance, the effect of the product size on the adsorption and desorption times of the resin was analyzed, and its service life was estimated. The present study provides a case for the establishment of an alkaline pretreatment scheme for high chemical oxygen demand (COD) wastewater and the recovery of effective substances.

2. Materials and Methods

2.1. Materials

Baotashan New Materials Co., Ltd. (Xianyang, China) generates 3 m^3 phenolic wastewater per day. Its pH is between 9 and 10, and the COD is 140,000 mgL^{-1}. The wastewater contains large molecules of phenols that are difficult to volatize. The aim was to discharge water that met the Integrated Wastewater Discharge Standard (GB8978-1996), with a COD of <500 mgL^{-1}.

2.2. Methods

2.2.1. Alkaline Prehydrolysis Experiment

Wastewater (100 mL) was used for the alkaline hydrolysis experiment. (1) After adding 10 mL of 1200 mgL^{-1} polyaluminum chloride (PAC) to the wastewater, it flocculated and settled for 12 h, and the COD of the supernatant was measured as a control. (2) The wastewater was hydrolyzed at pH 12 for 12 h; then, 10 mL of 1200 mgL^{-1} PAC was added to the mixture, which was flocculated and precipitated for 12 h, and the COD of the supernatant was measured. (3) The wastewater was hydrolyzed at pH 12 for 12 h; then, it was flocculated and precipitated for 12 h after adding 10 mL of 1200 mgL^{-1} PAC, after which its pH was adjusted to 8, and the COD of its supernatant was measured. (4) The wastewater was hydrolyzed at pH 12 for 4, 6, 8, 10, 12, and 24 h, after which its pH was adjusted to 8. Next, it was flocculated and precipitated for 12 h after adding 10 mL of 1200 mgL^{-1} PAC; then, the COD of its supernatant was measured.

2.2.2. Macroporous Resin Adsorption Experiment

The macroporous resin (XDA-1G, Sunresin New Materials Co., Ltd., Xi'an, China) was pretreated with 100 mL of anhydrous ethanol to fill the column. The column was washed with 2–3 times the volume of water until there was no alcohol odor at the outlet. The initial 80 mL effluent was discharged as waste; then, its COD was measured, and LC–MS (8050CL) and FTIR (FTIR-650) were carried out.

The measurement parameters of the LC–MS were as follows: ion source ESI (electron spray ionization) (negative), drying gas (temperature 325 °C, flow rate 10 L min^{-1}), atomization pressure 50 psi, capillary voltage 4000 V, and mass spectrometry scan range (m/z) 50~500.

The measurement parameters of the FTIR were as follows: spectral range (4000–400 cm^{-1}), resolution (0.1–0.5 cm^{-1}), light range (2–10 cm), and scanning speed (1–20 cm/s).

2.2.3. Regeneration Life Test of the Macroporous Resin Adsorption–Desorption

Wastewater (100 mL) flowed through the packed column at a rate of 0.1 L min^{-1}. After the adsorption test, the column was backwashed with heated 4% NaOH solution. Subsequently, the column was flushed with 2–3 times the volume of water for desorption and to remove the residual alcohol in the resin. The regeneration capacity of the macroporous resin was determined based on the COD of the effluent after desorption.

2.2.4. Liquid Chromatography–Mass Spectrometry (LC–MS) Detection

The effluent samples were collected and analyzed using LC–MS. The parameters used were ion source ESI (electron spray ionization) (negative), drying gas (temperature of 325 °C, flow rate of 10 L min^{-1}), atomization air pressure of 50 psi, capillary voltage of 4000 V, and mass spectrometry (MS) scan range (m/z) of 50 to 500.

The simulations used pseudo-first-order, pseudo-second-order, Elovich kinetic, and intra-particle diffusion models.

The effect of the alkaline hydrolysis on the degradation of the phenolic substances was investigated using the pseudo-first-order, pseudo-second-order, and Elovich kinetic models, and the equation for intra-particle diffusion. The simulation curve was drawn with Origin software (9.0).

Pseudo-first-order kinetic model:

$$\ln(Q_e - Q_t) = -k_1 t + \ln Q_e \quad (1)$$

Pseudo-second-order kinetic model:

$$t/Q_t = t/Q_e + 1/\left(k_2 {Q_e}^2\right) \quad (2)$$

Elovich kinetic model:

$$Q_t = (1/\beta)\ln t + (1/\beta)\ln(\alpha\beta) \quad (3)$$

The equation for intra-particle diffusion:

$$Q_t = k_3 t^{1/2} \quad (4)$$

where Q_e and Q_t refer to the amount of COD removed at equilibrium and at any time t (h), respectively; k_1, k_2, β, and k_3 are the rate constants for the pseudo-first-order kinetic, pseudo-second-order kinetic, Elovich kinetic, and intra-particle diffusion models, respectively, in units of $L \cdot min^{-1}$, $g \cdot mg^{-1} \cdot min^{-1}$, $L \cdot min^{-1}$, and $mg \cdot g^{-1} \cdot min^{-1}$, respectively, for the reaction process; α is the parameter for the initial reaction rate of the Elovich kinetic equation, $mg \cdot g^{-1} \cdot min^{-1}$; and β is the Elovich kinetic equation constant, $mg \cdot g^{-1}$.

Each data point was measured in triplicate.

3. Results

3.1. Effect of the Hydrolysis pH on the Flocculation and Phenol Removal

The COD of the raw water was 100,000 mgL^{-1}, and the pH was between 8 and 10. As shown in Figure 1, when the PAC was added directly, the residual COD after flocculation reached 80,000 mgL^{-1}. After 12 h of alkaline hydrolysis at pH 12, the PAC was added to the wastewater, and the COD of the supernatant after flocculation decreased to 68,000 mgL^{-1}. After hydrolysis, the pH was adjusted to 8, and the COD of the supernatant after flocculation was 62,000 mgL^{-1}, which was significantly lower than the previous result. After alkaline hydrolysis of the phenol-containing wastewater, the PAC was first added for flocculation, and then the pH was adjusted to 8. The COD in the supernatant was 66,400 mgL^{-1}. This result was between the treatment effects of the two abovementioned methods.

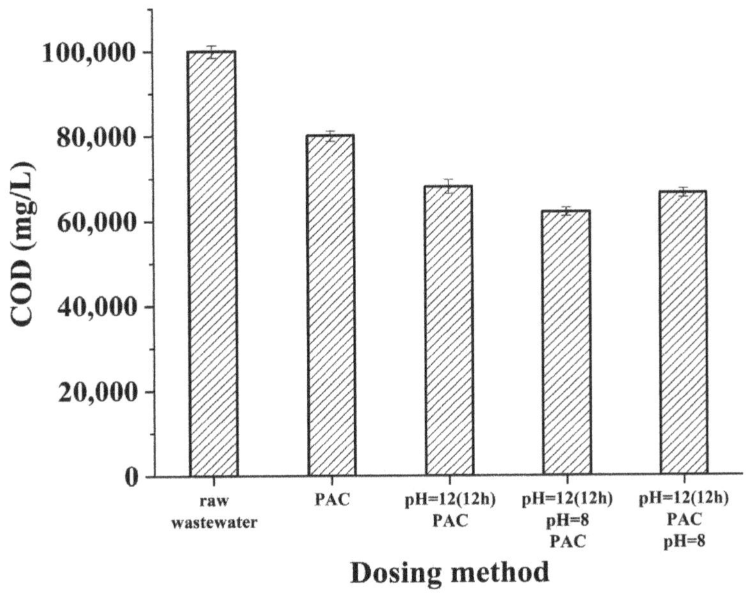

Figure 1. Effect of the hydrolysis pH on the COD removal by flocculation.

Therefore, alkaline hydrolysis was necessary for the pretreatment of the phenolic wastewater. This was because the phenol structure in the wastewater loosened, and the C-O bond gradually broke due to the catalysis of the alkaline reagents [11,12]. Macromolecular phenolic substances were disassembled into small molecules, which were then wrapped and precipitated by the PAC and removed as precipitates, resulting in a substantial decrease in the COD. After the alkaline hydrolysis, the wastewater pH was adjusted to 8, which resulted in the best flocculation performance of the PAC. This was because the PAC was hydrolyzed to stable and strongly bridged Al_{13} at pH 8 [13]. When the pH was 12, the hydrolytic precipitation of the PAC was converted into soluble ions, such as AlO_2, with poor electric neutralization and sweep ability. This led to a poor flocculation effect [14]. In addition, different pH values led to different forms of phenolic compounds, which led to different reaction processes. Most phenolics are hydrophilic and have a constant to measure the degree of acid–base dissociation. When the wastewater pH is too high and greater than this constant, the phenolic substances are in ionic form [5]. However, the phenolic compounds were more vulnerable to be attacked by alkaline reagents when they existed in molecular form, causing their structure to be destroyed. Finally, after hydrolysis and flocculation, the effect of the pH adjustment on the COD reduction was not obvious. This was because the PAC formed a more stable structure with the phenolic substances and their derivatives (Section 3.3); therefore, it did not participate in the subsequent reaction at pH 8.

3.2. *Effect of the Alkaline Hydrolysis Time on the COD Removal*

The effect of the hydrolysis time on the COD removal of the wastewater was studied, and the results are shown in Figure 2A. The initial COD reached 100,000 mgL^{-1}, and the residue COD was reduced to 47,000 mgL^{-1} after 24 h, with a 53% removal efficiency. With increasing time, the COD significantly decreased from 56,000 to 48,800 mg L^{-1} within 4–8 h, and the cumulative removal efficiency rapidly increased from 44% to 51.2%. With increasing hydrolysis time, the C-O and O-O bonds of the hydroxyl groups were broken more completely, as the alkaline reagent and the phenolics in the wastewater came into full contact [12]. In this case, more macromolecules were broken into smaller macromolecules, resulting in a significant decrease in the COD. However, when the alkaline hydrolysis exceeded a duration of 12 h, the COD slightly decreased from 47,200 to 47,000 mgL^{-1}, and the COD removal efficiency curve gradually flattened to 52.9–53%. The results showed that the system reached equilibrium after 12 h, and the removal of the COD by hydrolysis did not only depend on hydrolysis time. All the original phenolic substances in the wastewater were destroyed in the structure, the C-O and O-O bonds were completely broken, and the disassembled substances combined with the PAC to form new substances with a stable structure. Therefore, the residual COD did not decrease significantly, and there was almost no increase in the cumulative removal efficiency.

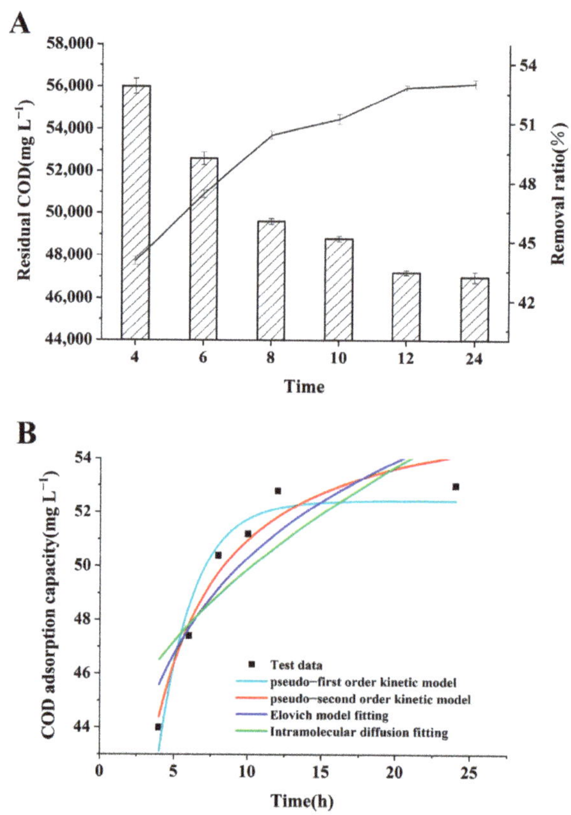

Figure 2. Effect of the alkali hydrolysis time on the COD removal (**A**) and its fitting curve (**B**).

In contrast to the oxidation methods for the treatment of phenol-containing wastewater [15,16], alkaline hydrolysis uses only alkaline reagents and facilitates the extraction and recovery of useful substances from the intermediate products obtained from the treatment. In contrast to hydrolytic acidification, the phenolic wastewater is hydrolyzed from insoluble organic matter to soluble material by the action of a large number of hydrolytic bacteria, which converts insoluble substances into easily biodegradable ones, both of which are released into the mixture, so that the COD concentration of the reactor effluent is still very high, with a COD removal rate of only 30% [17]. However, the alkaline reagent of alkaline hydrolysis acts directly on the C-O bond of the phenols, resulting in an increase in the COD removal of up to 50%.

Notably, because the hydrolysis time was related to the molecular size of the phenols, and the molecular size of phenols was related to the removal effect of the PAC adsorption, the hydrolysis time was indirectly related to the PAC adsorption. Therefore, an adsorption kinetic model was used for fitting. With increasing alkaline hydrolysis time, the decline in the COD in the water was simulated using pseudo-first order, pseudo-second order, Elovich kinetic, and intra-particle diffusion models, as shown in Figure 2B; the fitting parameters are listed in Table 1.

Table 1. Kinetic parameters.

Models	Parameters	
Pseudo-first-order kinetic model	K_1 R^2	0.432 0.93453
Pseudo-second-order dynamics model	K_2 R^2	0.01616 0.93947
Elovich kinetic model	α β R^2	8891.15855 0.1939 0.78161
Intra-particle diffusion model	K_3 R^2	2.90067 0.6252

Table 1 shows that the R^2 values of the alkaline hydrolysis in the four kinetic models were 0.93453, 0.93947, 0.78161, and 0.6252, and both the pseudo-first-order and pseudo-second-order kinetic models were applicable. However, the R^2 value of the pseudo-second-order dynamics model was higher than that of the pseudo-first-order kinetic model, indicating that the alkaline hydrolysis process was most consistent with the proposed secondary kinetic model, and the reaction rate was mainly controlled by the chemical reaction. The pseudo-first-order kinetic model was generally suitable for faster rate reaction processes, while the pseudo-second-order dynamics model was more suitable for complex multi-step reaction processes [18,19]. This showed that the alkaline hydrolysis of this phenol-containing wastewater was a complex multistep process.

3.3. The Putative Metabolism Pathway of the BHT

Based on the known main production reaction of the factory, it was speculated that the abundant substances present in the initial wastewater were 2,6-di-tert-butyl-4-methylphenol (BHT) (220.35), 4,6-di-tert-butyl-m-cresol (220.35), and m-cresol (108.14). The substances obtained after different treatments are shown in Table 2. The m-cresol was prone to oxidize to 3-methylbenzaldehyde by O_2 at 25 °C and standard pressure [20]. The Gibbs free energy of the reaction was −32.69 KJ, which indicated that it could occur spontaneously at room temperature and standard pressure. Therefore, 3−methylbenzaldehyde (120.15) could replace m−cresol when the wastewater was stored in the collector for a long time.

Table 2. The composition of the raw sewage and after different treatments.

Different Treatments	Composition
Raw wastewater	m-cresol; 3-methylbenzaldehyde
Alkaline hydrolysis	m-tolylmethanol
Alkaline hydrolysis + PAC	3-hydroxy-5-methylbenzyl chloride; 1-methyl-3-[(3-methylphenyl) methylperoxymethyl]benzene
Alkaline hydrolysis + PAC + resin adsorption	6-{[3,5-bis(tert-butyl)-4-hydroxyphenyl]methyl}-2-(tert-butyl)-4-chlorophenol

Figure 3 shows the changes in the residual substances detected in the wastewater after the successive treatment of the alkaline hydrolysis, the PAC flocculation–sedimentation, and the resin adsorption. 3-Methylbenzaldehyde can undergo the Cannizzaro reaction to form m-tolylmethanol (122.17), which requires adjusting the wastewater to a strong alkaline condition [21]. When the PAC was added, it was hydrolyzed to produce HCl, which replaced the methane H in water to produce dichloromethane [22]. The dichloromethane and m-tolylmethanol underwent Fourier alkylation in the alkaline environment in which $AlCl_3$ existed [23], and the H on the benzene ring was replaced to form 3-hydroxy-5-methylbenzyl chloride (156.61). Meanwhile, two molecules of m-tolylmethanol formed 1-methyl-3-[(3-methylphenyl)methylperoxymethyl]benzene (242.31). Both 3-hydroxy-5-methylbenzyl chloride and 1-methyl-3-[(3-methylphenyl) methylperoxymethyl]benzene showed 10^6 levels of strength in the mass spectrum, and neither of these substances was detected after the resin adsorption, indicating that they were removed by the adsorption. The mass spectrum of the resin adsorbed effluent showed that all substances were at a level of 10^6. It was speculated that the tert-butyl group of 4,6-di-tert-butyl-m-cresol in the raw water was replaced by Cl^-, which was generated by the hydrolysis of the PAC [22], after which it reacted with 2,6-di-tert-butyl-4-methylphenol (BHT) to generate 6-{[3,5-bis(tert-butyl)-4-hydroxyphenyl]methyl}-2-(tert-butyl)-4-chlorophenol (403). Its molecule was larger than the pore size of the resin; therefore, it was not absorbed and removed by the resin.

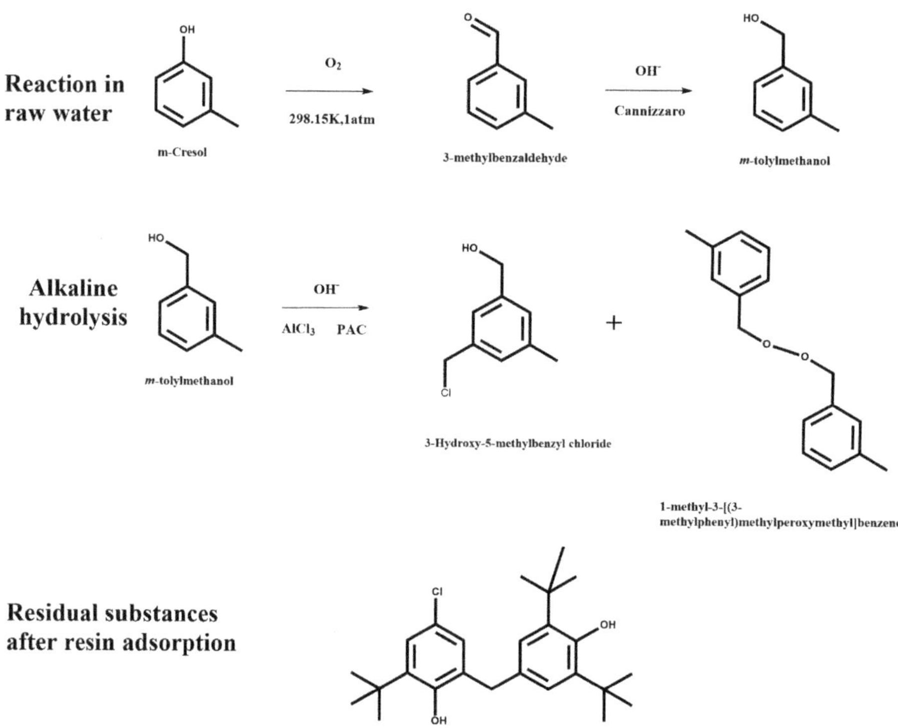

Figure 3. Inferred residual material structure in the wastewater treatment.

In short, although m-cresol was processed and produced by the factory, the real pollutant in the wastewater after exposure to O_2 was a low water-soluble aldehyde. Alkaline hydrolysis was conducive to the conversion of the aldehydes into m-toluene methanol dissolved in water, which was conducive to the water solubility of phenol and its capture by the PAC. Macromolecules that could not be adsorbed by the PAC were also excluded from the pore size of the resin. This reduced the inflow concentration pressure of the resin and assisted in the purification of valuable m-toluene methanol and resin regeneration.

3.4. FTIR Characteristic Analysis of the Functional Groups on the Surface of the Flocs after Hydrolysis

The FTIR spectra of the effluents from the alkaline hydrolysis, pre-oxidation, alkaline hydrolysis + resin adsorption, and pre-oxidation + resin adsorption are shown in Figure 4. The results of the different treatments showed no significant difference between 400 and 1500 cm^{-1}. The peak at 1500–1800 cm^{-1} was the stretching of the carbonyl C=O and the contraction of the benzene ring C=C bond [24,25]. Based on the MS analysis, it was found that the effluent from the four treatments formed similar products, which comprised 3-methylbenzaldehyde and 1-methyl-3-[(3-methylphenyl)methylperoxymethyl]benzene (see Figure 3) and contained numerous C=O bonds. However, the peaks formed by the other three treatments were comparable, except that the alkaline hydrolytic effluent had the lowest absorption peak. This may be because the H on the phenol hydroxyl was easily ionized with alkali to form negative oxygen ions [26]; thereby, the absorption peak shifted to the right, and the absorption intensity weakened.

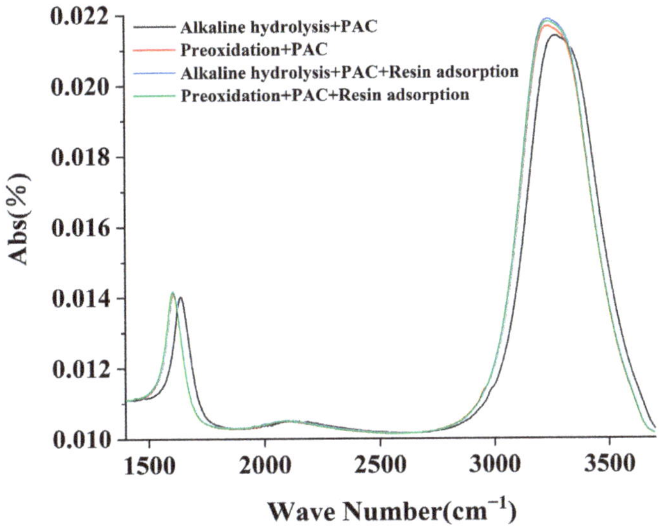

Figure 4. The FTIR characteristics of the wastewater after different treatments.

The peak at 3000–3500 cm^{-1} should be the contraction vibration of OH [27,28]. The order of the peak strength was the alkaline hydrolysis < pre-oxidation < alkaline hydrolysis + resin adsorption < pre-oxidation + resin adsorption. The OH bond vibration without the resin was weaker than that with the resin. This may be due to the alkaline dehydrogenation of two m-tolylmethanol to form a 1-methyl-3-[(3-methylphenyl)methylperoxymethyl]benzene molecule, which was equivalent to the reduction of two phenol hydroxyl molecules [29]. After the resin adsorption, 1-methyl-3-[(3-methylphenyl)methylperoxymethyl]benzene was removed, and only the 6-{[3,5-bis(tert-butyl)-4-hydroxyphenyl]methyl}-2-(tert-butyl)-4-chlorophenol contraction vibration was reflected using the FTIR (see Figure 3).

3.5. Effect of the Prehydrolysis on Improving the Service Life of the Resin Adsorption

The raw water and the alkaline-hydrolyzed wastewater were adsorbed by macroporous resin. Next, the hot lye was backwashed three times for regeneration. The phenol hydroxyl group produced sodium phenol in the alkaline environment, and the adsorption balance between it and the resin adsorption site was broken; therefore, the adsorbed molecules were dissolved back into the water phase [30]. As shown in Figure 5, the removal capacity of the resin to COD in the raw water or alkali-hydrolyzed water decreased with the increasing number of elutions. The two types of wastewater were treated by passing the resin at the same flow rate. After being eluted five times, the adsorption capacity of the resin for the COD in the raw water stabilized at approximately 45,000 mg L^{-1}, which was 89.79% of its initial capacity. However, the adsorption capacity of the resin for the COD in the alkali-hydrolyzed water stabilized at approximately 47,000 mg L^{-1}, reaching 93% of its initial capacity. The regenerated adsorption COD of the resin for the alkali hydrolysis was significantly higher than that of the raw water, and the prehydrolysis prolonged the service life of the resin adsorption.

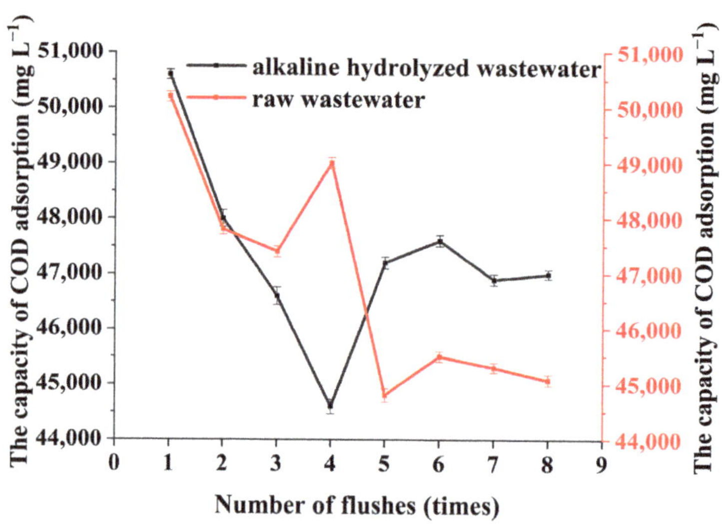

Figure 5. Effect of the prehydrolysis on improving the resin adsorption capacity.

The sizes of the substances obtained from the different treatments are shown in Table 3. The mass spectrum analysis results showed that the alkaline hydrolytic effluent contained [3-(chloromethyl)-5-methylphenyl] methanol and 3,3′-[peroxybis(methylene)]bis(toluene) with diameters of 6.4 and 8.3 nm, respectively (see Figure 3). The raw water contained m-cresol, 4,6-di-tert-butyl-m-cresol, and 2,6-di-tert-butyl-4-methylphenol (BHT) with diameters of 6.3, 8.3, and 8.8 nm, respectively. However, the average pore size of the resin used was fixed at 8.5–9.5 nm. The diameters of the substances in the raw water and the hydrolysis water were smaller than the average pore size of the resin, which proves that they can be captured by the macropores. However, it was assumed from the molecular size that the substances produced by the alkaline hydrolysis were smaller; therefore, the hydrolyzed effluent was more easily adsorbed by the resin to reduce more COD. It was also determined that the resin was more selective in adsorbing organic compounds with smaller molecular sizes. The pore and particle size decreased along with the increase in the number of flushes [31].

Table 3. The molecular size of the substances produced.

Material	Long	Wide	High	Volume	Diameter
m-Cresol	8.653	7.189	4.018	249.95	6.3
4,6-di-tert-butyl-m-cresol	10.752	8.196	6.505	573.24	8.306
2,6-di-tert-butyl-4-methylphenol (BHT)	11.453	8.990	6.505	669.77	8.75
3-Methylbenzaldehyde	9.022	7.316	4.018	265.21	6.424
m-tolylmethanol	8.868	7.736	4.019	275.71	6.508
3-hydroxy-5-methylbenzyl chloride	11.116	8.087	4.019	361.29	8.322
1-methyl-3-[(3-methylphenyl) methylperoxymethyl]benzene	15.773	8.040	4.545	576.37	8.255
6-{[3,5-bis(tert-butyl)-4-hydroxyphenyl]methyl}-2-(tert-butyl)-4-chlorophenol	15.372	10.212	8.312	1304.8	10.928

4. Conclusions

The wastewater produced during petroleum production contains phenolic compounds, which are organic substances that are highly toxic and difficult to degrade. In the present study, a novel technology combined with alkaline hydrolysis and resin adsorption was used to treat phenol-containing wastewater.

The hydrolysis pH was controlled at 12 and maintained for 12 h. Phenolic compounds existed in molecular form. They were more vulnerable to attack and destruction by alkaline reagents. After the pH was adjusted to 8, phenols were wrapped with the PAC and precipitated, resulting in a COD reduction of 38%. With the increase in the hydrolysis time, the COD removal rate increased rapidly from 44% to 51.2%. After 12 h, the COD removal rate gradually stabilized at 52.9–53%. Based on the kinetic model simulation, the hydrolysis + flocculation results were the most consistent with the secondary kinetic model. It was a complex multistep process, and the COD removal rate was mainly controlled by the prehydrolysis. The metabolic pathway revealed that although m-cresol was produced in the factory, it was oxidized to a low water-soluble aldehyde. Alkaline hydrolysis converted aldehydes into m-toluene-methanol, which was conducive to the dissolution of phenol and for capturing the PAC. The FTIR showed that the effluents from the alkali hydrolysis, pre-oxidation, and resin adsorption had similar products, containing numerous C=O bonds. Because the alkaline dehydrogenation of two m-methylphenyl carbinols forms 1-methyl-3-[(3-methylphenyl) methylperoxymethyl] benzene molecules, which were removed by the resin, the OH bond vibration of the remaining substances after the resin adsorption was stronger than that without the resin. The particle size of the material produced by the alkali hydrolysis was 6.4–8.3 nm, which was smaller than that of the raw water (6.3–8.8 nm). The adsorption capacity of the resin for the COD in the raw water was 45,000 mg L^{-1}, with 89.79% renewability, while the adsorption capacity for the alkaline-hydrolyzed water was 47,000 mg L^{-1}, with 93% renewability.

In conclusion, the alkali prehydrolysis before the resin adsorption can enhance the treatment effect of high-COD wastewater, reduce the concentration burden of the resin inflow, facilitate the control of the molecular size of adsorbed substances for repurification, and provide innovative ideas to help resin elution and extend its lifespan.

Author Contributions: C.G.: Conceptualization; Methodology; Supervision; K.Y.: Methodology; Writing—Original Draft; L.B.: Formal analysis; Validation; X.Z.: Data Curation; Investigation; Y.H.: Resources; Project administration; J.F.: Resources; Project administration; J.Q.: Resources; Project administration. All authors have read and agreed to the published version of the manuscript.

Funding: This work was financially supported by the Natural Science Foundation of China (Grant No. 51808044), the Key R&D Program of Shaanxi Province, China (2022KXJ-119), the Shendong Coal Branch Technology Innovation Project of China Shenhua Energy Co., Ltd. (Grant No. CEZB210304069), and the Fundamental Research Funds for the Central Universities, CHD (Grant No. 300102281502).

Data Availability Statement: The datasets used or analyzed during the current study are available from the corresponding author on reasonable request.

Conflicts of Interest: The authors declare that they have no known competing financial interests or personal relationships that could have appeared to influence the work reported in this paper.

References

1. Junyi, L.; Yu, Z.; Lei, Z. Research progress in the treatment technologies of industrial wastewater containing phenol. *Ind. Water Treat.* **2018**, *38*, 12–16.
2. González-Muñoz, M.; Luque, S.; Álvarez, J.; Coca, J. Recovery of phenol from aqueous solutions using hollow fibre contactors. *J. Membr. Sci.* **2003**, *213*, 181–193. [CrossRef]
3. Ronggui, F.; Haijuan, G.; Xian, L. New techniques and research progress on phenol-containing wastewater treatment. *Technol. Water Treat.* **2013**, *39*, 5–8.
4. Busca, G.; Berardinelli, S.; Resini, C.; Arrighi, L. Technologies for the removal of phenol from fluid streams: A short review of recent developments. *J. Hazard. Mater.* **2008**, *160*, 265–288. [CrossRef]
5. Zhang, M.; Zhang, Z.; Liu, S.; Peng, Y.; Chen, J.; Ki, S.Y. Ultrasound-assisted electrochemical treatment for phenolic wastewater. *Ultrason. Sonochemistry* **2020**, *65*, 105058. [CrossRef]
6. Zhang, C.; Wang, X.; Ma, Z.; Luan, Z.; Wang, Y.; Wang, Z.; Wang, L. Removal of phenolic substances from wastewater by algae. A review. *Environ. Chem. Lett.* **2019**, *18*, 377–392. [CrossRef]
7. Wang, W.; Kannan, K. Inventory, loading and discharge of synthetic phenolic antioxidants and their metabolites in wastewater treatment plants. *Water Res.* **2018**, *129*, 413–418. [CrossRef]
8. Brillas, E.; Garcia-Segura, S. Benchmarking recent advances and innovative technology approaches of Fenton, photo-Fenton, electro-Fenton, and related processes: A review on the relevance of phenol as model molecule—ScienceDirect. *Sep. Purif. Technol.* **2020**, *237*, 116337. [CrossRef]
9. Xu, Z.; Xiong, X.; Zeng, Q.; He, S.; Yuan, Y.; Wang, Y.; Wang, Y.; Yang, X.; Su, D. Alterations in structural and functional properties of insoluble dietary fibers-bound phenolic complexes derived from lychee pulp by alkaline hydrolysis treatment. *LWT* **2020**, *127*, 109335. [CrossRef]
10. Guzmán, J.; Mosteo, R.; Sarasa, J.; Alba, J.A.; Ovelleiro, J.L. Evaluation of solar photo-Fenton and ozone based processes as citrus wastewater pre-treatments. *Sep. Purif. Technol.* **2016**, *164*, 155–162. [CrossRef]
11. Cao, S.; Jiang, W.; Zhao, M.; Liu, A.; Wang, M.; Wu, Q.; Sun, Y. Pretreatment Hydrolysis Acidification/Two-Stage AO Combination Process to Treat High-Concentration Resin Production Wastewater. *Water* **2022**, *14*, 2949. [CrossRef]
12. Valsania, M.C.; Fasano, F.; Richardson, S.D.; Vincenti, M. Investigation of the degradation of cresols in the treatments with ozone. *Water Res.* **2012**, *46*, 2795–2804. [CrossRef] [PubMed]
13. Hu, C.-Z.; Liu, H.-J.; Qu, J.-H. Coagulation behavior of Al13 species. *Huan Jing Ke Xue* **2007**, *27*, 2467–2471.
14. Wei, N.; Zhang, Z.; Liu, D.; Wu, Y.; Wang, J.; Wang, Q. Coagulation behavior of polyaluminum chloride: Effects of pH and coagulant dosage. *Chin. J. Chem. Eng.* **2015**, *23*, 1041–1046. [CrossRef]
15. Zhaochun, T.; Quanjin, W.; Weiquan, Y. Chlorine oxidation test for phenol-containing wastewater. *China Water Wastewater* **2002**, *18*, 42–43.
16. Yabin, Q.; Pengfei, X.; Jiacheng, R. study on advanced treatment of coal chemical industry phenolic wastewater by Fenton oxidation. *Energy Chem. Ind.* **2019**, *40*, 59–62.
17. Ranran, S. *Research on Pretreatment of Phenolic Wasterwater with the Integrated Hydrolysis Acidification-Iron-Carbon Microelectrolysis Reactor*; Harbin Institute of Technology: Harbin, China, 2010.
18. Qiu, H.; Lv, L.; Pan, B.-C.; Zhang, Q.-J.; Zhang, W.-M.; Zhang, Q.-X. Critical review in adsorption kinetic models. *J. Zhejiang Univ. A* **2009**, *10*, 716–724. [CrossRef]
19. Deng, S.; Zhang, Q.; Nie, Y.; Wei, H.; Wang, B.; Huang, J.; Yu, G.; Xing, B. Sorption mechanisms of perfluorinated compounds on carbon nanotubes. *Environ. Pollut.* **2012**, *168*, 138–144. [CrossRef]
20. Wang, F.; Xu, J.; Liao, S.-J. One-step heterogeneously catalytic oxidation of o-cresol by oxygen to salicylaldehyde. *Chem. Commun.* **2002**, *6*, 626–627. [CrossRef]
21. Mei, X.; Guo, Z.; Liu, J.; Bi, S.; Li, P.; Wang, Y.; Shen, W.; Yang, Y.; Wang, Y.; Xiao, Y.; et al. Treatment of formaldehyde wastewater by a membrane-aerated biofilm reactor (MABR): The degradation of formaldehyde in the presence of the cosubstrate methanol. *Chem. Eng. J.* **2019**, *372*, 673–683. [CrossRef]
22. Degirmenci, V.; Uner, D.; Yilmaz, A. Methane to higher hydrocarbons via halogenation. *Catal. Today* **2005**, *106*, 252–255. [CrossRef]
23. Wei, X.; Liu, M.; Lu, K.; Wu, H.; Wu, J. Friedel-Crafts alkylation modification and hydrophilic soft finishing of meta aramid. *J. Eng. Fibers Fabr.* **2021**, *16*, 1558925021999061. [CrossRef]
24. Li, Y.-M.; Sun, S.-Q.; Zhou, Q.; Qin, Z.; Tao, J.-X.; Wang, J.; Fang, X. Identification of American ginseng from different regions using FT-IR and two-dimensional correlation IR spectroscopy. *Vib. Spectrosc.* **2004**, *36*, 227–232. [CrossRef]
25. Liu, H.-X.; Sun, S.-Q.; Lv, G.-H.; Chan, K.K. Study on Angelica and its different extracts by Fourier transform infrared spectroscopy and two-dimensional correlation IR spectroscopy. *Spectrochim. Acta Part A Mol. Biomol. Spectrosc.* **2006**, *64*, 321–326. [CrossRef]
26. Chen, L.; Wei, X.; Wang, H.; Yao, M.; Zhang, L.; Gellerstedt, G.; Lindström, M.E.; Ek, M.; Wang, S.; Min, D. A modified ionization difference UV–vis method for fast quantitation of guaiacyl-type phenolic hydroxyl groups in lignin. *Int. J. Biol. Macromol.* **2022**, *201*, 330–337. [CrossRef] [PubMed]

27. Vijayalakshmi, R.; Ravindhran, R. Comparative fingerprint and extraction yield of Diospyrus ferrea (willd.) Bakh. root with phenol compounds (gallic acid), as determined by uv–vis and ft–ir spectroscopy. *Asian Pac. J. Trop. Biomed.* **2012**, *2*, S1367–S1371. [CrossRef]
28. Tiainen, E.; Drakenberg, T.; Tamminen, T.; Kataja, K.; Hase, A. Determination of Phenolic Hydroxyl Groups in Lignin by Combined Use of 1H NMR and UV Spectroscopy. *Holzforschung* **1999**, *53*, 529–533. [CrossRef]
29. Wang, H.; Cao, X.; Xiao, F.; Liu, S.; Deng, G.-J. Iron-Catalyzed One-Pot 2,3-Diarylquinazolinone Formation from 2-Nitrobenzamides and Alcohols. *Org. Lett.* **2013**, *15*, 4900–4903. [CrossRef]
30. Hong, C.; Zhang, W.; Pan, B.; Lv, L.; Han, Y.; Zhang, Q. Adsorption and desorption hysteresis of 4-nitrophenol on a hyper-cross-linked polymer resin NDA-701. *J. Hazard. Mater.* **2009**, *168*, 1217–1222. [CrossRef]
31. Liu, X. *The Preliminary Study on the Effects of the Using-Period of the Extracted Liquids of Chinese Herbal Medice on Absorption and the Regeneration of Fouled Macroporous Absorbing Resin*; Nanjing University of Chinese Medicine: Nanjing, China, 2008.

water

MDPI

Article

Influences of Anthropogenic Pollution on the Dynamics of Sedimentary Fulvic Acid Fractions as Revealed via Spectroscopic Techniques Combined with Two-Dimensional Correlation Spectroscopy

Yanchun Xiao [1], Huibin Yu [2,*] and Yonghui Song [2,*]

1 Institute of Agricultural Engineering Technology, Fujian Academy of Agricultural Sciences, Fuzhou 350003, China; FAASHJKXGroup@163.com

2 Watershed Research Center for Comprehensive Treatment of Water Environmental Pollution, Chinese Research Academy of Environmental Sciences, Beijing 100012, China

* Correspondence: yhbybx@163.com (H.Y.); songyh@craes.org.cn (Y.S.)

Abstract: To identify the influences of anthropogenic activities on the composition, spatial distribution, sources, and transformation mechanism of sedimentary fulvic acid (FA) fractions from different reaches of an urban river were tracked via excitation-emission matrix (EEM) fluorescence spectroscopy with parallel factor (PARAFAC) analysis and two-dimensional correlation spectroscopy (2D-COS). Sediment samples were collected from Baitapu River (BR) along gradients with human activities (e.g., rural, town, and urban sections) in Shenyang, northeast China, from which FA fractions were extracted and then determined via EEM fluorescence spectroscopy. According to optical indices, the autochthonous sources of sedimentary FA fractions in BR were more significant than the terrestrial sources. Among the sections, the contribution from autochthonous sources decreased in the following order: Rural > Urban > Town. Six components of sedimentary FA fractions were identified via EEM–PARAFAC: C1 comprised tryptophan-like (TRL) compounds; C2 was associated with microbial humic-like (MHL) compounds; C3, C4, and C5 were associated with FA-like (FAL) compounds; and C6 comprised humic acid-like (HAL) compounds. The proportion of sedimentary FA fractions decreased in the following order: MHL + FAL + HAL (humus, 77.37–88.90%) > TRL (protein, 11.10–22.63%) for the three sections, showing that humus dominated. The town section exhibited the highest sedimentary FA fractions (5328.87 ± 1315.82 Raman unit [R.U.]), followed by the urban (4146.49 ± 535.75 R.U.) and rural (2510.56 ± 611.00 R.U.) sections. Three pollution sources were determined via principal component analysis (i.e., the dominant industrial source, domestic wastewater, and agricultural effluent). Additionally, the results from 2D-COS revealed that sedimentary FA fractions tended to stabilize as the protein-like component was transformed into the HAL component. Furthermore, we used the structural equation model to validate the critical environmental variables affecting the FA fraction transformation. The results can elucidate the influences of human activities on the dynamics of sedimentary FA fractions in urban rivers.

Keywords: river sediment; fulvic acid fractions; variation; parallel factor analysis; structural equation model

Citation: Xiao, Y.; Yu, H.; Song, Y. Influences of Anthropogenic Pollution on the Dynamics of Sedimentary Fulvic Acid Fractions as Revealed via Spectroscopic Techniques Combined with Two-Dimensional Correlation Spectroscopy. *Water* **2023**, *15*, 2256. https://doi.org/10.3390/w15122256

Academic Editor: Luisa Bergamin

Received: 22 May 2023
Revised: 12 June 2023
Accepted: 14 June 2023
Published: 16 June 2023

1. Introduction

River sediments are an important part of the river ecosystem and function as the main carriers and significant heterogeneous adsorbents for migrating, transforming, and accumulating environmental pollutants [1,2]. Thus, they can significantly influence the water quality and ecological functions of river ecosystems [3,4]. Fulvic acid (FA) fractions are black substances enriched in sediments, which present a strong adsorption capacity with contaminants [5,6]. Previous studies have reported that sedimentary FA fractions play an important role in governing the environmental geochemistry behavior of heavy metal pollutants, which is attributed to the active functional groups such as phenolic, carboxyl, and nitrogen-containing groups [7,8]. Nevertheless, few studies have been conducted on the dynamics of sedimentary FA fractions. The external input of sedimentary FA fractions is mainly influenced by anthropogenic activities (e.g., agriculture practices, industrialization, and urbanization) [9]. They usually show spatial variations, and the dynamics of the fractions are closely linked to environmental variables [10,11]. Therefore, it is necessary to further elucidate the effects of environmental variables and transformation mechanisms of sedimentary FA fractions, which is crucial for managing urban rivers.

The interactive mechanism of FA fractions and fluorophore groups can be described via excitation–emission matrix (EEM) fluorescence spectroscopy [12]. Furthermore, parallel factor (PARAFAC) analysis can mathematically decompose an EEM spectrogram into individual fluorescent components and quantify its relative content by weakening the interference from overlapping fluorophores [13,14]. Two-dimensional correlation spectroscopy (2D-COS) proved an effective technique in spectrum analysis, and it was generally used to improve the resolution of spectra, establish unambiguous assignments, and determine the sequence of the spectra [15]. Cui et al. [7] identified three components from the FA EEM, each component with two ligands, and assessed the interaction of different FA ligands and heavy metals via EEM-PARAFAC and 2D-COS combined technology; that is, high humification FA ligand with more metal-binding sites had the strong metal-binding ability and low metal-binding speed, whereas, low humification FA ligand presented high metal-binding speed. Structural equation models (SEM) can be used to identify underlying processes or unmeasured "latent" variables and explore the theoretical and empirical relationships of variables in a network form [16]. In this study, we employed EEM-PARAFAC, 2D-COS, and SEM to reveal the dynamic changes and transformation mechanisms of sedimentary FA fractions affected by multiple anthropogenic activities.

Considering the unique geological features of the Baitapu River (BR) in Shenyang, northeastern China, and the related human activities, we investigated the effects of anthropogenic activities on the transformation of sedimentary FA fractions. Numerous farmlands, industrial parks, and residential communities exist along the banks of BR, which can be divided into different ecosystem compartments (i.e., rural, town, and urban sections). Thus, it is ideal for studying the dynamic changes in sedimentary FA fractions derived from different anthropogenic activities. In recent years, sewage emission caused by increased agricultural activities, industrial manufacturing, and population has deteriorated water quality [17]. Moreover, the Chinese Environmental Quality Standards for Surface Water (GB-3838-2002) has classified the water environment of BR to be below grade V, which necessitates urgent treatment. Consequently, identifying pollution sources and tracking the variation of sedimentary FA fractions to facilitate pollution control is vital. The main objectives of this study are: (1) to investigate the occurrence and spatial distributions of sedimentary FA fractions in different ecosystem compartments of BR; (2) to identify potentially anthropogenic pollution sources of sedimentary FA fractions through combined statistical analysis models; (3) to compare the variations in FA fractions in different ecosystem compartments and establish the differences and similarities; and (4) to explore the potential mechanisms of transformation of sedimentary FA fractions from urban rivers, which is important to reveal the fate of sedimentary FA fractions in aquatic systems. The results can provide a theoretical basis and support for preventing anthropogenic pollutants in urban rivers.

2. Materials and Methods

2.1. Sampling and Processing

BR (41°37′–41°43′ N, 123°39′–123°20′ E) is a level-one tributary of the Hunhe River in Shenyang, Liaoning, China, with a total length of 51.5 km and a basin area of 182 km^2. BR belongs to a continental semi-humid climate zone with a mean annual average precipitation of 600 to 800 mm and a mean annual average temperature of 6.2 °C to 9.7 °C. It comprises the central area of Shenyang City, including Hunnan District and Heping District, and flows from southeast to northwest through different geological units (Figure 1). The runoff is associated with agriculture; livestock and poultry breeding; and industrial, domestic, and natural activities. In addition, its annual mean runoff volume is 227.2 million m^3/year, which is its main source of water supply [17]. Owing to the low flow rate in BR, which plays a vital role in pollutant retention, BR is divided into three ecosystem compartments: agricultural, industrial, and urban.

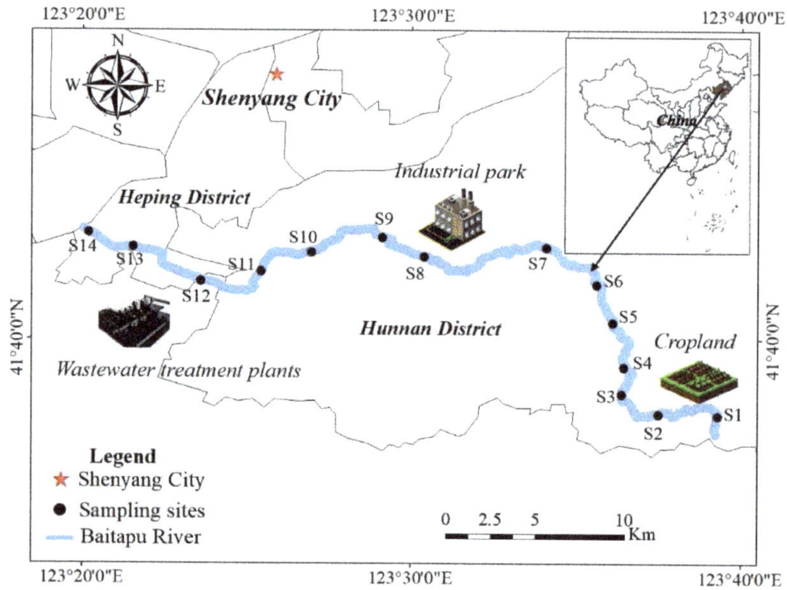

Figure 1. Distribution of sediment samples from BR taken in April 2016.

To cover the three major ecosystem compartments, potential pollution sources, and hydraulic characteristics, 14 sampling sites along BR were selected (Figure 1). Among them, sites S1 to S4 are located in the upper reach covering the rural region, where rural domestic sewage, effluent from livestock and poultry farms, and agriculture sewage are discharged. S5 to S8 is situated in the middle reach covering the town region, where industrial wastewater and domestic sewage are discharged through sewage pipes. S9 to S14 are located in the lower reach covering the urban region, which is affected by effluent from wastewater treatment plants. In April 2016, surface sediments of 0–10 cm were collected using a Peterson grab sampler (ETC200, Shanghai, China). To avoid potential bias caused by accidental sampling, five surface sediments (~1.0 kg) were collected from the center point of the sampling site and four positive directions toward the east, west, south, and north according to the BR section width. Seventy sediment samples were collected from these sites and refrigerated for transport to the laboratory within 12 h of sampling. All sediment samples were freeze-dried, crushed, and sieved through a 0.2 mm mesh sieve before analysis.

2.2. *Sedimentary FA Fraction Extraction*

Sedimentary FA fraction extraction was performed following the International Humic Substances Society (IHSS) recommendations. In short, 10-g freeze-dried, crushed, and sieved sediment was suspended in 100-mL mixed solution (50 mL of 0.1 M $Na_4P_2O_7 \cdot 10H_2O$ and 50 mL of 0.1 M NaOH) to maintain the dry sediment to liquid ratio at 1:10. Subsequently, the suspension was shaken for 24 h (200 rpm) at 25 °C in the absence of light. The supernatant was collected by settling the sediment after centrifugation at 8500 rpm for

10 min. This process was repeated in triplicate, and the supernatant was pooled together to filter through a 0.45-μm Millipore membrane to obtain a humus solution. Then, 6 M HCl was added into the humus solution to adjust the pH value to 1–2, and the supernatant was collected after precipitation at 4 °C for 12 h to obtain the FA fractions.

2.3. Physicochemical Analysis

The pH and electrical conductivity (EC) of the sediment were determined in a sediment/water suspension (sediment/water ratio = 1:2.5) using a multifunctional water analyzer (Hydrolab DS5, HACH, USA). The DO concentration of overlying bottom water (5 ± 1 cm above the sediment) was recorded at the time of sampling in the field using the Hydrolab DS5 water analyzer. The total nitrogen (TN) content of the sediment was measured with a continuous-flow automated analyzer (FUTURA, Alliance, France) after digestion with H_2SO_4-H_2O_2 [18]. The total phosphorus (TP) content of the sediment was analyzed using the Murphy-Riley method after digestion with H_2SO_4-$HClO_4$ [19]. The combustion oxidation nondispersive infrared absorption method (HJ 501-2009) was used to measure the total organic carbon (TOC) content of sedimentary FA fractions with a TOC analyzer (TOC-LCPH, Shimadzu, Japan).

2.4. Fluorescence Measurements and PARAFAC Analysis

The fluorescence EEM spectra of sedimentary FA fractions were obtained using a fluorescence spectrometer (Hitachi F-7000, Tokyo, Japan). Wavelengths of the excitation (E_x) and emission (E_m) were at 200–450 (step 5 nm) and 260–550 (step 5 nm), respectively, with a scan rate of 2400 nm·min^{-1} [20]. Each measured EEM was corrected for the inner-filter effect according to the absorption spectra of the same sample, and the blank signals of Milli-Q water were subtracted to eliminate the influence of Raman scattering. Rayleigh scattering was eliminated using an interpolation technique. The Rayleigh scattering and spectrum data of the upper side were set to zero to eliminate the Rayleigh scattering effect [21]. Then, the corrected EEM was modeled with PARAFAC using MATLAB R2019a software with the DOMFluor toolbox (www.models.life.ku.dk, version 1.7) [22]. The correct number of identified fluorescent components was determined via residual analysis, split-half analysis, and visual inspection. Fluorescence intensity was normalized to Raman units (R.U.), and the relative concentrations of each component were estimated using maximum fluorescence intensity (F_{max}). The contribution of each component to the total fluorescence (%C1–%C6) was used as indices for the composition of sedimentary FA fractions.

Three optical indices, namely, fluorescence index (FI), biological index (BIX), and humification index (HIX) were used to infer the aromaticity, source, and humification degree of sedimentary FA fractions, respectively, which could be calculated using EEM data [23]. Higher (>1.9) and lower (<1.4) values of FI indicate weak and strong aromaticity, respectively [24]. Higher BIX (>1) and HIX (>10) are associated with a stronger biological/aquatic bacterial contribution and a higher humic content. In comparison, lower BIX (<0.8) and HIX (<4) are associated with a weaker contribution from autochthonous sources and a lower humification degree [25]. The three indices are calculated and interpreted using the following formulas:

$$FI = \frac{I_{370/450}}{I_{370/500}}, \quad (1)$$

$$BIX = \frac{I_{310/380}}{I_{310/430}}, \quad (2)$$

$$HIX = \frac{\sum I_{254/435\rightarrow480}}{\sum I_{254/300\rightarrow345}}, \quad (3)$$

where I indicates the fluorescence intensity at each specific E_x/E_m wavelength pair or range.

2.5. Two-Dimensional Correlation Spectroscopy (2D-COS) Analysis

Two-dimensional COS analysis using 2D Shige software version 1.3 (Kwansei–Gakuin University, XiGong, Japan) can provide specific orders of component variation information under external perturbations. The variation in sedimentary FA fractions can be investigated via 2D-COS analysis according to the *Ex* loadings of the PARAFAC components [26]. Noda et al. [27] summarized the rules for analyzing synchronous and asynchronous maps. If the signs of the two maps are the same in a given wavelength range, the event at the x-coordinate is precedent to that at the y-axis, while if the signs are opposite, the preferential changes are reversed with respect to the coordinates [28].

2.6. Statistical Analysis

Principal component analysis (PCA) is an excellent statistical tool for handling multivariate data with minimum loss of information [29]. PCA was performed to identify the possible sources and compositions of sedimentary FA fractions from BR using pH, EC, DO, TN, TP, TOC, FI, BIX, HIX, and the F_{max} of C1–C6.

Path analysis based on a structural equation model (SEM) was used to predict interactive relationships between observed and latent variables and causality among latent variables [30]. The applicability and overall goodness-of-fit test for SEM were determined using the ratio of Chi-square, degrees of freedom, and significance coefficient (i.e., Chi-square/df is less than 5, and the *p-value* is less than 0.05) [31].

3. Results and Discussion

3.1. Physicochemical Parameters

The three sections exhibited variations in physicochemical parameters (Table S1 and Figure S1). The average pH of the sediments gradually increased from the rural (7.62 $\pm$ 0.28) to town (7.85 $\pm$ 0.35) sections, and the pH of the urban section (6.77 $\pm$ 0.69) was significantly lower than those of the rural and town sections. Zhang et al. [12] reported that the relative abundance of *Anaerolineae* in sediment from the urban section was higher than those in sediments from the rural and town sections, which facilitated protein degradation, accompanied by the production of a large amount of organic acid, resulting in a weakly acidic environment in the sediment from the urban section. The urban section exhibited the highest mean EC value (511.50 $\pm$ 330.95 $\mu S \cdot cm^{-1}$), followed by the rural (400.44 $\pm$ 58.87 $\mu S \cdot cm^{-1}$) and town (357.47 $\pm$ 98.34 $\mu S \cdot cm^{-1}$) sections, consistent with the results of Mahabeer et al. [32]. The urban dominant zones exhibited the highest EC value, and agricultural activities increased the EC value through the enrichment of pollutants from the associated anthropogenic activities. The DO concentration of the overlying bottom water from BR varied significantly from 0.15 to 1.67 $mg \cdot L^{-1}$, which was lower than the monitoring threshold (DO < 2 $mg \cdot L^{-1}$) according to the guidelines for the treatment of rural black and black-odorous water (trial) (China 2019). This indicates that the sediments in the whole river were all in a hypoxic state [33].

Additionally, the average content of TN increased from the rural (2283.33 $\pm$ 880.83 $mg \cdot kg^{-1}$) to town (3176.80 $\pm$ 547.94 $mg \cdot kg^{-1}$) sections and then decreased in the urban (2554.68 $\pm$ 1089.90 $mg \cdot kg^{-1}$) section. The town section exhibited the highest TN, which may be related to the industrial activities in this region. A large fertilizer factory, identified in the industrial park along the town section, was possibly one of the leading contributors to the elevated TN level in this section. The average TN content in the rural section was comparable to that in the urban section, which may be related to non-point sources from the predominant farming across the rural region, where nitrogen-containing fertilizers could be leached into BR owing to surface runoff. Furthermore, the mean TP content in the rural (842.22 $\pm$ 436.98 $mg \cdot kg^{-1}$) section was approximately half of those in the town (1777.62 $\pm$ 643.39 $mg \cdot kg^{-1}$) and urban (1870.12 $\pm$ 1086.19 $mg \cdot kg^{-1}$) sections. Phosphorous is relatively immobile in soils, and given its application as fertilizer in farming activities in rural areas, phosphorus uptake by biota increases its retention in the area [32]. In contrast, in the town and urban areas, phosphorus could enter the river system through impervious structures in the anthropogenic zones. According to the sediment pollution evaluation criteria proposed by the United States Environmental Protection Agency (EPA), over 90% of the 14 sampling sites were heavily contaminated (TN > 2000 mg·kg^{-1}, TP > 600 mg·kg^{-1}), indicating that river ecosystems were destroyed by human activities, leading to eutrophication. Moreover, the average TOC content of sedimentary FA fractions decreased in the following order: Town (65.18 ± 31.36 mg·L^{-1}) > Urban (30.98 ± 13.20 mg·L^{-1}) > Rural (20.57 ± 6.38 mg·L^{-1}), which showed that the town section exhibited the highest TOC, related to the industrial and residential activities. The effluent from the industrial park with 20,000 m^3·days^{-1} entered into the river at site S8; consequently, it exhibited a higher TOC content than the other sampling sites. Specifically, the TOC content significantly varied throughout the sedimentary FA fractions, with a coefficient of variation (CV) of 69.38% (Table S1), indicating that the distributions of sedimentary FA fractions were strongly affected by multiple anthropogenic inputs and that significant spatial differences occurred.

3.2. *Fluorescence Spectroscopy Characteristics of Sedimentary FA Fractions*

Through visual inspection, three representative pollution source types were obtained in sedimentary FA fractions with six fluorescence peaks (Figure S2): peak tryptophan-like (TRL, E_x/E_m = 225–237 nm/340–381 nm); peak microbial humic-like (MHL, E_x/E_m = 290–310 nm/370–410 nm); peak soil-derived FA-like (FAL_1, E_x/E_m = 260–300 nm/400–500 nm); peak ultraviolet FA-like (FAL_2, E_x/E_m = 237–260 nm/400–500 nm); peak visible FA-like (FAL_3, E_x/E_m = 300–370 nm/400–500 nm); and peak terrestrial humic acid-like (HAL, E_x/E_m = 370–390 nm/480–500 nm) [20]. Sediments affected by different pollution sources were measured and plotted as an EEM spectrum with various advantageous peaks. In the spectrum of the sediments affected by agricultural effluent, peak FAL_2 was the predominant fluorophore of FA fractions (Figure S2a). The sedimentary FA fractions dominated by peak MHL, FAL_2, and FAL_3 were mainly affected by industrial effluent (Figure S2b). Untreated sewage effluent was raw sewage in municipal pipes, which flowed into BR owing to the overloading of wastewater treatment plants that increased the fluorescence intensity of peak FAL_2 and MHL (Figure S2c).

The FI values of sedimentary FA fractions from BR were 1.55–2.03, with an average value of 1.76 ± 0.09 (Figure S3a). Generally, FI values greater than 1.55 were associated with microbial-derived sources [34]. This shows that the biological source of sedimentary FA fractions was more significant than the terrestrial source in BR. The FI mean value in the rural section (1.85 ± 0.07) was significantly higher than those in the urban (1.74 ± 0.05) and town (1.69 ± 0.07) sections, indicating that the sedimentary FA fractions in the town and urban sections exhibited stronger aromaticity than those in the rural section.

The BIX values were positively correlated with FA fractions formed from biological activity [30]. The BIX values of sedimentary FA fractions from BR ranged from 0.67 to 1.23, with an average value of 0.93 ± 0.11, indicating that most sediment samples of BR were dominated by FA fractions from mixed sources (Figure S3a). The rural section exhibited a higher mean BIX value (1.04 ± 0.10) than the urban (0.92 ± 0.06) and town (0.84 ± 0.09) sections. This indicates that the contribution from autochthonous sources decreased in the following order: Rural > Urban > Town.

The HIX values of sedimentary FA fractions from BR ranged from 2.14 to 9.23, with a mean value of 5.93 ± 1.91, exhibiting a wide range of variation (Figure S3b). The town section exhibited a slightly higher mean HIX value (7.00 ± 1.75) than the urban section (6.44 ± 1.45). The rural section (4.08 ± 1.25) exhibited the lowest average value of HIX. This indicates that the sedimentary FA fractions from the town and urban sections contained higher amounts of condensed polyaromatic structures than that from the rural section. Additionally, several sediment samples from the rural section exhibited low humification levels, which may be associated with relatively low human and strong microbial activities; thus, more endogenous FA fractions were released [23]. Furthermore, all sedimentary FA fractions from the town and urban sections exhibited moderate humification levels

(4 < HIX < 10), indicating that they were partly from exogenous input and biogeochemical processes.

3.3. Identification of PARAFAC Analysis

To characterize the composition and distribution of sedimentary FA fractions, all EEM data from the three sections were modeled using PARAFAC, and six independent components were analyzed (Figure 2 and Table S2). The six components can be grouped into two categories: one protein-like (C1) and five humus-like (C2, C3, C4, C5, and C6) components. C1 (E_x/E_m of 225 nm/380 nm) was identified as tryptophan-like (peak TRL) associated with plankton production [30]. C2 (E_x/E_m of 285 nm/390 nm) was identified as microbial humic-like (peak MHL), related to biological products from the release of algae and bacteria leachate [35]. C3 (E_x/E_m of 260 nm/470 nm) was similar to soil-derived FA-like (peak FAL_1) [36]. C4 (E_x/E_m of 240 nm/435 nm) represented the UV FA-like (peak FAL_2), which was usually derived from the breakdown of plant substances [35]. C5 (E_x/E_m of 335 nm/420 nm) was identified as visible FA-like (peak FAL_3), which was usually derived from the breakdown of animal substances [35]. The broad and long wavelength peak of C6 (E_x/E_m of 360 nm/480 nm) was categorized as terrestrial HAL compounds (peak HAL) with a high molecular weight [14].

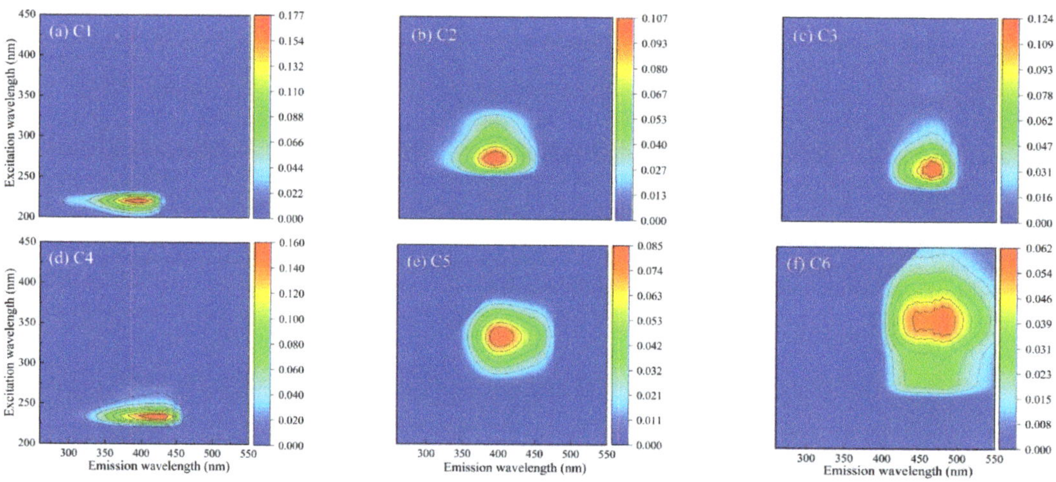

Figure 2. Six components determined by a PARAFAC model fit the EEM of sedimentary FA fractions. **(a)** C1; **(b)** C2; **(c)** C3; **(d)** C4; **(e)** C5; **(f)** C6.

3.4. Distribution in Fluorescence Components

To investigate the variations in sedimentary FA fractions from different sections, the variations in fluorescence intensity and proportion (%) of each component of the six PARAFAC components were evaluated in detail (Figure 3). The total fluorescence intensities in the town (5328.87 ± 1315.82 R.U.) and urban (4146.49 ± 535.75 R.U.) sections were significantly higher than that in the rural (2510.56 ± 611.00 R.U.) section (Figure 3a). This indicates that the town and urban sections faced heavier pollution of sedimentary FA fractions than the rural section, which may be related to the large amounts of industrial wastewater and urban domestic sewage occurring in the town and urban sections, respectively. The rural section was surrounded by sparse villages and farmlands, associated with a low intensity of anthropogenic activities, resulting in a low fluorescence intensity of sedimentary FA fractions. The proportion of each component varied owing to the different pollution source inputs (Figure 3b). For the rural section, %C4 (26.35 ± 1.36%) exhibited the highest mean value of F_{max} proportion, followed by %C1 (22.63 ± 2.47%),

%C2 (17.61 ± 0.19%), %C3 (16.82 ± 2.33%), %C5 (11.85 ± 0.82%), and %C6 (4.74 ± 0.95%). Peak FAL_2 (C4) exhibited the strongest signal of sedimentary FA fractions in the rural section, which is likely associated with the agricultural catchments in that region [25]. Peak TRL (C1) exhibited the second-strongest signal of sedimentary FA fraction, which was a predominant biologically derived component associated with high primary production. The mean values of F_{max} proportion in the town section decreased in the following order: %C3 (21.05 ± 1.34%) > %C4 (19.16 ± 3.03%) > %C2 (18.79 ± 1.18%) > %C5 (17.92 ± 3.33%) > %C6 (11.98 ± 3.14%) > %C1 (11.10 ± 4.84%). The mean values of F_{max} in the urban section decreased as follows: %C4 (23.81 ± 0.89%) > %C3 (19.85 ± 0.94%) > %C1 (18.03 ± 1.90%) > %C2 (17.67 ± 0.75%) > %C5 (13.42 ± 0.94%) > %C6 (7.22 ± 0.75%). A high protein tryptophan-like (C1) abundance of sedimentary FA fractions was found in the rural and urban sections, consistent with previous studies [37,38]. The percentages of sedimentary FA fractions for the three sections decreased as follows: %C2 +%C3 +%C4 +%C5 +%C6 (humus-like, 77.37–88.90%) > %C1 (protein-like, 11.10–22.63%). As described, humus-like compounds dominated the FA fraction because the terrestrial substances were the dominant inputs, and labile protein-like compounds were rapidly consumed by microorganisms, thus contributing a lower amount of protein-like fluorescence for sedimentary FA fractions. In contrast, relatively stable humus-like compounds were preferentially preserved in the sediments under hypoxic or anoxic conditions.

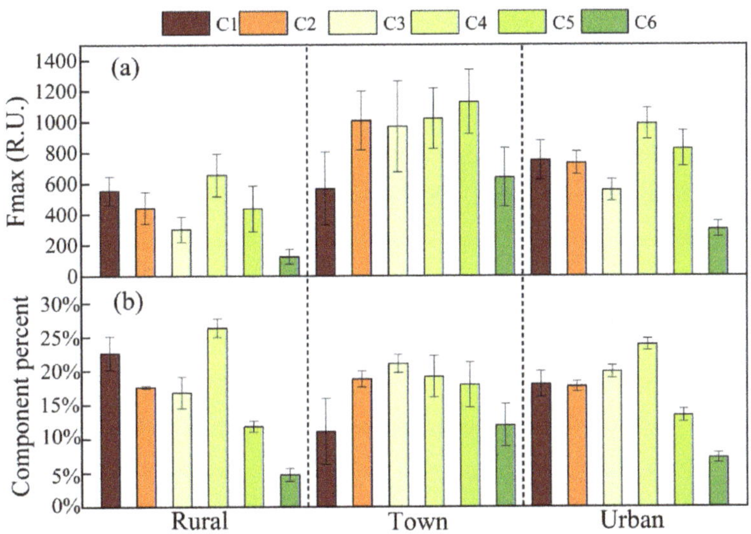

Figure 3. Plots of maximum fluorescence intensity (**a**) and proportion of six components (**b**) from the PARAFAC model in different sections.

3.5. Identification and Apportionment of Sedimentary FA Fractions Sources

The variations in sedimentary FA fractions and the influences of human activities were further investigated via PCA. Figure 4 summarizes the PCA results, including the factor loading and score plot. The loading whose absolute value was greater than 0.6 of the total variance was highlighted. The first two principal components explained 62.15% (PC1) and 23.84% (PC2) of the total variance. PC1 was positively correlated with TOC (r = 0.868), C2 (r = 0.818), C3 (r = 0.877), C5 (r = 0.756), and C6 (r = 0.908), but PC1 was negatively correlated with FI (r = −0.610). This indicates that higher PC1 was related to the terrestrial sources with condensed polyaromatic structures and humus content associated with industrial sources [29]. PC2 exhibited a strong positive correlation with TP (r = 0.910) and C4 (r = 0.653), indicating that PC2 was mainly associated with the emission of domestic sewage and agricultural effluent [39].

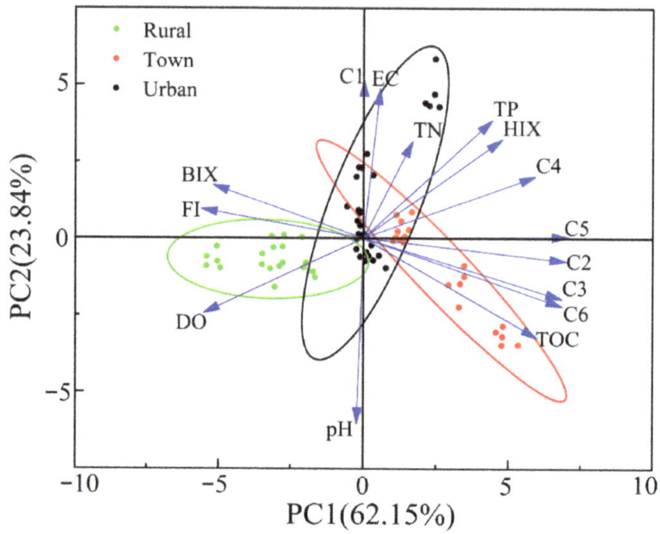

Figure 4. PCA results show the interrelation between physicochemical parameters, optical indices, and PARAFAC components. Samples are color-coded to represent the rural (green), town (red), and urban (black) groups; The green ellipse is the sedimentary FA fractions with a 95% confidence level in rural section; The red ellipse is the sedimentary FA fractions with a 95% confidence level in town section; The black ellipse is the sedimentary FA fractions with a 95% confidence level in urban section; Arrows (blue) represent the factor loadings.

The sedimentary FA fractions with a 95% confidence level formed a cluster for each ellipse in the score plot. The rural and town groups were assigned to different sides of PC1 and had different PC1 values. Sedimentary FA fraction signals of the rural group were identified by FI and BIX. In contrast, those of the town group were identified by a greater number of humus-like compounds (C2, C3, C5, and C6), which indicates a higher autochthonous feature in the rural section and a higher humification degree in the town section [40–42]. However, the urban group was located on the positive loading of PC2. It was positively correlated with C1, which may be related to the stimulated growth of phytoplankton owing to the increased nutrient level and decreased pH level in the urban section sediments. The combined results reveal that the properties of sedimentary FA fractions were affected by long-term external input, which possibly promoted autochthonous production owing to the decomposition and mineralization of terrestrial material by microbial activity and phytoplankton production.

3.6. Dynamic Variations in Sedimentary FA Fractions from Different Sections

Sedimentary FA fractions are heterogeneous and influenced by environmental factors such as pH, EC, and DO. These factors play a vital role in the dynamic variations of sedimentary FA fractions. The transformation order of the sedimentary FA fractions from different sections was further explored via 2D COS analysis.

Peak TRL was negatively correlated with FAL_1 in the synchronous and asynchronous maps (Figure 5a1,a2), while peak FAL_1 was positively correlated with FAL_2 in the two maps (Figure 5b1,b2). Moreover, peak FAL_3 was positively correlated with FAL_2 in the synchronous map but negatively correlated with FAL_2 in the asynchronous map (Figure 5c1,c2). The correlations between peak MHL and peak FAL_3 were similar to those between peak FAL_3 and peak FAL_2 (Figure 5d1,d2). According to Noda's rule [27], the sequential changes followed the order: TRL$\rightarrow$$FAL_1$$\rightarrow$ $FAL_2$$\rightarrow$$FAL_3$$\rightarrow$MHL. Studies have shown a competitive relationship between biodegradation and photodegradation when the primary source was endogenous [43]. Consequently, labile tryptophan-like substances in the rural section may be first transformed into relatively high concentrations of nitrogenous precursors for the formation of the FAL materials [39]. Under conditions of light radiation, microorganisms, particularly Actinobacteria, participated in the utilization of FAL components, and several complex FAL components were degraded into simple compounds, followed by the production of MHL components [44]. Particularly, the degradation order of FAL components (FAL_1, FAL_2, and FAL_3) was arranged by decreasing E_m wavelength, possibly because light destroyed and degraded the more resistant structure of FAL components into usable substrates. Peak MHL was positively correlated with peak HAL in both the synchronous and asynchronous spectra (Figure 5e1,e2), showing the order of MHL→HAL, which indicates that MAL compounds could form a stable HAL compound through microbial polymerization [45]. The sequential changes followed the order of TRL→FAL→MHL→HAL for sedimentary FA fractions in the rural section.

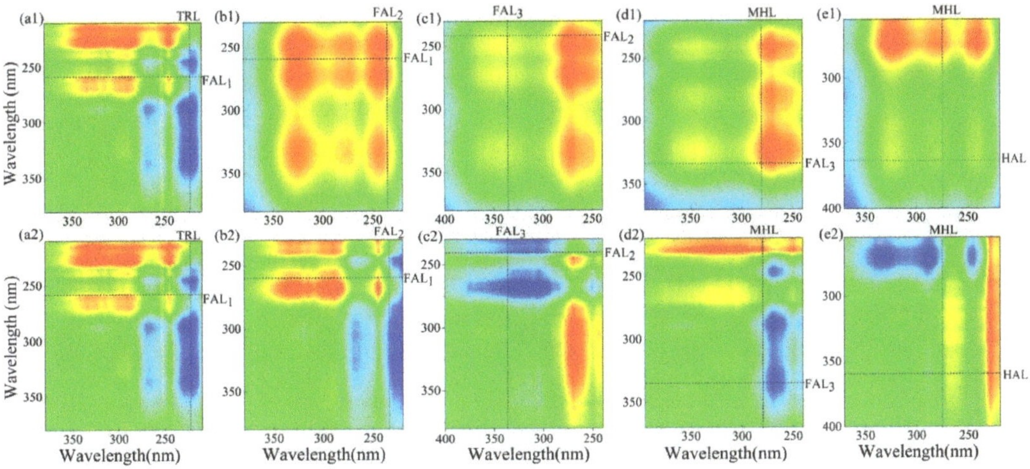

Figure 5. Synchronous and asynchronous maps from 2D-COS analysis in the rural section. Red and yellow represent positive correlations, and blue and green represent negative correlations. (**a1**) Synchronous map, C1 vs. C3; (**a2**) asynchronous map, C1 vs. C3; (**b1**) synchronous map, C3 vs. C4; (**b2**) asynchronous map, C3 vs. C4; (**c1**) synchronous map, C5 vs. C4; (**c2**) asynchronous map, C5 vs. C4; (**d1**) synchronous map, C2 vs. C5; (**d2**) asynchronous map, C2 vs. C5; (**e1**) synchronous map, C2 vs. C6; and (**e2**) asynchronous map, C2 vs. C6.

The synchronous and asynchronous maps for the town section are shown in Figure 6. Peaks TRL and FAL_2 exhibited the same relationship (Figure 6a1,a2), while peaks FAL_3 and FAL_2 exhibited opposite relationships (Figure 6b1,b2) in the two maps, indicating that the change order was TRL→FAL_2→FAL_3. The correlation between peaks FAL_3 and FAL_1 (Figure 6c1,c2) showed the same signals, while peaks MHL and FAL_1 exhibited opposite signals (Figure 6d1,d2) in the two maps, which shows that the variation order was FAL_3→FAL_1→MHL. Peaks MHL and HAL exhibited the same relationships (Figure 6e1,e2) in the two maps, indicating that the variation order was MHL→HAL. Overall, the sequence variation of sedimentary FA fractions for the town section followed the order of TRL→FAL_2→FAL_3→FAL_1→MHL→HAL, similar to the variation trend of the rural section. The variation is probably related to the same controlling factors in the rural section, such as pH (Section 3.7). However, FAL_1 degraded later than FAL_2 and FAL_3, attributable to two factors, according to the reports by Yao et al. [46]: The first is physicochemical factors (e.g., pH, DO, TN, and TP), which were the most significant contributor to the river sediment bacterial community (49.50%). The second factor is the occurrence of heavy metals, which significantly contributed to the sediment fungal community (48.00%). The highest TN concentration in the town section could significantly increase enzyme activity, facilitating the conversion of active lignin and cellulose and promoting the degradation process of FAL_2 derived from the breakdown of plant substances, consistent with the findings of Saiya-Cork et al. [47], who observed that N addition increased βG (cellulolytic enzyme) activity. In our previous study of heavy metals, the Naoxiao River near the town section of BR was proven to undergo heavy metal pollution, particularly by Cd and Zn [48]. The presence of heavy metals in the town section caused by exogenous input, such as through industrial sewage and domestic sewage discharge, may alter the fungal community structure and properties. For example, the heavy metals Cd and Zn can significantly affect the distribution and microbial functions of saprotrophs, which were the dominant trophic mode of the fungal community in sediment, mainly responsible for the decomposition of plant and animal residues (FAL_2 and FAL_3) [49].

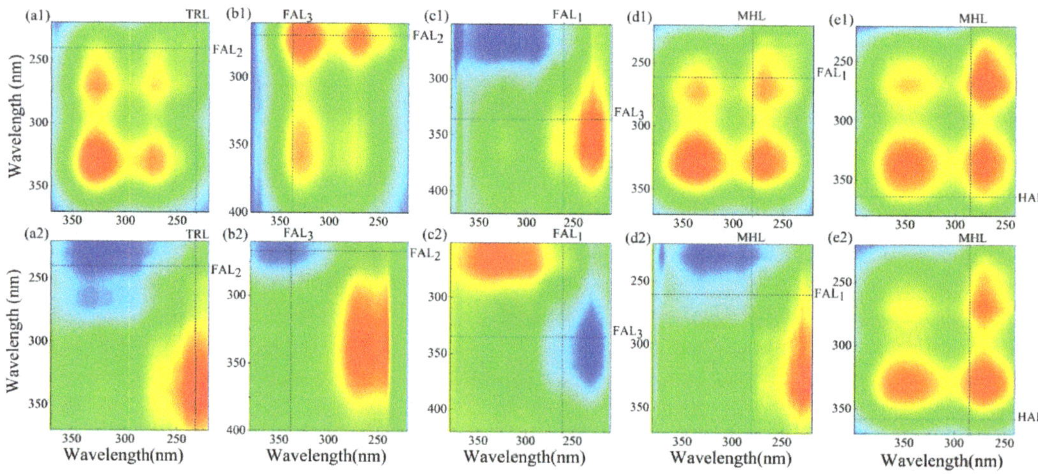

Figure 6. Synchronous and asynchronous maps from 2D-COS analysis in the town section. Red and yellow represent positive correlations, and blue and green represent negative correlations. (**a1**) Synchronous map, C1 vs. C4; (**a2**) asynchronous map, C1 vs. C4; (**b1**) synchronous map, C5 vs. C4; (**b2**) asynchronous map, C5 vs. C4; (**c1**) synchronous map, C5 vs. C3; (**c2**) asynchronous map, C5 vs. C3; (**d1**) synchronous map, C2 vs. C3; (**d2**) asynchronous map, C2 vs. C3; (**e1**) synchronous map, C2 vs. C6; and (**e2**) asynchronous map, C2 vs. C6.

Figure 7 shows the synchronous and asynchronous maps for the urban section. The relationship between peaks MHL and TRL (Figure 7a1,a2) exhibited opposite signals in the two maps. In contrast, peaks MHL and FAL_3 (Figure 7b1,b2), FAL_3 and FAL_2 (Figure 7c1,c2) and FAL_2 and FAL_1 (Figure 7d1,d2) exhibited the same signals, indicating that the variation followed the order of $TRL \rightarrow MHL \rightarrow FAL_3 \rightarrow FAL_2 \rightarrow FAL_1$. The correlation between peaks HAL and FAL_1 (Figure 7e1,e2) showed opposite signals in the two maps, indicating that the variation order was $FAL_1 \rightarrow HAL$. Hence, the sequential change order was $TRL \rightarrow MHL \rightarrow FAL_3 \rightarrow FAL_2 \rightarrow FAL_1 \rightarrow$ HAL for the urban section, which follows increasing E_m wavelength. Generally, it has been reported that the fluorescence features at longer E_m could be linked with larger-size and higher-density structures [22]. This indicates that the formation and transformation of sedimentary FA fractions tend to be more complex and stable [50–52], consistent with the normal composting owing to the lowest DO concentration (close to anoxia state) in the urban section (Table S1). The sequential change order of sedimentary FA fractions differed among the three sections, and the effects of environmental variables and terrestrial inputs on the decomposition and polymerization of sedimentary FA fractions remain unclear. Thus, further exploring the influencing mechanism is vital.

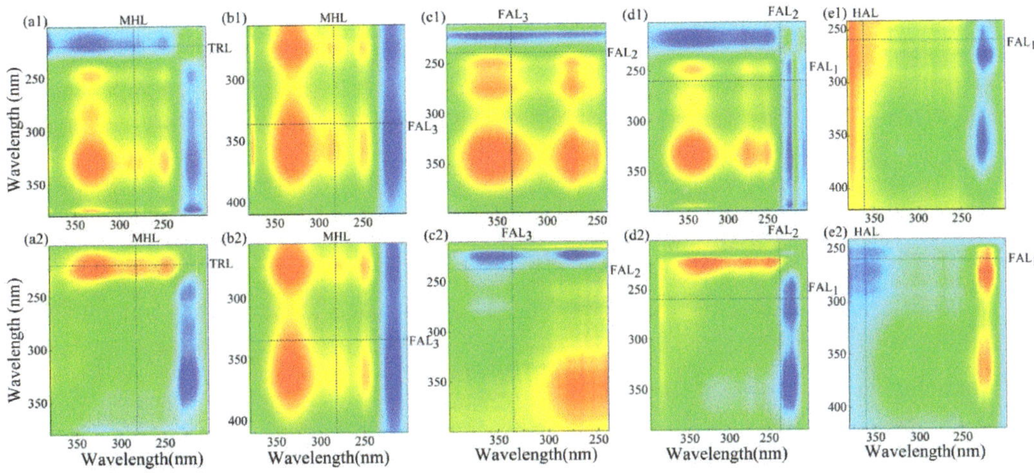

Figure 7. Synchronous and asynchronous maps from 2D-COS analysis in the urban section. Red and yellow represent positive correlations, and blue and green represent negative correlations. (**a1**) Synchronous map, C2 vs. C1; (**a2**) asynchronous map, C2 vs. C1; (**b1**) synchronous map, C2 vs. C5; (**b2**) asynchronous map, C2 vs. C5; (**c1**) synchronous map, C5 vs. C4; (**c2**) asynchronous map, C5 vs. C4; (**d1**) synchronous map, C4 vs. C3; (**d2**) asynchronous map, C4 vs. C3; (**e1**) synchronous map, C6 vs. C3; (**e2**) asynchronous map, C6 vs. C3.

3.7. Environmental Factors Affecting the Transformation of Sedimentary FA Fractions

IBM SPSS AMOS 22 was employed to further explore the causal relationships between environmental factors, nutrients, and sedimentary FA fractions from BR [53]. For Chi-square/df = 4.647 and $p = 0.000$, the SEM model revealed the correlation of environmental factors with the transformation of sedimentary FA fractions [30]. "Protein" and "Humus" were latent variables (in ovals) (Figure 8), while the others were observed variables (in rectangles). The path coefficients of TOC with pH, DO, and EC were 0.78, −0.64, and 0.28, respectively, which may be related to the conducive effect of the environmental variables on the biological processes of sedimentary FA fractions through the influence of the variables on microbial communities and functions. Particularly, pH showed the most significant direct correlation with nutrients, which was consistent with the previous reports that pH could significantly influence the environmental behavior of the sedimentary FA fractions by altering nitrogen-related functions [4,46]. Zhang et al. [12] found that phylum Proteobacteria and Chloroflexi were the dominant phyla in the river sediment, preferring the degradation of aromatic compounds and protein-like substances. Moreover, Wang et al. [54] reported that Actinomycetes and Planctomycetes in the sediments could be potentially involved in organic matter degradation and methane and nitrogen cycling processes, and the higher abundance of Limnohabitans was related to the lower pH and its sensitivity to acid. The path coefficients of TOC with protein and humus were −0.59 and 0.65, respectively, indicating that TRL materials were mainly derived from autochthonous sources, and humic-like materials were mainly derived from allochthonous sources [21]. Moreover, protein exhibited a significant negative effect (−0.21) on the humus, and the factor loading decreased in the following order: FI (1.00) > BIX (0.98) > C1 (−0.05). Thus, FI and BIX exhibited significant indirect negative effects on humus, mediated by the latent variable of protein, particularly FI. This can be attributed to the transformation of large amounts of allochthonous FA fractions in the sediments into autochthonous FA fractions through microbial metabolism, which indicates that the autochthonous origins were dominant in BR. Additionally, the path coefficient (0.37) of TP with humus was positive, possibly because the nutrients from terrestrial input could facilitate microbial transformation or humus production [55]. All TN/TP values were <20 (Table S1), which confirms previous reports that for TN/TP values less than 20, TN is deficient, and TP is the limiting factor [56]. TP as the prominent driver of the bacterial community, is considered closely related to chemoheterotrophy and fermentation [57]. In the humus measurement model, the factor loadings decreased in the following order: C5 (1.00) > C2 (0.99) > C3 (0.98) > C6 (0.94) > C4 (0.80). This indicates that the C5 and C2 exhibited a significant direct correlation with the latent variable of humus, related to the exogenous input and the diverse microorganisms existing in the sediment. With a higher TOC concentration and lower DO content, the bacteria in the sediment became more active, which caused a release of bacterial metabolizing FA fractions [58]. In Figure 8, doubled-headed arrows indicate correlations between variables (e.g., C1 and C2; C1 and C4), revealing that sedimentary FA fractions tended to stabilize as protein-like components were transformed into humus-like components [59], which was consistent with the results of Section 3.6.

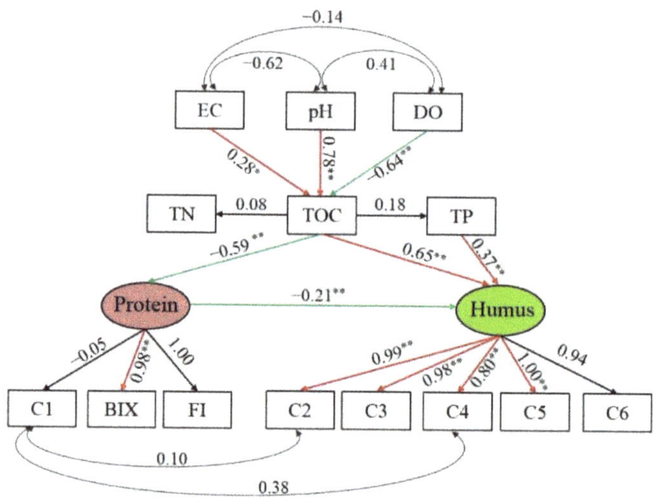

Figure 8. Predicted path model for the sedimentary FA fractions; solid red lines indicate significant positive effects, and solid green lines indicate significantly negative effects; solid black lines indicate slightly significant effects; number adjacent to the lines are the standardized coefficients; single-headed arrows represent unidirectional causal relationships; and double-headed arrows represent non-causal covariance. ——— Positive correlation; ——— Negative correlation; * $p < 0.05$; ** $p < 0.01$.

4. Conclusions

Variations in sedimentary FA fractions in three sections of an urban river were explored. In the sediments, the biological source of FA fractions was more significant than the terrestrial source, particularly in the rural section. The results of PCA revealed that the sedimentary FA fractions were mainly affected by the discharge of industrial wastewater, domestic wastewater, and agricultural effluent. Six components were identified from the sedimentary FA fractions: C1 comprised TRL compounds; C2 was associated with MHL compounds; C3, C4, and C5 were related to FAL compounds; and C6 comprised HAL compounds. The sequential variation of sedimentary FA fractions for the rural and town sections followed the order of TRL→FAL→MHL→HAL. In contrast, that of sedimentary FA fractions for the urban section followed the order of TRL→MHL→FAL→HAL. According to SEM, the pH and TP were the critical environmental variables that influenced the transformation of sedimentary FA fractions, labile protein-like component (C1) was transformed into a stable HAL component (C6), and the stable HAL component might persist for a long time in the environment and promote the internal release. Consequently, the strict control of external source discharge and the adoption of remedial measures to improve the river ecosystem is highly recommended.

Supplementary Materials: The following supporting information can be downloaded at: https://www.mdpi.com/article/10.3390/w15122256/s1, Table S1: Physicochemical properties of sediments, overlying bottom water, and sedimentary FA fractions from BR, China; Table S2. Description of the six PARAFAC model components from the sedimentary FA fractions; Figure S1: Boxplots display the spatial variation of physicochemical parameters. White circles indicate the mean values, boxes show standard error, and whiskers delimit maximum and minimum; Figure S2: Representative EEM of sedimentary FA fractions with various pollution types and sources: (a) agricultural effluent (S1); (b) industrial effluent (S8); and (c) untreated sewage effluent (S13); Figure S3: Spatial variations of the values of FI, BIX and HIX.

Author Contributions: Y.X.: Methodology, Software, Data curation, and Writing—original draft. H.Y.: Investigation, Resources, Writing—review & editing, and Data curation. Y.S.: Writing—review & editing and Supervision. All authors have read and agreed to the published version of the manuscript.

Funding: This work was supported by the National Key R&D Program of China (No. 2021YFC3201500), the Natural Science Foundation of Fujian Province (Nos. 2020J011374 and 2021J01500), the Fujian Provincial Department of Science and Technology (Nos. 2020R1032002 and 2020R1032006), and the Foreign Cooperation Project of Fujian Academy of Agricultural Sciences (No. DWHZ-2022-17).

Institutional Review Board Statement: Not applicable.

Informed Consent Statement: Not applicable.

Data Availability Statement: Not applicable.

Acknowledgments: The authors are grateful to the editors and the anonymous reviewers for their insightful comments and suggestions.

Conflicts of Interest: The authors declare no conflict of interest.

References

1. Nargis, A.; Habib, A.; Harun, O.R.; Harun, H.B.; Islam Sarker, M.S.; Jin, R.; Cai, M. Status of multielement in water of the river Buriganga, Bangladesh: Aquatic chemistry of metal ions in polluted river water. *Emerg. Contam.* **2021**, *7*, 99–115. [CrossRef]
2. Ali, J.; Li, Y.; Wang, X.J.; Zhao, J.; Xi, N.N.; Zhang, Z.R.; Xia, X.H. Climate-zone-dependent effect mechanism of humic acid and fulvic acid extracted from riverine sediments on aggregation behavior of graphene oxide. *Sci. Total Environ.* **2020**, *721*, 137682. [CrossRef] [PubMed]
3. Gontijo, E.S.J.; Herzsprung, P.; Lechtenfeld, O.J.; de CBueno, C.; ACBarth, J.; Rosa, A.H.; Friese, K. Multi-proxy approach involving ultrahigh resolution mass spectrometry and self-organising maps to investigate the origin and quality of sedimentary organic matter across a subtropical reservoir. *Org. Geochem.* **2021**, *151*, 104165. [CrossRef]
4. Song, F.H.; Wu, F.C.; Guo, F.; Wang, H.; Feng, W.Y.; Zhou, M.; Deng, Y.H.; Bai, Y.C.; Xing, B.S.; Giesy, J.P. Interactions between stepwise-eluted sub-fractions of fulvic acids and protons revealed by fluorescence titration combined with EEM-PARAFAC. *Sci. Total Environ.* **2017**, *605–606*, 58–65. [CrossRef] [PubMed]
5. Fallah, R.E.; Rouillon, R.; Vouvé, F. Spectral characterization of the fluorescent components present in humic substances, fulvic acid and humic acid mixed with pure benzo (a) pyrene solution. *Spectrochim. Acta Part A Mol. Biomol. Spectrosc.* **2018**, *199*, 71–79. [CrossRef]
6. Qian, G.R.; Xu, L.; Li, N.; Wang, K.L.; Qu, Y.W.; Xu, Y.F. Enhanced arsenic migration in tailings soil with the addition of humic acid, fulvic acid and thiol-modified humic acid. *Chemosphere* **2022**, *286*, 131784. [CrossRef]
7. Cui, H.Y.; Wen, X.; Wu, Z.H.; Zhao, Y.; Lu, Q.; Wei, Z. Insight into complexation of Cd(II) and Cu(II) to fulvic acid based on feature recognition of PARAFAC combined with 2DCOS. *J. Hazard. Mater.* **2022**, *440*, 129758. [CrossRef]
8. Li, H.; Wang, J.H.; Zhao, B.Y.; Gao, M.S.; Shi, W.J.; Zhou, H.J.; Xie, Z.L.; Zhou, B.; Lü, C.W.; He, J. The role of major functional groups: Multi-evidence from the binding experiments of heavy metals on natural fulvic acids extracted from lake sediments. *Ecotoxicol. Environ. Saf.* **2018**, *162*, 514–520. [CrossRef]
9. Yu, H.B.; Song, Y.H.; Du, E.; Yang, N.; Peng, J.F.; Liu, R.X. Comparison of PARAFAC components of fluorescent dissolved and particular organic matter from two urbanized rivers. *Environ. Sci. Pollut. Res.* **2016**, *23*, 10644–10655. [CrossRef]
10. Aranganathan, L.; Radhika Rajasree, S.R.; Govindaraju, K.; Sivarathna kumar, S.; Gayathri, S.; Remya, R.R.; Suman, T.Y. Spectral and microscopic analysis of fulvic acids isolated from marine fish waste and sugarcane bagasse co-compost. *Biocatal. Agric. Biotechnol.* **2020**, *29*, 101762. [CrossRef]
11. Wen, Z.D.; Song, K.S.; Shang, Y.X.; Lyu, L.L.; Tao, H.; Liu, G. Natural and anthropogenic impacts on the DOC characteristics in the Yellow River continuum. *Environ. Pollut.* **2021**, *287*, 117231. [CrossRef] [PubMed]
12. Zhang, L.; Sun, Q.X.; Dou, Q.H.; Lan, S.; Peng, Y.Z.; Yang, J.C. The molecular characteristics of dissolved organic matter in urbanized riverine sediments and their environmental impact under the action of microorganisms. *Sci. Total Environ.* **2022**, *827*, 154289. [CrossRef] [PubMed]

13. Feng, L.K.; Zhang, J.; Fan, J.L.; Wei, L.L.; He, S.F.; Wu, H.M. Tracing dissolved organic matter in inflowing rivers of Nansi Lake as a storage reservoir: Implications for water-quality control. *Chemosphere* **2021**, *286*, 131624. [CrossRef] [PubMed]
14. Xu, X.T.; Kang, J.; Shen, J.M.; Zhao, S.X.; Wang, B.Y.; Zhang, X.X.; Chen, Z.L. EEM-PARAFAC characterization of dissolved organic matter and its relationship with disinfection by-products formation potential in drinking water sources of northeastern China. *Sci. Total Environ.* **2021**, *774*, 145297. [CrossRef] [PubMed]
15. Fan, T.T.; Yao, X.; Ren, H.Y.; Liu, L.; Deng, H.G.; Shao, K.Q. Regional-scale investigation of the molecular weight distribution and metal-binding behavior of dissolved organic matter from a shallow macrophytic lake using multispectral techniques. *J. Hazard. Mater.* **2022**, *439*, 129532. [CrossRef]
16. Lupi, L.; Bertrand, L.; Monferrán, M.V.; Amé, M.V.; Diaz, M.P. Multilevel and structural equation modeling approach to identify spatiotemporal patterns and source characterization of metals and metalloids in surface water and sediment of the Ctalamochita River in Pampa region, Argentina. *J. Hydrol.* **2019**, *572*, 403–413. [CrossRef]
17. Liu, D.P.; Lu, K.T.; Yu, H.B.; Gao, H.J.; Xu, W.N. Applying synchronous fluorescence spectroscopy conjunct second derivative and two-dimensional correlation to analyze the interactions of copper (II) with dissolved organic matter from an urbanized river. *Talanta* **2021**, *235*, 122738. [CrossRef]
18. Soon, Y.K.; Kalra, Y.P. Short communication: A comparison of plant tissue digestion methods for nitrogen and phosphorus analyses. *Can. J. Soil Sci.* **1995**, *75*, 243–245. [CrossRef]
19. Bao, S. *Soil Agrochemical Analysis*; China Agricultural Press: Beijing, China, 2000; Volume 30.
20. Guo, X.; Yu, H.; Yan, Z.; Gao, H.; Zhang, Y. Tracking variations of fluorescent dissolved organic matter during wastewater treatment by accumulative fluorescence emission spectroscopy combined with principal component, second derivative and canonical correlation analyses. *Chemosphere* **2018**, *194*, 463–470. [CrossRef]
21. Liu, D.P.; Yu, H.B.; Gao, H.J.; Liu, X.Y.; Xu, W.N.; Yang, F. Insight into structural composition of dissolved organic matter in saline alkali soil by fluorescence spectroscopy coupled with self-organizing map and structural equation modeling. *Spectrochim. Acta Part A Mol. Biomol. Spectrosc.* **2022**, *279*, 121311. [CrossRef]
22. Aftab, B.; Shin, H.S.; Hur, J. Exploring the fate and oxidation behaviors of different organic constituents in landfill leachate upon Fenton oxidation processes using EEM-PARAFAC and 2D-COS-FTIR. *J. Hazard. Mater.* **2018**, *354*, 33–41. [CrossRef]
23. Zhang, H.; Cui, K.P.; Guo, Z.; Li, X.Y.; Chen, J.; Qi, Z.G.; Xu, S.Y. Spatiotemporal variations of spectral characteristics of dissolved organic matter in river flowing into a key drinking water source in China. *Sci. Total Environ.* **2020**, *700*, 10. [CrossRef] [PubMed]
24. Niloy, N.M.; Haque, M.M.; Tareq, S.M. Characterization of dissolved organic matter at urban and industrial rainwater of Bangladesh by fluorescence spectroscopy and EEM-PARAFAC modeling. *Environ. Chall.* **2021**, *5*, 100250. [CrossRef]
25. Duplá, M.V. Characterization of cDOM in the Elkhorn Slough estuary using EEM spectroscopy and its potential for macrophyte monitoring. *J. Mar. Syst.* **2022**, *226*, 103661. [CrossRef]
26. Wu, J.Q.; Yao, W.K.; Zhao, L.; Zhao, Y.; Qi, H.S.; Zhang, R.J.; Song, C.H.; Wei, Z.M. Estimating the synergistic formation of humus by abiotic and biotic pathways during composting. *J. Clean. Prod.* **2022**, *363*, 132470. [CrossRef]
27. Noda, I. Generalized two-dimensional correlation method applicable to infrared, raman, and other types of spectroscopy. *Appl. Spectrosc.* **1993**, *47*, 1329–1336. [CrossRef]
28. Maqbool, T.; Hur, J. Changes in fluorescent dissolved organic matter upon interaction with anionic surfactant as revealed by EEM-PARAFAC and two-dimensional correlation spectroscopy. *Chemosphere* **2016**, *161*, 190–199. [CrossRef] [PubMed]
29. Shafiquzzaman, M.; Ahmed, A.T.; Shafiul Azam, M.; Razzak, A.; Askri, B.; Hassan, H.F.; Ravikumar, B.N.; Okuda, T. Identification and characterization of dissolved organic matter sources in Kushiro river impacted by a wetland. *Ecol. Eng.* **2014**, *70*, 459–464. [CrossRef]
30. Lu, K.T.; Gao, H.J.; Yu, H.B.; Liu, D.P.; Zhu, N.M.; Wan, K.L. Insight into variations of DOM fractions in different latitudinal rural black-odor waterbodies of eastern China using flfluorescence spectroscopy coupled with structure equation model. *Sci. Total Environ.* **2022**, *816*, 151531. [CrossRef]
31. Hooper, D.; Coughlan, J.; Mullen, M.R. Structural equation modelling: Guidelines for determining model fit. Electron. *Electron. J. Bus. Res. Methods* **2008**, *6*, 53–60.
32. Mahabeer, P.; Tekere, M. Anthropogenic pollution influences on the physical and chemical quality of water and sediments of the umdloti river system, Kwazulu-Natal. *Phys. Chem. Earth* **2021**, *123*, 103030. [CrossRef]
33. Chakraborty, P.; Chakraborty, S.; Jayachandran, S.; Madan, R.; Sarkar, A.; Linsy, P.; Nath, B.N. Effects of bottom water dissolved oxygen variability on copper and lead fractionation in the sediments across the oxygen minimum zone, western continental margin of India. *Sci. Total Environ.* **2016**, *556*, 1052–1061. [CrossRef] [PubMed]
34. Cory, R.M.; Miller, M.P.; McKnight, D.M.; Guerard, J.J.; Miller, P.L. Effect of instrument-specific response on the analysis of fulvic acid fluorescence spectra. *Limnol. Oceanogr. Methods* **2010**, *8*, 67–78.
35. Pitta, E.; Zeri, C. The impact of combining data sets of fluorescence excitation–emission matrices of dissolved organic matter from various aquatic sources on the information retrieved by PARAFAC modeling. *Spectrochim. Acta Part A Mol. Biomol. Spectrosc.* **2021**, *258*, 119800. [CrossRef]
36. Osburn, C.L.; Boyd, T.J.; Montgomery, M.T.; Bianchi, T.S.; Coffifin, R.B.; Paerl, H.W. Optical proxies for terrestrial dissolved organic matter in estuaries and coastal waters. *Front. Mar. Sci.* **2016**, *2*, 127. [CrossRef]
37. Ma, Y.M.; Mao, R.; Li, S.Y. Hydrological seasonality largely contributes to riverine dissolved organic matter chemical composition: Insights from EEM-PARAFAC and optical indicators. *J. Hydrol.* **2021**, *595*, 125993. [CrossRef]

38. Wilson, H.F.; Xenopoulos, M.A. Effects of agricultural land use on the composition of fluvial dissolved organic matter. *Nat. Geosci.* **2009**, *2*, 37–41. [CrossRef]
39. Yu, Z.; Liu, X.M.; Zhao, M.H.; Zhao, W.Q.; Liu, J.; Tang, J.; Liao, H.P.; Chen, Z.; Zhou, S.G. Hyperthermophilic composting accelerates the humification process of sewage sludge: Molecular characterization of dissolved organic matter using EEM-PARAFAC and two-dimensional correlation spectroscopy. *Bioresour. Technol.* **2019**, *274*, 198–206. [CrossRef]
40. Asaoka, S.; Jadoon, W.A.; Umehara, A.; Takeda, K.; Otani, S.; Ohno, M.; Fujitake, N.; Sakugawa, H.; Okamura, H. Organic matter degradation characteristics of coastal marine sediments collected from the Seto Inland Sea, Japan. *Mar. Chem.* **2020**, *225*, 103854. [CrossRef]
41. Shen, D.L.; Huang, S.H.; Zhang, Y.P.; Zhou, Y.C. The source apportionment of N and P pollution in the surface waters of lowland urban area based on EEM-PARAFAC and PCA-APCS-MLR. *Environ. Res.* **2021**, *197*, 111022. [CrossRef]
42. Zhou, Y.Q.; Zhang, Y.L.; Jeppesen, E.; Murphy, K.R.; Shi, K.; Liu, M.L.; Liu, X.H.; Zhu, G.W. Inflow rate-driven changes in the composition and dynamics of chromophoric dissolved organic matter in a large drinking water lake. *Water Res.* **2016**, *100*, 211–221. [CrossRef] [PubMed]
43. Tranvik, L.J.; Bertilsson, S. Contrasting effects of solar UV radiation on dissolved organic sources for bacterial growth. *Ecol. Lett.* **2008**, *4*, 458–463. [CrossRef]
44. Song, X.Y.; Zhao, M.Y.; Chen, A.Q.; Xie, X.Y.; Yang, H.Y.; Zhang, S.B.; Wei, Z.M.; Zhao, Y. Effects of input of terrestrial materials on photodegradation and biodegradation of DOM in rivers: The case of Heilongjiang River. *J. Hydrol.* **2022**, *609*, 127792. [CrossRef]
45. Zhang, J.; Lv, B.Y.; Xing, M.Y.; Yang, J. Tracking the composition and transformation of humic and fulvic acids during vermicomposting of sewage sludge by elemental analysis and fluorescence excitation- emission matrix. *Waste Manag.* **2015**, *39*, 111–118. [CrossRef] [PubMed]
46. Yao, Y.; Zhao, J.Q.; Miao, L.Z.; Hou, J. Effects of sediment physicochemical factors and heavy metals on the diversity, structure, and functions of bacterial and fungal communities from a eutrophic river. *Environ. Pollut.* **2022**, *303*, 119129. [CrossRef]
47. Saiya-Cork, K.R.; Sinsabaugh, R.L.; Zak, D.R. The effects of long-term nitrogen deposition on extracellular enzyme activity in an Acer saccharum forest soil. *Soil Biol. Biochem.* **2002**, *34*, 1309–1315. [CrossRef]
48. Xiao, Y.C.; Yu, H.B.; Song, Y.H. Spatial differentiation characteristics and response relationship of DOM, nutrients, and heavy metals in river sediments. *Environ. Sci.* **2022**, *43*, 226–236. [CrossRef]
49. Wang, J.; Yuan, S.; Tang, L.; Pan, X.; Pu, X.; Li, R.; Shen, C. Contribution of heavy metal in driving microbial distribution in a eutrophic river. *Sci. Total Environ.* **2020**, *712*, 136295. [CrossRef]
50. Bai, Y.C.; Wu, F.C.; Xing, B.S.; Meng, W.; Shi, G.L.; Ma, Y.; Giesy, J.P. Isolation and characterization of Chinese standard fulvic acid sub-fractions separated from forest soil by stepwise elution with pyrophosphate buffer. *Sci. Rep.* **2015**, *5*, 8723. [CrossRef]
51. Carvalho, S.I.; Otero, M.; Duarte, A.C.; Santos, E.B. Spectroscopic changes on fulvic acids from a kraft pulp mill effluent caused by sun irradiation. *Chemosphere* **2008**, *73*, 1845–1852. [CrossRef]
52. Łomińska-Płatek, D.; Anielak, A.M. Quantitative balance and analysis of fulvic acids changes in the process of municipal sewage treatment. *Water Resour. Ind.* **2021**, *26*, 100155. [CrossRef]
53. Igolkina, A.A.; Meshcheryakov, G. semopy: A python package for structural equation modeling. *Struct. Equ. Model. Multidiscip. J.* **2020**, *27*, 952–963. [CrossRef]
54. Wang, H.J.; Liu, X.C.; Wang, Y.L.; Zhang, S.Q.; Zhang, G.M.; Han, Y.Y.; Li, M.X.; Liu, L. Spatial and temporal dynamics of microbial community composition and factors influencing the surface water and sediments of urban rivers. *J. Environ. Sci.* **2023**, *124*, 187–197. [CrossRef]
55. Williams, C.J.; Yamashita, Y.; Wilson, H.F.; Jaffé, R.; Xenopoulos, M.A. Unraveling the role of land use and microbial activity in shaping dissolved organic matter characteristics in stream ecosystems. *Limnol. Oceanogr.* **2010**, *55*, 1159–1171. [CrossRef]
56. Guildford, S.J.; Hecky, R.E. Total nitrogen, total phosphorus, and nutrient limitation in lakes and oceans: Is there a common relationship. *Limnol. Oceanogr.* **2000**, *45*, 1213–1223. [CrossRef]
57. Jiao, C.; Zhao, D.; Zeng, J.; Guo, L.; Yu, Z. Disentangling the seasonal co-occurrence patterns and ecological stochasticity of planktonic and benthic bacterial communities within multiple lakes. *Sci. Total Environ.* **2020**, *740*, 140010. [CrossRef]
58. Cao, J.X.; Sun, Q.; Zhao, D.H.; Xu, M.Y.; Shen, Q.S.; Wang, D.; Wang, Y.; Ding, S.M. A critical review of the appearance of black-odorous waterbodies in China and treatment methods. *J. Hazard. Mater.* **2020**, *385*, 18. [CrossRef]
59. Zhu, L.J.; Zhao, Y.; Bai, S.C.; Zhou, H.X.; Chen, X.M.; Wei, Z.M. New insights into the variation of dissolved organic matter components in different latitudinal lakes of northeast China. *Limnol. Oceanogr.* **2020**, *65*, 471–481. [CrossRef]

MDPI AG
Grosspeteranlage 5
4052 Basel
Switzerland
Tel.: +41 61 683 77 34

Water Editorial Office
E-mail: water@mdpi.com
www.mdpi.com/journal/water